the
UNIVERSITY
of
GREENWICH

BIOLOGICAL CONTROL
WITH EGG PARASITOIDS

CAB INTERNATIONAL is an intergovernmental organization providing services worldwide to agriculture, forestry, human health and the management of natural resources.

The information services maintain a computerized database containing over 2.7 million abstracts on agricultural and related research with 150,000 records added each year. This information is disseminated in 47 abstract journals, and also on CD-ROM and online. Other services include supporting development and training projects, and publishing a wide range of academic and professional books.

The four scientific institutes are centres of excellence for research and identification of organisms of agricultural and economic importance: they provide annual identifications of over 30,000 insect and microorganism specimens to scientists worldwide, and conduct international biological control projects.

International Institute of Biological Control
An Institute of CAB INTERNATIONAL

The Institute:
- provides research, training and information in support of biological pest control as the foundation of sustainable pest management to all countries;
- operates from bases in the UK, Switzerland, Trinidad and Tobago, Benin, Kenya, Pakistan and Malaysia;
- supports programmes to control exotic weed and insect pests of agriculture and conservation areas through the introduction of carefully selected and specific natural enemies, and provides associated services in exploration, safety testing and quarantine;
- supports the development of safe and effective biopesticides as economical commercial alternatives to chemical pesticides, particularly in developing countries;
- supports the implementation of integrated pest management (IPM) programmes through assistance to policy making, training of trainers and on-farm, participatory research;
- provides information on all aspects of biological control and IPM to researchers, extensionists, policy makers and others through its quarterly journal, *Biocontrol News and Information*, and other products;
- offers training courses and in-service training throughout the world.

International Institute of Biological Control   *Tel*: Ascot (0344) 872999
Silwood Park   *Telex:* 9312102255 BC G
Buckhurst Road, Ascot   *E-mail:* cabi@cabi.org
Berkshire SL5 7TA, UK   *Fax:* (0344) 875007

For further details of the services provided by the Institute, please contact the Institute Director.

BIOLOGICAL CONTROL WITH EGG PARASITOIDS

Edited by

E. Wajnberg

INRA, Antibes
France

and

S.A. Hassan

BBA, Darmstadt
Germany

CAB INTERNATIONAL

on behalf of the

International Organization for Biological Control
of Noxious Animals and Plants (IOBC)

CAB INTERNATIONAL
Wallingford
Oxon OX10 8DE
UK

Tel: Wallingford (0491) 832111
Telex: 847964 (COMAGG G)
E-mail: cabi@ cabi.org
Fax: (0491) 833508

A catalogue entry for this book is available from the British Library.

ISBN 0 85198 896 2

Typeset by Solidus (Bristol) Limited

Printed and bound in Great Britain by
Biddles Ltd, Guildford and King's Lynn

Contents

Contributors

BIGLER, FRANZ, *Swiss Federal Research Station for Agronomy, Reckenholtzst. 191, 8046 Zürich, Switzerland*

BIN, FERDINANDO, *Agricultural Entomology Institute, University of Perugia, Borgo XX Giugno, 06121 Perugia, Italy*

BOIVIN, GUY, *Research Station, Agriculture Canada, 430 Boul. Gouin, Saint-Jean-sur-Richelieu, Quebec, Canada J3B 3E6*

GRENIER, SIMON, *Laboratoire de Biologie Appliquée, Bât. 406 INSA, LA INRA 227, 20 avenue Albert Einstein, 69621 Villeurbanne Cedex, France*

HASSAN, SHERIF A., *Institut für Biologische Schädlingsbekämpfung, BBA, Heinrichst. 243, D-6100 Darmstadt, Germany*

LI, LI-YING, *Guangdong Entomological Institute, 106 Xingang Road West, Guangzhou 510260, China*

NORDLUND, DONALD A., *USDA-ARS, Subtropical Agriculture Research Laboratory, Biological Control of Pests Research Unit, 2413 East Highway 83, Weslaco, Texas 78596, USA*

PINTO, JOHN D., *Department of Entomology, University of California, Riverside, California 92521, USA*

SCHMIDT, JONATHAN M., *Department of Environmental Biology, University of Guelph, Guelph, Ontario, Canada N1G 2W1*

SMITH, SANDY M., *Faculty of Forestry, University of Toronto, 33 Willcocks Street, Toronto, Ontario, Canada M5S 3B3*

STOUTHAMER, RICHARD, *Department of Entomology, Agricultural University, P.O. Box 8031, 6700 EH Wageningen, The Netherlands*

VINSON, S. BRADLEIGH, *Department of Entomology, Texas A&M University, College Station, Texas 77843-2475, USA*

WAJNBERG, ERIC, *Laboratoire de Biologie des Invertébrés, Unité de Biologie des Populations, INRA, 37 Bld. du Cap, 06600 Antibes, France*

Preface

The idea for publishing this book was suggested on several occasions by members of the Working Group '*Trichogramma* and other egg parasitoids' of the International Organization for Biological Control (IOBC). Therefore, we are pleased to complete it at a time when a sound knowledge of egg parasitoids is urgently needed.

Plant protection remains a fundamental issue in agricultural science. It should not only lead to efficient interventions but also be conducted with concern for the environment. The use of biological control agents for the control of pests has long been an essential part of pest management strategies in crop protection. The importance and unique advantages of the method are now well recognized.

Numerous egg parasitoids are effective natural enemies of important agricultural and forestry pests. The impact of these egg parasitoids on pest populations in nature, as well as the potential of their use in biological pest control by augmentation, has long drawn the attention of research entomologists.

The wide distribution of such parasitoid species throughout the world, especially those of the genus *Trichogramma* (Hymenoptera, Chalcidoidea, Trichogrammatidae) indicates their importance as natural enemies of pests. In the genus *Trichogramma* alone, more than one hundred species are known to occur on a large number of crops in very different types of agroecosystems throughout the world. For the purposes of biological control, the advantage of using these egg parasitoids, compared to larval parasitoids, is that they prevent the hatching of the larvae which are usually, for the crop, more damaging at the early instar stage. Moreover, due to their usual low host specificity, egg parasitoids can be mass-reared more easily in large numbers, on different natural or factitious hosts.

We believe that there are various groups of people that need practical knowledge of biological control with egg parasitoids. These vary from students who may never specialize in the subject, but need it as part of an agricultural education, to research entomologists working on egg parasitoids or commercial operators in biological control programmes. The present book, which has been designed to provide up-to-date information, is therefore offered to fulfil the needs of all these groups. It can be used by those taking an introductory course, but also by research entomologists and mass-producers of egg parasitoids who need detailed information concerning these fascinating wasps.

The different topics are arranged into 12 chapters and were chosen in order to cover almost all the scientific aspects involved in the use of these insects in biological control. Chapter 1 begins with a consideration of Trichogrammatidae systematics, with a special emphasis on the *Trichogramma* genus. The world-wide distribution of the different species is also considered. The following Chapter provides a survey of the use of *Trichogramma* for biological control of several pests, all over the world. Then, in Chapter 3, the different strategies available to choose the optimal wasp species are presented, while the next Chapter discusses mass-production on artificial diets. Quality control in mass-rearing systems is considered in Chapter 5 while Chapter 6 provides the key points necessary to adjust properly release methods (including timing) in the field. Finally, Chapter 7 gives a summary of biological control programmes developed with egg parasitoids other than *Trichogramma*.

The optimal utilization of egg parasitoids in biological control programmes cannot be developed without a sound knowledge of their ecology and biology. Thus the mechanisms used by parasitoid females to locate the habitat of their hosts and the way they recognize and attack them are discussed in Chapters 8 and 9, respectively. Physiological interactions between the parasites and their hosts are considered in Chapter 10, while Chapter 11 examines the overwintering strategies of egg parasitoids. Finally, in order to improve wasp efficiency through artificial selection programmes, we need to estimate accurately the genetic variability in the population of these tiny hymenopterous parasitoids. Chapter 12 discusses this issue, with a special emphasis on intra-population levels of genetic variability.

Throughout this book, the reader will be aware that both fundamental and applied research has been conducted on these egg parasitic wasps. We hope this book will serve as a stimulus to expand our knowledge of egg parasitoids and therefore to increase their use and efficiency as biological control agents.

Finally we wish to thank IOBC for support, and several anonymous reviewers who greatly improved the final quality of the book. Miss Ch. Curty is also thanked for designing the 'flip book' of drawings in the lower corner of the pages.

Eric Wajnberg
Sherif A. Hassan

Systematics of the Trichogrammatidae with Emphasis on *Trichogramma* [1]

JOHN D. PINTO[1] & RICHARD STOUTHAMER[2]

[1]*Department of Entomology, University of California, Riverside, California 92521, USA;* [2]*Department of Entomology, Agricultural University, P.O. Box 8031, 6700 EH Wageningen, The Netherlands*

Abstract

The Trichogrammatidae (Chalcidoidea) consists of over 600 species and about 80 genera of insect egg parasitoids. The systematics of the group is summarized. The genus *Trichogramma* has received the most attention because of its importance in biological control. *Trichogramma* is worldwide in distribution and consists of 145 described species. The systematics of the group depends heavily on male genitalia. Because of general morphological homogeneity among species, investigators have assumed that the genus consists of numerous cryptic species. For this reason reproductive and allozymic characters have been utilized extensively in an attempt to better distinguish and relate species. Our analysis indicates that a typological approach has characterized much of the systematics research in *Trichogramma*. We attribute this, at least in part, to a conflict between the goals of biological control and systematics.

Systematics of Trichogrammatidae

The Trichogrammatidae is a cosmopolitan family of Chalcidoidea of approximately 620 species and 80 genera. This group of insect egg parasitoids includes the smallest of insects, ranging in length from 0.2 to 1.5 mm. The three-segmented tarsus separates the family from all other chalcidoids. Although this chapter is concerned primarily with *Trichogramma*, brief attention is given to the systematics of the entire family. *Trichogramma* may be the most important from an applied perspective but it is not necessarily the most typical nor the most common genus.

Historical resume

Most of the descriptive work on the family occurred before 1960 and is attributable to Arsene Girault who authored 45% of the nominal genera. Early family classifications include those by Girault (1912, 1918), Kryger (1918) and Blood (1923). The most comprehensive study of the Trichogrammatidae was by Doutt & Viggiani (1968). This is a useful work of synthesis. It examined original material of most of the type species, illustrated much of it, synonymized many names, and provided a key to genera, checklists of nominal taxa and generic diagnoses. A classification by Viggiani followed in 1971 dividing the family into two subfamilies, each with two tribes (Trichogrammatinae: Trichogrammatini, Paracentrobini; Oligositinae: Oligositini, Chaetostrichini).

Viggiani's classification is based almost exclusively on male genitalia. Consequently genera whose males are unknown (e.g. *Xenufens*) cannot be confidently placed. An alternate classification stemming from Girault (1918) is based on antennal structure, grouping genera with a funicle in the Trichogrammatinae and those without in the Lathromerinae (Peck *et al.*, 1964; Gauld & Bolton, 1988). Tribes were not proposed. Both classifications are basically monothetic and require testing with other characters. In general, relationships within the family remain poorly understood and require considerable study.

Several new genera have been added since 1968 (e.g. Doutt, 1974; Hayat & Husain, 1981; Lin, 1981, 1991; Viggiani, 1982, 1989; Pinto & Viggiani, 1987; Yousuf & Shafee, 1987). Although certain genera have been reviewed (Doutt, 1973; Nagaraja, 1978a; Pinto, 1990; Velten & Pinto, 1990; Pinto & Viggiani, 1991), not a single genus has undergone comprehensive revision. This is due largely to the absence of adequate collections of these minute and fragile insects, and the fact that most regions of the world are poorly sampled. Consequently, current generic concepts are tentative and subject to modification once sufficient material becomes available. Intermediate taxa are already coming to our attention that blur formerly distinct generic and tribal limits (e.g. Hayat & Subba Rao, 1985; Pinto, 1990).

Important taxonomic catalogues and checklists of the family include Doutt & Viggiani (1968) and Yousuf & Shafee (1986a) (genus group names), Yousuf & Shafee (1986b) (checklist of world species), Hayat & Viggiani (1984) (catalogue, Oriental region), Burks (1979) (catalogue, United States) and De Santis (1989) (catalogue, Latin America).

Size and distribution of genera

Most of the genera of Trichogrammatidae are small; 55 include five or fewer species and 30 of these are monotypic. Several appear to represent derived or primitive elements of other genera and probably should be synonymized. Only 14 genera have more than ten species (Fig. 1.1). The two largest, *Tricho-gramma* and *Oligosita*, with ca. 145 and 110 species, respectively, contain 40% of the family's diversity. The fact that *Trichogramma* has the most described species is probably a result of its importance to biological control.

The distribution of genera in the biogeographic regions of the world was summarized by Yousuf & Shafee (1987). Virtually all larger genera are cosmopolitan with the exception of *Trichogrammatoidea* which either is not present or is poorly represented in the Palaearctic and Nearctic. Several of the smaller genera also are widely distributed. The total number of genera (followed by the number of endemic genera) in the six biogeographic regions,

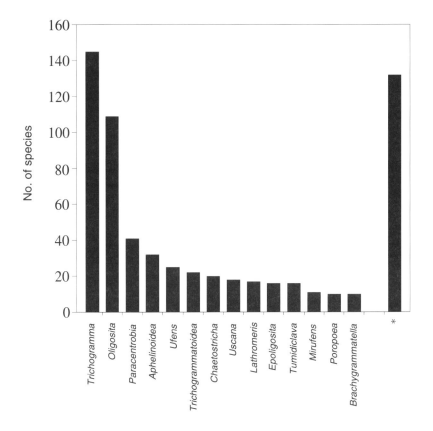

Fig. 1.1. Graph indicating number of species of the 14 largest genera of Trichogrammatidae. Asterisk indicates all remaining (ca. 65) genera.

based on a summary of the literature and recent collections, are as follows: Palaearctic (37, 7), Nearctic (36, 2), Australian (32, 10), Oriental (32, 2), Neotropical (27, 5), Afrotropical (15, 0).

Intensified sampling efforts in the last few years have rapidly increased our knowledge of generic distributions. For example, as recently as 1979 Burks reported only 17 genera in America north of Mexico, and only 19 were listed as Nearctic by Yousuf & Shafee in 1987. Recent collecting has increased the number of Nearctic genera to 36 (J.D. Pinto, unpubl.). Certain previously unrecorded taxa such as *Chaetostricha*, *Mirufens* and *Haeckeliania* are represented by several undescribed species (unpubl.). Similarly, only two genera were known from Costa Rica previously. Intensive collecting in that country has now increased the number to 22 (Pinto, 1994). Faunal studies stressing generic diversity have been published for New Zealand (Noyes & Valentine, 1989), Taiwan (Lin, 1981), Peru (Ruiz & Korytkowski, 1979) and India (Hayat & Subba Rao, 1985; Yousuf & Shafee, 1987).

Hosts

The Trichogrammatidae consists of solitary or gregarious endoparasitoids of insect eggs. There are also a few recorded exceptions to egg parasitism (e.g. puparia of Cecidomyiidae: Viggiani, 1981). A list of hosts has never been published for the family. However, reviews for Oriental taxa are given by Hayat & Viggiani (1984) and Yousuf & Shafee (1987). A few additional associations are available in Doutt & Viggiani (1968) but most must be gleaned from the primary literature. A recent survey of this literature located at least one reasonably reliable host record for 53 of the 80 recognized genera. Ten orders of Insecta have been recorded as hosts. The number of genera associated with each are as follows: Homoptera (29), Coleoptera (18), Hemiptera (9), Lepidoptera (8), Odonata (4), Orthoptera (4), Diptera (3), Hymenoptera (1), Thysanoptera (1) and Neuroptera (1).

Most of the better known genera of Trichogrammatidae have a broad range of hosts, and several have been recorded from eggs of more than one order. For example, *Oligosita* has been recorded from Homoptera, Hemiptera, Orthoptera and Coleoptera, and, although *Trichogramma* is usually associated with Lepidoptera, there also are records from eggs of Coleoptera, Diptera, Hemiptera, Homoptera, Hymenoptera (Symphyta) and Neuroptera. Early indications suggest a greater fidelity to microhabitat than to taxon of host. Thus, members of *Lathromeroidea* have been associated with eggs of Odonata, Hemiptera and Homoptera that occur in aquatic habitats, and certain species of *Trichogramma* parasitize Lepidoptera, Neuroptera and Hymenoptera eggs that occur together on the same plants (J.D. Pinto, unpubl.). Certain small genera, however, appear to be host specialists. Probable examples include

Poropoea and *Ophioneurus*, which have only been retrieved from eggs of Curculionoidea.

Systematics of *Trichogramma*

Introduction

In this section we review the systematics of *Trichogramma*, a genus of major concern to biological control. Some attention is given to *Trichogrammatoidea*, its close relative, which is also of interest from an applied perspective, and has had a similar taxonomic history.

The importance of systematics in taxa of economic significance is obvious. This is particularly true in *Trichogramma* because all sampled terrestrial habitats harbour native species. Consequently, any well-conceived augmentation programme must identify this fauna, distinguish the various species from one another and from introduced exotics, and monitor the identity of mass cultures to avoid contamination. An appreciation of native species and their relationship to those in other faunas may actually preclude the need to introduce exotics in the first place.

The urgency of identifications for applied work coupled with considerable morphological homogeneity in *Trichogramma* has prompted use of non-traditional characters for species discrimination. These include reproductive compatibility, mode of reproduction and allozymes, as well as relatively minor morphological differences. Below we review the status of *Trichogramma* systematics with particular attention to the role played by these character sources and to problems that continue to retard systematic progress. Finally, because reproductive incompatibility and mode of reproduction have a prominent role in systematics studies in this genus, we close by summarizing what is known about their cause and speculate on their evolutionary significance.

Taxonomic review

Trichogramma was described by Westwood (1833) with *T. evanescens* as type species. There were several additions in the late 19th and early 20th centuries, and by 1940 approximately 50 nominal species had been added, the majority from Europe and North America. These were based largely on morphological differences such as colour and setation, traits eventually found to be correlated with body size, seasonality and host (Flanders, 1931; Oldroyd & Ribbands, 1936; Salt, 1937; Quednau, 1960). This and the unavailability of types led to numerous synonymies until, in 1960, Flanders & Quednau proposed that only six species were structurally identifiable.

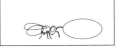

Because of morphological homogeneity, it was proposed relatively early that certain problems in *Trichogramma* systematics could be resolved only with biological and reproductive characters (Howard & Fisk, 1911; Harland & Atteck, 1933). This was re-emphasized by Flanders & Quednau (1960) and Quednau (1960) who argued that a biological assessment under controlled laboratory conditions was a prerequisite for valid species descriptions. They counted reproductive compatibility and mode of reproduction among the most important characters. Unfortunately, because these data stemmed from few cultures and could not be correlated with morphological differences they did little to improve the long-standing systematics problems in the genus.

Studies of male genitalia by Nagarkatti & Nagaraja (1968, 1971) were a breakthrough and ushered in a new era of *Trichogramma* taxonomy. They found that the genitalia varied considerably and that differences correlated well with reproductive data. For the first time, dependable morphological traits were available for identification. This resulted in the description of numerous species. Of the approximately 145 species of *Trichogramma* now recognized, 114 (almost 80%) were described after 1968. Male genitalia have also been important in *Trichogrammatoidea* (Nagaraja, 1978a). Other useful characters, but of secondary importance, include colour, antennal features (shape, sensilla and setation – primarily in males), wing venation and setal tracks, scutellar setae length and ovipositor length (Nagarkatti & Nagaraja, 1971; Voegelé *et al.*, 1975; Pinto *et al.*, 1978; Sugonyaev, 1986). Cuticular sculpturing, a useful character source in many chalcidoid groups, is relatively homogeneous (however, see Pinto (1992) for an exception). Unfortunately traits for identifying females with the same level of confidence as males are unavailable.

Taxonomic structure of the genus

A species catalogue for *Trichogramma* was published by Zerova & Fursov (1989). Features separating the genus were summarized by Pinto (1992). Briefly, *Trichogramma* is distinguished by the characteristic 'sigmoid' venation and RS_1 setal track on the forewing, and the dorsal lamina associated with the male genitalia (Figs 1.2, 1.8–1.10).

Trichogrammatoidea, its closest relative according to Nagaraja (1978a), lacks the RS_1 and dorsal lamina (cf. Figs 1.8–1.11). The female antenna is virtually identical in both genera (Fig. 1.3) consisting of two funicular segments followed by a relatively short, solid and somewhat asymmetrical clavus or club. Although specifics vary somewhat, antennae of basically similar shape and structure occur in females of other trichogrammatid genera known to parasitize Lepidoptera eggs (*Paratrichogramma, Trichogrammatomyia, Australufens, Xenufens*). It is unknown if these six genera represent a monophyletic lineage or if the similarity is homoplastic.

Males of the vast majority of *Trichogramma* are distinguished from those of other genera by the antenna which has a single, elongate flagellar segment representing the fusion of all funicular and claval elements (Fig. 1.4). The male antenna of *Trichogrammatoidea* consists of distinct funicular (two) and claval (three) segments (Fig. 1.7).

Trichogramma is currently divided into two subgenera, the nominate subgenus, worldwide in distribution, and *Trichogrammanza*, an Australian assemblage of three species (Carver, 1978; Oatman & Pinto, 1987). *Trichogrammanza* is characterized by a two-segmented funicle and one-segmented clavus in males (Fig. 1.6). This phenetic intermediacy suggested to Nagarkatti & Nagaraja (1977) that *Trichogrammanza* (known but unnamed in 1977) represented a transition between *Trichogrammatoidea* and *Trichogramma* (*Trichogramma*). Voegelé & Pintureau (1982) concurred and portrayed *Trichogrammanza* as the sister taxon of the nominate subgenus.

Kostadinov (1988) erected the genus *Nuniella* for a species entirely similar to nominate *Trichogramma* except for its highly derived male genitalia. Pinto & Oatman (1985) treated a related North American species as *Trichogramma*. *Nuniella* was recently synonymized with *Trichogramma* by Pintureau (1993).

Recently, Pinto (1992) described four unique *Trichogramma* from Central America and Australia that modified the generic diagnosis somewhat. In these species the male antenna has a two-segmented funicle and a three-segmented clavus (Fig. 1.5). The male genitalia retain the dorsal lamina found in all other *Trichogramma*. Because such antennae occur in other taxa (e.g. *Australufens*, *Trichogrammatomyia*) and are almost certainly plesiomorphic these species were not given generic or subgeneric status. Instead they were tentatively divided into two species groups and assigned to the nominate subgenus.

The nominate subgenus with approximately 140 recognized species (over 160 named) is divided into a variable number of species groups depending on authority (Nagarkatti & Nagaraja, 1977; Voegelé & Pintureau, 1982; Viggiani & Laudonia, 1989). Except for the recently described Lachesis and Primaevum groups (Pinto, 1992), all are remarkably similar to one another. Most are differentiated by male genitalia. Thus far these taxa have not been defined phylogenetically. Although Voegelé & Pintureau (1982) did place their 14 groups in the context of a phylogram, the rational for its topology was not fully explained.

Distributions and faunal relationships

The number of named and generally accepted species of *Trichogramma* from the six biogeographic regions of the world (based on published records only) are as follows: Palaearctic (50), Oriental (35), Nearctic (28), Neotropical (24), Afrotropical (8) and Australian (7). These numbers are highly correlated with collecting and agricultural application. For example, the relatively few species from Australia is largely explained by limited collecting primarily confined to

J.D. Pinto & R. Stouthamer

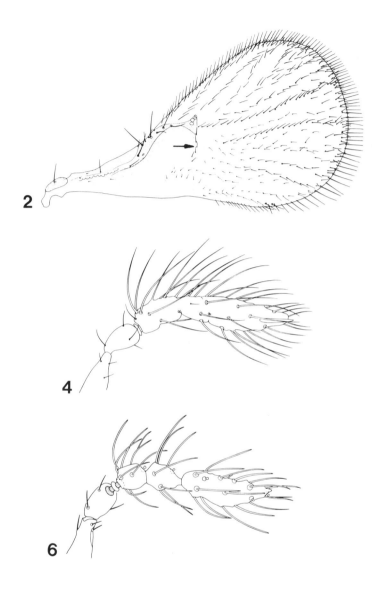

2

4

6

Figs 1.2–1.7. Forewing (*Trichogramma*) and antennae (*Trichogramma* and *Trichogrammatoidea*). 2, Forewing, arrow indicates RS$_1$ setal track (absent in *Trichogrammatoidea*); 3, antenna of female *Trichogramma nubilale* Ertle & Davis (typical of all *Trichogramma* and *Trichogrammatoidea*); 4, antenna of male *Trichogramma nubilale* (typical of most *Trichogramma*); 5, antenna of male *Trichogramma primaevum* Pinto (type restricted to Lachesis & Primaevum groups of *Trichogramma*); 6, antenna of male *Trichogramma funiculatum* Carver (typical of subgenus *Trichogrammanza*); 7, antenna of male *Trichogrammatoidea annulata* De Santis (typical of entire genus).

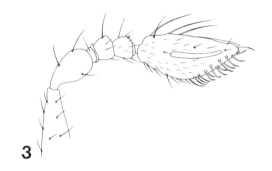

3

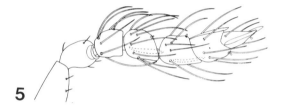

5

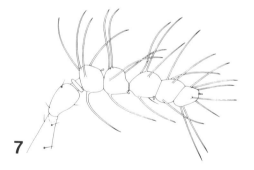

7

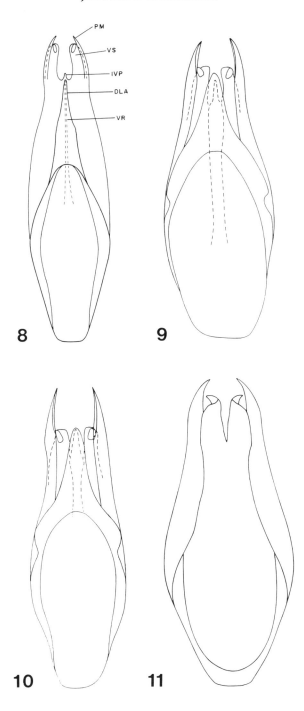

the eastern third of the country. In general, *Trichogramma* remains poorly known in most parts of the world. *Trichogrammatoidea*, a much smaller taxon of 22 species, occurs in the Oriental (10), Afrotropical (6), Neotropical (3) and Australian (3) regions. One Palaearctic species, *T. stammeri*, was described from a single female (Novicky, 1946). We are unaware of other records. *T. bactrae* Nagaraja, a widespread species in the Oriental and Australian regions, was released recently in the southwestern United States (Hutchison *et al.*, 1990) but establishment there has yet to be confirmed.

Thus far no terrestrial ice-free habitat is known to lack *Trichogramma*. The few Nearctic localities surveyed in some detail harbour ca. 5–10 species (Thorpe, 1984; Hung *et al.*, 1985; Pinto & Oatman, 1988; Pinto, unpubl.). In the Nearctic we have examined species from well north of the Arctic Circle, at elevations above 3000 m, and from the hottest deserts. The genus appears to be common in the tropics but it is too early to speculate how its diversity there compares to that in temperate zones. Regional studies of taxonomic importance include Sorokina (1984, 1992), Sugonyaev (1986) and Pintureau (1987) for the Palaearctic, Pintureau & Babault (1988) for the Afrotropical, Nagaraja (1973), Pang & Chen (1974) and Oatman *et al.* (1982) for the Oriental, Oatman & Pinto (1987) for the Australian, Ruiz & Korytkowski (1979), Brun *et al.* (1984) and Zucchi (1988) for the Neotropical, and Nagaraja & Nagarkatti (1973), Pinto *et al.* (1978) and Pinto & Oatman (1985) for the Nearctic. Several of these studies include keys to species for the geographic region covered. Keys of worldwide scope have been published by Voegelé & Pintureau (1982), and, very recently, by Sorokina (1993). The latter treats 127 of the currently recognized species of *Trichogramma*. Although considerably derivative of the literature, this work provides a useful character summary for most of the world's species.

Pending a needed phylogenetic classification and a greater collecting effort, we can offer few general statements regarding relationships among elements of the world fauna. Clearly, the genus is remarkably homogeneous everywhere. With the exception of *Trichogrammanza* there is no infrageneric grouping with significant morphological differences obviously associated with

---

Figs 1.8–1.11. (*Opposite*) Male genitalia (dorsal view) in *Trichogramma* and *Trichogrammatoidea* showing genital structure and examples of 'major' and 'minor' species differences in *Trichogramma* as discussed in text (DLA, dorsal lamina; IVP, intervolsellar process; PM, parameres; VR, ventral ridge; VS, volsellar digiti). 8, *Trichogramma carverae* Oatman & Pinto; 9, *Trichogramma exiguum*; 10, *Trichogramma fuentesi* Torre; 11, *Trichogrammatoidea bactrae*. The presence of the DLA distinguishes *Trichogramma* from *Trichogrammatoidea*. Genitalic characters in Fig. 1.8 vs. Figs 1.9 and 1.10 represent an example of 'major' differences; characters in Fig. 1.9 vs. Fig. 1.10 represent an example of 'minor' differences.

a major geographical area. This is reflected by the distribution of the species groups recognized by Voegelé & Pintureau (1982). Except for the monotypic groups, all are broadly distributed. One possible exception is an unnamed group of ca. ten species represented by *T. atopovirilia* Oatman & Platner and *T. drepanophorum* Pinto & Oatman, known so far only from the southern Nearctic and the Neotropical regions (unpubl.).

Faunal overlap appears to be considerable at the species-group level but there is little evidence that individual species commonly occur in non-contiguous biogeographical regions. However, the Evanescens, Minutum, Chilonis, and Parkeri groups (*sensu* Voegelé & Pintureau, 1982; Pintureau, 1987) are represented in the Palaearctic and Nearctic by close relatives, and we suggest that a careful comparison of faunas would reveal some conspecificity. A similarly close relationship occurs between the Oriental and Australian faunas (unpubl.).

Certain species, especially those well adapted to agricultural and other disturbed habitats, are widely distributed. For example, *T. evanescens* is reported throughout the Palaearctic and probably occurs in the Oriental region as well. *T. chilonis* Ishii occurs in the Oriental region, in eastern Australia and on many Pacific islands including Hawaii. *T. exiguum* Pinto & Platner and *T. pretiosum* Riley are distributed throughout much of the New World. In general, however, records suggesting widespread distribution of *Trichogramma* species are suspect unless confirmed by authoritative identifications. Clearly, most if not all records prior to 1968 require verification.

Biogeographical comparisons are complicated further by widespread introductions for at least the last 100 years. Considering this long history it is surprising how little evidence we have that releases actually result in permanent additions to regional faunas. This is because faunas are poorly understood and pre- and post-release surveys are rarely, if ever, done. Also, in some cases such as the recent introduction of the Oriental species *T. ostriniae* Pang & Chen to North America, the exotic is so similar to native species (*T. pretiosum* in this case) that its presence could easily be overlooked. The only release that we are aware of suggesting successful establishment of a non-native species is *T. pretiosum* in Australia (Pinto *et al.*, 1993). Material from Europe identified as *T. evanescens* has been introduced into North America on several occasions but there is no evidence that it is established. The species was released in southern California by Oatman *et al.* (1968) and in Missouri by Parker (1970). Recent efforts to collect it at both sites were unsuccessful (unpubl.).

Summary and critique of recent studies

The value of male genitalia in *Trichogramma* systematics has been mentioned. This character source has not been a complete solution to taxonomic problems however, as morphologically similar forms frequently display biological differences, and it is assumed that the genus contains numerous cryptic species that can only be recognized by studying reproductive, allozymic and morphological traits under controlled laboratory conditions (Nagarkatti & Nagaraja, 1977; Nagaraja, 1987). Numerous studies have utilized this approach but, in the aggregate, they have not solved many problems. Continuing concerns revolve around two major issues. First, inadequate attention to potential intraspecific variation of all characters has fostered a typological approach to *Trichogramma* systematics. Secondly, problems associated with type specimens have added to the confusion over species definitions. Below, we discuss current problems associated with types and the use of morphological, allozymic and reproductive characters.

Type specimens

Several types of *Trichogramma* have either been lost, ignored or are unusable. This breech in the historical record resulted in such a severe nomenclatural dislocation in North America that virtually all species names had come to be applied inconsistently and incorrectly (Pinto *et al.*, 1978). Because of this and the absence of voucher specimens, practically nothing of the considerable early literature on North American *Trichogramma* is of use at the species level. Pinto *et al.* (1978) designated neotypes for those species lacking types and drew attention to the characteristics of existing types. Although the North American fauna remains poorly known, this at least stabilized names of the most common species.

Parallel problems, however, continue to prevent name stabilization especially in the Old World (Sugonyaev, 1986). Types of several important species are unknown. This includes *T. dendrolimi* Matsumura, *T. cacoeciae* Marchal, *T. brassicae* Bezdenko, *T. pini* Meyer, *T. turkestanica* Meyer and *T. pallida* Meyer (Sugonyaev, 1986; Sorokina, 1992). The result has been considerable disagreement among European specialists as to species identity and synonymy. For example, *T. cacoeciae* is treated as a synonym of *T. embryophagum* (Hartig) by Pintureau (1987) but as a synonym of *T. dendrolimi* by Sorokina (1992). Pintureau (1987) treats *T. turkestanica* as a synonym of *T. meyeri* Sorokina and *T. maidis* Pintureau & Voegelé as a synonym of *T. brassicae* Bezdenko. Sorokina (1992) disagrees in both cases pointing out that these synonymies can not be substantiated because the types are absent. Even the identity of the type species, *T. evanescens*, is exceptionally controversial (Sugonyaev, 1986; Kostadinov & Pintureau, 1991). A primary type exists but, as a damaged female, it is of limited value. A similar problem concerns the identity of *T. brasiliense* Ashmead. The original description is based on the female type from Brazil in the US National Museum. Because the type is female, confident application of *brasiliense* is not possible. Yet, recent studies continue using the name (see De Santis, 1989).

Controversy over the identity of many common and important species will continue until neotypes are designated and existing types redescribed. Also, a petition to the International Commission of Zoological Nomenclature should request invalidation of the existing female type of *T. evanescens* and replacement with a male. Until useful types are in place almost any reasonable definition of a given species will be no more or no less justified than any other. The need for neotype designations also was voiced by Sugonyaev (1988). We find it curious that confusion over the names of Old World species in particular has been tolerated as long as it has.

Character studies

MORPHOLOGY

It is generally agreed that there is no *a priori* level of morphological differentiation required for species recognition. Many examples exist of good biological species with virtually no structural differences at all (White, 1978). However, species hypotheses require certain assumptions, namely that the differences used are genetic and that they consistently separate virtually all populations of a species from close relatives. When major structural differences occur they usually are tacitly treated as genetic and universality is assumed even if the form is known from few specimens. As differences become smaller however, concerns about genetic control and potential variation increase. To base species on relatively minor morphological differences requires some knowledge of character plasticity and levels of intraspecific variation. Of course, the definition of 'major' and 'minor' differences varies with the taxon and can be appreciated only after a group has undergone considerable study. Figures 1.8–1.10 provide examples of the two character classes in *Trichogramma*.

As already indicated, many of the characters used early on for distinguishing *Trichogramma* species fell into disrepute because of their plasticity. Documented character plasticity notwithstanding, species descriptions have continued to utilize differentia that provide minimal separation. This is reflected in keys to species which depend on subtle differences in setal length and number, and minor variances in ratios of anatomical parts (e.g. Nagaraja & Nagarkatti, 1973; Voegelé & Pintureau, 1982). In many cases, original descriptions were derived from small series providing little confidence that intraspecific variation was considered.

Studies by Sorokina (1987a) and Pinto *et al.* (1989) showed that several taxonomic characters are subject to ecophenotypic variation. The latter study indicated that in certain cases the magnitude of difference commonly attributed to species was within the range of variation of isofemale strains reared at different treatments. Also, allometric relationships were suggested for certain characters. For example, the ratio of ovipositor to hind tibial length, commonly used to separate females, is negatively correlated with body size. Number of antennal setae, ratio of forewing fringe to wing width, and number of antennal basiconic sensilla, all used to distinguish *Trichogramma* species, also are size sensitive. This is not to say that variation in these traits is totally inappropriate for species. However, to reduce the chance that it is intraspecific or ecophenotypic, consistency should be verified by examining relatively large series from different locales and preferably from several hosts. Although the effect of host on morphology has yet to receive adequate attention in this genus, the extreme host-dependent dimorphism reported by Salt (1937) for *T. semblidis* does not appear to be common.

The use of multivariate statistics to separate anatomically similar species shows promise in parasitic Hymenoptera (Woolley & Browning, 1987). So far, in *Trichogramma*, these procedures have been restricted to separating laboratory cultures of various species (Russo & Pintureau, 1981; Voegelé & Pintureau, 1982; Kostadinov & Pintureau, 1991). Again, such studies need to be expanded to ensure that intraspecific variation is adequately represented.

ALLOZYMES

Trichogramma has been the subject of numerous electrophoretic studies in the last 20 years. Most have focused on the Palaearctic fauna and have stressed putative esterase loci (e.g. Voegelé & Bergé, 1976; Jardak *et al.*, 1979; Pintureau & Babault, 1981, 1982; Pintureau, 1987, 1993; Pintureau & Keita, 1989). In these studies allozymic differences are used to separate species, characterize species groups and hypothesize intrageneric relationships. Relatively little work has concentrated on New World (Hung, 1982; Hung & Huo, 1985; Pinto *et al.*, 1992, 1993; Pintureau, 1993) or Oriental species (Cao *et al.*, 1988).

It is difficult to evaluate the considerable electrophoretic work on the Palaearctic fauna. First of all we are troubled that much of it is based on putative esterase loci that have not been confirmed by crossing experiments. Interpreting esterases is notoriously problematic (e.g. Richardson *et al.*, 1986; Heckel, 1993). Also, in several cases, characterization is based on limited samples (often a single culture) and should be considered preliminary. Furthermore, several of the species examined, especially those in the relatively homogeneous Evanescens and Minutum groups, remain incompletely studied morphologically and in certain instances are incompletely isolated reproductively (Pintureau, 1991). As stressed by Wright (1980), data from electrophoretic analysis can be most productive after morphological study either clearly outlines species differences or identifies the problems that prove intractable with morphology. In our opinion, the systematics of Palaearctic *Trichogramma* has not yet reached that stage.

Their premature employment notwithstanding, allozymic data have considerable promise for *Trichogramma* systematics. Because of the dearth of morphological traits, it represents an alternative character source so badly

needed for testing hypotheses of relationship that currently rely almost exclusively on male genitalia. In general, most of the allozymic work shows a reasonable correlation with morphology. However, intraspecific variation can be considerable, and genetic distances between homospecific samples can surpass those between heterospecifics (Pintureau, 1987; Pinto *et al.*, 1991, 1993). This is not an uncommon finding with electrophoretic data (Ayala, 1982) and argues further for adequate morphological analysis prior to electrophoretic work, and for basing that work on a reasonable sampling of the species range.

Molecular methods comparing nucleotide sequences have not yet contributed to the systematics of *Trichogramma*. Exploratory work in this area is underway, however (Vanlerberghe, 1991).

REPRODUCTIVE CHARACTERS

REPRODUCTIVE COMPATIBILITY: Hybridization experiments have been a frequent source of appeal for solving species problems in *Trichogramma* (e.g. Nagarkatti & Nagaraja, 1968, 1977; Oatman *et al.*, 1970; Fazaluddin & Nagarkatti, 1971; Nagarkatti & Fazaluddin, 1973; Oatman & Platner, 1973; Pinto *et al.*, 1978; Jardak *et al.*, 1979; Pintureau & Voegelé, 1980; Voegelé & Russo, 1981; Babi *et al.*, 1984; Nagaraja, 1987). Although crosses are often used to simply corroborate morphological differences, in certain cases they have been the primary basis for species descriptions. For example, the North American *Trichogramma platneri* Nagarkatti (Nagarkatti, 1975) was distinguished from *T. minutum* Riley after two morphologically identical cultures failed to cross. In Asia, *T. poliae* Nagaraja, *T. hesperidis* Nagaraja and *T. pallidiventris* Nagaraja were based largely on reproductive disjunction (Nagaraja, 1973). Recognition of Palaearctic species in the Evanescens group also was largely justified by reproductive incompatibility (Pintureau & Voegelé, 1980; Voegelé & Russo, 1981; Pintureau *et al.*, 1982; Pintureau, 1987). Compatibility data played a similar role in systematic decisions in *Trichogrammatoidea* (Nagaraja, 1978b).

Complete incompatibility is not a prerequisite for species separation in these studies as several report only partial compatibility in one or both directions. Also, the incompatibility, once discovered, initiates a search for differences in morphology, allozymes or host preferences. Just as incompatibility is stressed where morphological differences are slight or lacking, in *T. dendrolimi*, compatible allopatric forms with morphological and allozymic differences that would normally justify species recognition were instead treated as subspecies (Babi *et al.*, 1984). Reproductive data also have been used to estimate the strength of sexual isolation in *Trichogramma* as well as species relationships as measured by levels of interspecific hybridization (Nagarkatti & Fazaluddin, 1973).

MODE OF REPRODUCTION: Thelytokous forms of *Trichogramma* have been a taxonomic enigma because of the paucity of morphological features for distinguishing females. The occurrence of rare males, especially when reared at high temperatures, has allowed investigators to ally several of these populations with known arrhenotokous species (e.g. Sorokina, 1987b). Yet in some cases reproductive mode alone, often coupled with other biological traits, has prompted species recognition. Three thelytokous species, all Palaearctic, have been recognized – *T. oleae* (Voegelé & Pointel, 1979), *T. cordubensis* (Vargas & Cabello, 1985) and *T. telengai* (Sorokina, 1987b).

JUSTIFICATION FOR REPRODUCTIVE DATA: Reproductive incompatibility has been treated as apodictic evidence for species in *Trichogramma* in several cases (e.g. Oatman *et al.*, 1970; Nagaraja, 1973; Nagarkatti, 1975; Nagarkatti & Nagaraja, 1977; Pintureau & Voegelé, 1980). Morphological and sometimes life history and allozymic traits are used to further support species recognition but these are typically minor and, on their own, would not prompt taxonomic action. The tacit justification for weighting reproductive data is the biological species concept (Nagarkatti & Nagaraja, 1968) which treats genetically isolated populations as species. Our concern with much of this work is not necessarily the species concept adopted but that it is explicitly typological in allowing a small number of crosses (in some cases only one) to dictate species limits.

Studies of other organisms have shown that genetic exchangeability can vary considerably within species, and that incompatible forms are frequently linked through intermediates (e.g. Dobzhansky & Spassky, 1959; Mayr, 1963; Miller & Westphal, 1967; Raven, 1986; Coyne & Orr, 1989a; Verrell & Arnold, 1989). The possibility of populations intermediate to reproductively isolated cultures of *Trichogramma* is acknowledged by some authors (Nagarkatti & Fazaluddin, 1973), yet species recognition has generally followed the discovery of incompatible cultures without corroboration by additional sampling.

The unfortunate result of 'species' based primarily on reproductive incompatibility is that they can be defined only relative to another taxon. They lack characters of their own. Without even so much as adequate geographic definitions in most cases, they can, at best, be identified only as long as 'type cultures' or collections compatible with type cultures exist. Once these are lost the 'species' based on them are also irretrievably lost. It comes as no surprise that considerable taxonomic confusion persists in assemblages such as the Evanescens group where reproductive compatibility has contributed significantly to species hypotheses.

VARIATION IN COMPATIBILITY: We are not suggesting that reproductive data are not potentially useful. The argument is with how they are used. Estimates of variation are required before any character source can contribute

meaningfully to classification. For morphological characters this usually can be ascertained rapidly once adequate collections are in place. For non-traditional characters such as reproductive data, which require live material, considerably more effort is required. Yet variation estimates are no less important if a typological approach is to be avoided.

To better evaluate reproductive data in *Trichogramma* we briefly review the hundreds of crossing experiments performed in our laboratory at Riverside, California. These experiments can be divided into two categories: (i) crosses between cultures of different morphospecies, and (ii) crosses between cultures of the same morphospecies. Morphospecies of *Trichogramma* are defined as forms with consistently separable morphology regardless of environmental treatment. The crossing protocol utilized is detailed in Pinto *et al.* (1991). Briefly, compatibility between two cultures is estimated after examining progenies of 20 replicates of each heterogamic combination and ten concurrent replicates of each homogamic combination as controls (no-choice tests). A replicate consists of an isolated pair. Virgin females of each culture are individually isolated concurrently to verify arrhenotokous reproduction. Most homogamic crosses result in 60–80% females. Heterogamic crosses are arbitrarily considered partially compatible if the sex ratio (% females) in either direction is <75% that of the corresponding homogamic control.

Crosses between morphospecies: In almost all cases, morphological traits used to delimit species of North American *Trichogramma* are good predictors of reproductive incompatibility. Of 51 heterospecific crosses among 23 species performed in our laboratory, only three (6%) resulted in hybridization (unpubl.). All three were non-reciprocal and between extremely close relatives. These results differ considerably from those of Nagarkatti & Fazaluddin (1973) reporting that distinct morphospecies hybridize frequently in the laboratory. They performed 56 crosses among eight New World *Trichogramma*, and reported hybrids in almost 30% of these. For the most part, hybridization percentages were low (<33% females) and non-reciprocal. In all cases, hybrids were fertile and morphologically indistinguishable from the maternal parent. They also noted that the likelihood of hybridization was not correlated with morphological distance.

Although both studies found general correlation between morphology and levels of hybridization, the differences are significant. Our work indicates that species of *Trichogramma*, including close relatives, rarely hybridize. Nagarkatti & Fazaluddin (1973) suggest that hybridization is common, even among morphologically distinct and presumably distantly related species. A study of Palaearctic species by Pintureau (1991) reports intermediate results: hybridization was found to be rare among morphologically distant species but not uncommon among close relatives. How can these different results be rectified? In our opinion, at least some of the claims of interspecific hybridization are artifacts of experimental design and different species concepts.

Because crossing experiments by Nagarkatti and co-workers involved pooled individuals both during mating and oviposition (Nagarkatti & Nagaraja, 1968), it was impossible to determine how many females produced 'hybrids'. Our procedure, and presumably that of others (e.g. Pintureau, 1991), utilized isolated females. The latter allows the percentage of mothers producing putative hybrid F_1 females in a cross to be determined. This is important because it is difficult to ensure the virginity of all females prior to exposure to males. Although one attempts to only use females for these experiments that emerged from host eggs that did not also issue males, occasionally a male sibling, especially a runt, escapes detection. Most cases of low-level interspecific hybridization in our laboratory are not due to hybridization at all, but instead to one or two parental females that had previously mated with an undetected sibling. When such crosses are repeated they almost always verify complete incompatibility. Thus, claims of interspecific hybridization must be based on isolated pairs not pooled individuals. We also suggest that they be based on female progeny production in at least 25% of the replicates. If produced in fewer, then the cross should be repeated to confirm the results. Electrophoretic and morphometric analysis of putative hybrids also is recommended.

Thus, we suggest that at least some of the interspecific hybridization reported by Nagarkatti & Fazaluddin (1973) was due to incorporating a small number of non-virgin females in their experiments. The low percentage of hybrids, the fact that they typically occurred only in one of the two mating trials performed, their fertility and consistent resemblance to the maternal species, as well as the poor correlation between likelihood of hybridization and morphological distance, are consistent with this suggestion.

The results of Pintureau (1991) are probably similar to ours. Although the report of considerable hybridization among closely related species appears to differ, we note that most of the partial compatibility reported is among three morphologically similar species of the Evanescens group. Although we do not know these taxa well, it is possible that these results, with levels of partial compatibility similar to that which we record within morphospecies (see below), merely reflect a less conservative species concept.

Crosses within morphospecies: We are finding varying levels of compatibility within several morphospecies of *Trichogramma*. This variation includes complete disjunction, partial reciprocal and partial to complete non-reciprocal compatibility and, of course, complete compatibility. The complexity of relationships that can occur has recently been reported in two morphospecies, *T. deion* Pinto & Oatman, and what we tentatively call the *T. minutum* complex (Pinto *et al.*, 1991).

Trichogramma deion is a widespread species in western North America

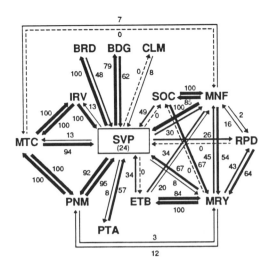

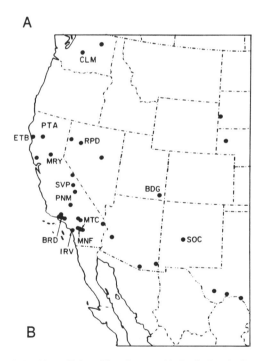

Fig. 1.12. Crossing relationships in *T. deion* (A) and geographic distribution of cultures crossed (B). Arrows in A point to parental female; numbers on arrows indicate level of reproductive compatibility of heterogamic cross relative to corresponding homogamic check cross (e.g. 34 indicates percentage female progeny produced in heterogamic cross is ca. one-third the percentage resulting from homogamic cross). See Pinto *et al.* (1991) for data on cultures. Modified from Pinto *et al.* (1991).

(Fig. 1.12B) (Pinto *et al.*, 1986). Results of 47 crosses among cultures of this morphospecies showed considerable variation from almost complete disjunction to complete compatibility (Fig. 1.12A). In certain instances incompatible cultures each were partially compatible with a common third culture. There is no geographic correlate to incompatibility in *T. deion*. Compatible and significantly disjunct samples can occur in adjacent locales or even at the same locale. It was confirmed that progeny from at least one of the non-reciprocal crosses were fertile, compatible with both parentals, and represented true hybrids (Stouthamer, 1989; Pinto *et al.*, 1991).

The *Trichogramma minutum* complex includes what are perhaps the most widely used biological control agents in North America (e.g. Oatman &

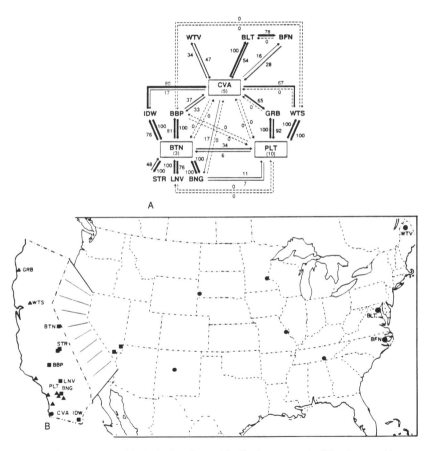

Fig. 1.13. Crossing relationships in the three forms of the *T. minutum* complex (A) and geographic distribution of cultures crossed in the USA (B). For B: ●, *T. minutum*; ▲, *T. platneri* (coastal form); ■, *T. platneri* (interior form). See Fig. 1.12 for further explanation and Pinto *et al.* (1991) for data on cultures. Modified from Pinto *et al.* (1991).

Platner, 1985; Smith *et al.*, 1990). As in *T. deion*, crossing relationships in this morphologically uniform complex are anything but straightforward (Fig. 1.13A), except in this case they are correlated with geography (Fig. 1.13B). Three reproductive groupings, referred to as forms, occur in North America. Although correlated with geography, reproductive variation in the complex includes considerable interform compatibility as well as intraform incompatibility. As in other morphospecies studied, reproductively disjunct cultures were often partially compatible to common third cultures. Also, as found in *T. deion*, progeny from at least one non-reciprocal cross between two different forms were fertile and compatible with both parentals. Electrophoretic studies showed them to be true hybrids (Pinto *et al.*, 1991, 1992).

Allozymic variation at 14 loci in 22 cultures of the *T. minutum* complex correlates well with these crossing data (Pinto *et al.*, 1992) and is consistent with the tentative recognition of three allopatric forms. Interestingly, the genetic distances reported among cultures of this complex were in many cases greater than those between cultures of *T. pretiosum* and *T. deion*, two close but reproductively isolated and morphologically distinct species (Pinto *et al.*, 1993). The taxonomic status of the forms of the *T. minutum* complex remains questionable and, in part, will depend on reproductive relationships in areas of geographic intermediacy which have yet to be adequately studied.

These results point to the danger of basing *Trichogramma* species on reproductive data especially if only a few crosses are considered. As in other organisms studied, reproductive relationships are complex and this complexity is known to occur intraspecifically. Extensive sampling throughout the range of a morphospecies is necessary before using crossing for species taxonomy. Taxonomic decisions that seem straightforward after examining a few cultures, often turn out to be anything but, once additional representatives are studied.

CAUSES OF INCOMPATIBILITY AND THELYTOKY IN *TRICHOGRAMMA*:

Incompatibility: Early stages of speciation are expected to be associated with reduced compatibility between different geographic isolates of a species. In most diploid–diploid species such reduced compatibility expresses itself in the sterility or low viability of the heterogametic sex formed in one of the two crosses between isolates. This so-called Haldane's Rule, the topic of intense research in *Drosophila* (Coyne & Orr, 1989b), obviously does not apply to haplo-diploid Hymenoptera. However, as already indicated, many crosses between sympatric and allopatric conspecifics of *Trichogramma* result in incompatibility including complete bidirectional incompatibility and non-reciprocal cross incompatibility (NRCI). NRCI is typical of cytoplasmic incompatibility caused by microbial infection in many insect species (e.g. Richardson *et al.*, 1987). However, in the few cases studied in *Trichogramma*, no evidence for microbe-induced incompatibility was found. Cytoplasmic factors were excluded as a cause for NRCI in two isolates of *T. deion*

(Stouthamer, 1989). Instead, NRCI in this species appears to be caused by genes on nuclear chromosomes that somehow code for a factor involved in sperm recognition. Eggs fertilized with incompatible sperm die in the early stages of development. As indicated above, taxonomic separation of forms with NRCI is not justified without additional differences.

Thelytoky: At least two different causes of thelytokous reproduction have been described in *Trichogramma*: hybridization and microbial infection. Species with thelytokous forms are listed in Table 1.1. Hybridization-caused thelytoky has been reported twice. Pintureau & Babault (1981) report a case of transient thelytoky resulting from a cross of *T. voegelei* Pintureau and *T. brassicae* Bezdenko. The hybrid F_1 females from this cross produced female offspring without fertilization. However, their daughters were arrhenotokous. The F_1 females were true hybrids and their thelytoky was likely caused by an epistatic interaction between parental genomes. The cytogenetic mechanism that caused thelytoky in these hybrids apparently leads to complete homozygosity in their offspring. The arrhenotokous reproduction of the F_2 females

Table 1.1. *Trichogramma* species with thelytokous forms.

Species	References
brasiliensis[a]	Kfir (1982)
brevicapillum	Stouthamer (1990)
cocoeciae	Marchal (1936)
chilonis	Stouthamer *et al.* (1990a, b)
cordubensis	Vargas & Cabello (1985), Stouthamer *et al.* (1990a, b)
deion[b]	Stouthamer *et al.* (1990a, b)
embryophagum	Birova (1970)
evanescens?	Marchal (1936)
oleae	Voegelé & Pointel (1979)
pintoi	Wang & Zhang (1988)
platneri	Stouthamer *et al.* (1990a, b)
pretiosum	Stouthamer *et al.* (1990a, b)
evanescens (as *rhenana*)	Pintureau (1987)
telengai	Sorokina (1987b)
n. sp.	(unpubl.)
sp.	Hase (1925)
sp.	Howard & Fiske (1911)

[a] Of unknown identity. *T. brasiliensis* is known only from its female holotype.
[b] The thelytokous record for *T. semifumatum* by Bowen & Stern (1966) probably refers to this species.

may be the result of the reshuffling of two grandparental genomes breaking up the original epistatic interactions. Crosses between other cultures of *T. voegelei* and *T. brassicae* failed to produce thelytokous F_1 females (R. Stouthamer, unpubl.).

The second description of hybridization-induced thelytoky was by Nagarkatti (1970). She established a thelytokous line with a female resulting from a cross between what may have been *T. exiguum* females (as *T. perkensi* Girault) and *T. californicum* Nagaraja & Nagarkatti males. Six sisters of this female also were tested and proved to be arrhenotokous.

Hybridization is suspected to be the cause of permanent thelytoky in a number of *Trichogramma*. This includes certain strains of the Old World *T. embryophagum*, *T. cacoeciae* and several unnamed forms in North America. These have been referred to as 'uncurable' forms of thelytokous *Trichogramma* by Stouthamer *et al.* (1990b). The cytogenetic mechanisms have been studied in some of these. Heterozygosity is maintained from generation to generation by the replacement of normal meiosis with a mitotic process (R. Stouthamer & D. Kazmer, unpubl.). Little is known about the origin of these forms, however, and their potential parental species have not been identified.

The taxonomic status of the 'uncurable' thelytokous forms is uncertain. For one thing, males, which are necessary for identification, are extremely rare (<1% of progeny) or non-existent (Stouthamer, 1991). Examination of males may not result in unequivocal identification in any event, since, if these forms do arise through hybridization, their males may have characters of one or both of the parental species. In the former case, a single isofemale line may produce two types of males. With the development of mitochondrial DNA markers it should be possible to identify the maternal species of such hybrid forms. Both parental species could be identified using other genetic markers (allozymes, restriction fragment polymorphisms, etc.). However, once the parental species of hybrid forms have been identified naming them is still not straightforward considering the debate now surrounding the taxonomy of hybridization-induced parthenogenetic lizards (Cole, 1985; Frost & Wright, 1988).

Microbe-associated thelytoky has now been found in ten *Trichogramma* species [*chilonis, deion, pretiosum, platneri, brevicapillum* Pinto & Platner, *oleae, evanescens* (as *rhenana* Voegelé & Russo), *cordubensis*, certain strains of *embryophagum*, and an undescribed species from California], and is associated with an infection of the wasps with *Wolbachia* bacteria (Rousset *et al.*, 1992; Stouthamer & Werren, 1993; Stouthamer *et al.*, 1993). By killing the *Wolbachia* by either temperature or antibiotic treatment such thelytokous forms of *Trichogramma* can be rendered permanently sexual (Stouthamer *et al.*, 1990a).

Microbe-induced thelytokous individuals occur with arrhenotokous conspecifics in many field populations. Generally its frequency does not reach high levels in populations. For example, only eight of 538 field-collected females (i.e. less than 2%) of *T. deion* tested were thelytokous (unpubl.). So far only one population (i.e. *T. pretiosum* from Kauai, Hawaii) is believed to have fixed thelytoky of this type. Because of the low frequency of thelytokous females in arrhenotokous populations they may be overlooked if laboratory cultures are started from several field-collected individuals. This probably explains why some laboratory cultures of *Trichogramma* become unisexual after reproducing bisexually for several generations.

Treating thelytokous females carrying microbes with the antibiotics tetracycline, rifampicin or sulfamethoxazole causes them to produce male progeny within two or three days (Stouthamer *et al.*, 1990a, b; Stouthamer, 1991). Arrhenotokous lines can be started by mating the males with thelytokous females and then treating these females with antibiotics. Daughters which they produce after three or more days of treatment can be used to initiate arrhenotokous lines. Such lines have been established for several species of *Trichogramma* (Stouthamer *et al.*, 1990a, b) and are permanently arrhenotokous. Similarly, microbe-associated thelytokous females can be converted to arrhenotoky by rearing them at 30°C. When such females are reared at temperatures of ca. 28°C, a high proportion of the offspring consist of gynandromorphs (i.e. individuals consisting of a mosaic of male and female tissues) (Bowen & Stern, 1966; Cabello & Vargas, 1985; R. Stouthamer, unpubl.).

The cytogenetic mechanisms that allow microbe-associated thelytokous females to produce daughters from unfertilized eggs is a form of gamete duplication (R. Stouthamer & D. Kazmer, unpubl.). Eggs undergo normal meiosis resulting in a haploid pronucleus. The diploid number of chromosomes is restored by fusion of the products of the first mitotic division. Genetically, this causes such eggs to develop into completely homozygous thelytokous females. If these females are exposed to conspecific males they are able to mate, store sperm, and fertilize their eggs just as their arrhenotokous counterparts do. Fertilized eggs of such females also follow the same course of development as in arrhenotokous forms. They develop into daughters whose genome consists of a set of chromosomes from each parent, but since they carry the microbe reproduction continues to be thelytokous. Consequently, thelytokous females occurring in primarily arrhenotokous conspecific populations are not genetically isolated.

Microbe-associated thelytokous females do not differ from their bisexual counterparts taxonomically. They should simply be considered as 'infected' forms of the same species. We do not know of any case in *Trichogramma* where a microbe-associated thelytokous form does not have a naturally occurring arrhenotokous conspecific with which it is capable of interbreeding. It is conceivable, however, that if this type of thelytoky becomes fixed in a population, females will eventually be incapable of interbreeding with the arrhenotokous species they were derived from. In such populations, mutations can be expected to accumulate in unexpressed parts of the genome that

code for female traits associated with mating and fertilization, and for male-limited traits. These females, and any males they eventually produce, may then be incapable of sexual reproduction. This scenario is unknown in *Trichogramma* but seems to have taken place in the thelytokous aphelinid *Encarsia formosa* Gahan (Zchori-Fein *et al.*, 1992). In this species only thelytokous populations are known in nature. Treatment with antibiotics results in male progeny but these males are unable to successfully inseminate females. Consequently, arrhenotokous lines cannot be started in this species. The time required to accumulate a sufficient number of mutations in populations fixed for thelytoky is unknown. However, after more than 300 generations in the laboratory as a thelytokous culture, a collection of *T. deion* from Sanderson, Texas, was still completely compatible with arrhenotokous conspecifics (Stouthamer *et al.*, 1990b).

Conclusion

We have been somewhat critical of the state of *Trichogramma* systematics in this review. Generally, we feel that there has been a typological approach to much of the recent research. The tendency to describe similar forms as species and to characterize them without first attempting to estimate limits of within-species variation has compounded problems in this taxonomically difficult group. We suggest that one reason for this approach is a basic conflict that has arisen between species-level systematics and biological control.

The species of most systematists is a monophyletic assemblage of populations which varies discontinuously from other such assemblages. Discontinuous variation among bisexual species is due to both reproductive and demographic disjunction (Templeton, 1989). However, each assemblage is expected to subsume considerable continuous variation, much of it of biological significance, but none the less, variation which cannot possibly be treated in a formal sense by taxonomy. An important task of biological control, on the other hand, is to identify differences among cognate populations because these may be critical to successful introductions. Whether this variation is continuous or discontinuous should be of little concern. Although the goals of biological control need not conflict with systematics, they do if it is assumed that variants important to biological control should have formal taxonomic treatment.

DeBach (1969) has argued that biological control is vitally interested in whether natural enemies differ from one another, regardless of our ability to distinguish them morphologically. He impugned systematists for often ignoring reproductively isolated but morphologically similar entities, or for designating them as strains or races rather than sibling species. Without formal designation as species, he believed, important populations may remain buried for possible use in biological control. We disagree with the protocol that this argument frequently fosters: the formal naming and characterizing of morphologically similar or identical forms as species following comparative studies of limited laboratory material. This practice, certainly not uncommon in *Trichogramma* research, is counterproductive and quickly leads to a taxonomy of cultures which cannot easily accommodate intraspecific variation and which can survive only as long as the laboratory cultures it is based on.

In our opinion, part of the reputation which *Trichogramma* has as a taxonomically difficult group is attributable to this conflict between biological control and systematics. Although their relationship should be symbiotic, and has been described as such (Knutson, 1980), we question whether this actually is the case. Systematics is a taxon-oriented endeavour and must be approached with global perspective. This is especially important in a group such as *Trichogramma* which has a relatively homogeneous world fauna. Unfortunately, much of the systematics of this genus reflects the problem-oriented and parochial concerns of biological control. In large part, this is attributable to the general lack of support for basic systematics, amazingly enough, even for the systematics of agriculturally significant taxa such as *Trichogramma*. Of course, systematics cannot be ignored by biological control and it is not. However, the absence of support for the area in its own right guarantees that taxonomic problems will be approached in an *ad hoc* fashion as an appendage of applied problems and often by investigators with little training or interest in systematics. In *Trichogramma* this is reflected by the unusually high number of entomologists recently involved in taxonomic studies. More than 40 authors, representing over 20 laboratories, are responsible for the species described since 1968 alone. With this many individuals, several with a passing interest in the systematics of the genus that is limited to local faunas and local problems, the global perspective that taxonomic stability requires has never materialized.

We realize that populations with unique attributes often are important to biological control and that these require recognition. However, formal taxonomic recognition may be inappropriate. Instead, we have suggested that a neutral term such as 'form' be used for interesting variants of a morpho-species until corroborative studies can resolve their taxonomic status (Pinto *et al.*, 1991). If voucher specimens are maintained, data tied to these forms will continue to be secure and useful in the future. What we are recommending is a global rather than a parochial approach to *Trichogramma* systematics. Only in this way can we begin to depart from the typology that has been responsible for much of the confusion in this group.

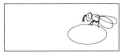

References

Ayala, F.J. (1982) The genetic structure of species. In: Milkman, R. (ed.) *Perspectives on Evolution.* Sinauer Assoc., Sunderland, Massachusetts, pp. 60–82.

Babi, A., Pintureau, B. & Voegelé, J. (1984) Etude de *Trichogramma dendrolimi* [Hym.: Trichogrammatidae], description d'une nouvelle sous-espèce. *Entomophaga* 29, 369–379.

Birova, H. (1970) A contribution to the knowledge of the reproduction of *Trichogramma embryophagum. Acta Entomologica Bohemoslovaca* 67, 70–82.

Blood, B.N. (1923) Notes on Trichogrammatinae taken around Bristol. *Proceedings of the Bristol Naturalists' Society (Annual Report)* 5, 253–258.

Bowen, W.R. & Stern, V.M. (1966) Effect of temperature on the production of males and sexual mosaics in uniparental race of *Trichogramma semifumatum. Annals of the Entomological Society of America* 59, 823–834.

Brun, P.G., Gomez de Moraes, G.W. & Soares, L.A. (1984) Três espécies novas de Trichogrammatidae parasitóides de Lepidópteros desfolhadores da mandioca e do eucalipto. *Pesquisa agropecuaria brasileira, Brasilia* 19, 805–810.

Burks, B.D. (1979) Chalcidoidea (part) and Cynipoidea. In: Krombein, K.V., Hurd, P.H., Smith, D.R. & Burks, B.D. (eds) *Catalog of Hymenoptera in America North of Mexico.* Smithsonian Institution Press, Washington, DC, pp. 768–1107.

Cabello, T. & Vargas, P. (1985) Temperatures as a factor influencing the form of reproduction of *Trichogramma cordubensis. Zeitschrift für Angewandte Entomologie* 100, 434–441.

Cao, G., Lu, W. & Long, S. (1988) Studies on comparison of esterase isozyme of different species of *Trichogramma.* In: Voegelé, J., Waage, J.K. & van Lenteren, J.C. (eds) Trichogramma *and Other Egg Parasites – Les Trichogrammes et autres parasitoïdes oophages. Les Colloques de l'INRA* 43, 35–44.

Carver, M. (1978) A new subgenus and species of *Trichogramma* Westwood (Hymenoptera: Chalcidoidea) from Australia. *Journal of the Australian Entomological Society* 17, 109–112.

Cole, C.J. (1985) Taxonomy of parthenogenetic species of hybrid origin. *Systematic Zoology* 34, 359–363.

Coyne, J.A. & Orr, H.A. (1989a) Patterns of speciation in *Drosophila. Evolution* 43, 362–381.

Coyne, J.A. & Orr, H.A. (1989b) Two rules of speciation. In: Otte, D. & Endler, J.A. (eds) *Speciation and Its Consequences.* Sinauer Assoc., Sunderland, Massachusetts, pp. 180–207.

DeBach P. (1969) Uniparental, sibling and semi-species in relation to taxonomy and biological control. *Israel Journal of Entomology* 4, 11–27.

De Santis, L. (1989) Catalogo de los himenopteros calcidoideos (Hymenoptera) al sur de los estados unidos. Segundo Suplemento. *Acta Entomologica Chilena* 15, 9–90.

Dobzhansky, T. & Spassky, B. (1959) *Drosophila paulistorum,* a cluster of species in statu nascendi. *Proceedings of the National Academy of Sciences* 45, 419–428.

Doutt, R.L. (1973) The genus *Paratrichogramma* Girault (Hymenoptera: Trichogrammatidae). *The Pan-Pacific Entomologist* 49, 192–196.

Doutt, R.L. (1974) *Chaetogramma,* a new genus of Trichogrammatidae (Hymenoptera: Chalcidoidea). *The Pan-Pacific Entomologist* 50, 238–242.

Doutt, R.L. & Viggiani, G. (1968) The classification of the Trichogrammatidae (Hymenoptera: Chalcidoidea). *Proceedings of the California Academy of Sciences* 35, 477–586.

Fazaluddin, M. & Nagarkatti, S. (1971) Reproductively incompatible crosses of *Trichogramma cacoeciae pallida* with *T. minutum* and *T. pretiosum* (Hymenoptera: Trichogrammatidae). *Annals of the Entomological Society of America* 64, 1470–1471.

Flanders, S.E. (1931) The temperature relationships of *Trichogramma minutum* as a basis for racial segregation. *Hilgardia* 5, 395–405.

Flanders, S.E. & Quednau, W. (1960) Taxonomy of the genus *Trichogramma*. *Entomophaga* 4, 285–294.

Frost, D.R. & Wright, J.W. (1988) The taxonomy of uniparental species, with special reference to parthenogenetic *Cnemidophorus*. *Systematic Zoology* 37, 200–209.

Gauld, I.D. & Bolton, B. (eds) (1988) *The Hymenoptera*. Oxford University Press, Oxford, 332 pp.

Girault, A.A. (1912) The chalcidoid family Trichogrammatidae. I. Tables of the subfamilies and genera and revised catalogue. *Bulletin of the Wisconsin Natural History Society* 10, 81–100.

Girault, A.A. (1918) *North American Hymenoptera Trichogrammatidae*. Privately published, Sydney, 11 pp. (Reprinted in Gordh, G., Menke, A.S., Dahms, E.C. & Hall, J.C. (eds) (1979) The privately printed papers of A.A. Girault. *Memoirs of the American Entomological Institute* 28, 142–152.)

Harland, S.C. & Atteck, O.M. (1933) Breeding experiments with biological races of *Trichogramma minutum*. *Zeitschrift Induktive Abstammungs und Vererbungslehre* 64, 54–76.

Hase, A. (1925) Beitrage der Lebensgeschichte der Schlupfwespe *Trichogramma evanescens*. *Arbeiten Biologische Reichsanstalt für Land und Forstwirtschaft* 14, 171–224.

Hayat, M. & Husain, T. (1981) A new genus of Trichogrammatidae from India. *Bollettino del Laboratorio di Entomologia Agraria 'Filippo Silvestri' di Portici* 38, 81–83.

Hayat, M. & Subba Rao, B.R. (1985) Chalcidoidea of India and adjacent countries. Family Trichogrammatidae. *Oriental Insects* 19, 239–245, 304–308.

Hayat, M. & Viggiani, G. (1984) A preliminary catalogue of the Oriental Trichogrammatidae (Hym.: Chalcidoidea). *Bollettino del Laboratorio di Entomologia Agraria 'Filippo Silvestri' di Portici* 41, 23–52.

Heckel, D.G. (1993) Comparative genetic linkage mapping in insects. *Annual Review of Entomology* 38, 381–408.

Howard, L.O. & Fiske, W.F. (1911) The importation into the United States of the parasites of the gypsy moth and the brown-tail moth. *USDA Bulletin* No. 91, 344 pp.

Hung, A.C.F. (1982) Chromosome and isozyme studies in *Trichogramma* (Hymenoptera: Trichogrammatidae). *Proceedings of the Entomological Society of Washington* 84, 791–796.

Hung, A.C.F. & Huo, S. (1985) Malic enzyme, phosphoglucomutase and phosphoglucose isomerase isozymes in *Trichogramma* [Hym.: Trichogrammatidae]. *Entomophaga* 30, 143–149.

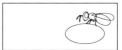

Hung, A.C.F., Vincent, D.L., Lopez, J.D. & King, E.G. (1985) *Trichogramma* (Hymenoptera: Trichogrammatidae) fauna in certain areas of Arkansas and North Carolina. *Southwestern Entomologist* 8, 11–20.

Hutchison, W.M., Moratorio, M. & Martin, J.M. (1990) Morphology and biology of *Trichogrammatoidea bactrae* (Hymenoptera: Trichogrammatidae), imported from Australia as a parasitoid of pink bollworm (Lepidoptera: Gelechiidae) eggs. *Annals of the Entomological Society of America* 83, 46–54.

Jardak, T.B., Pintureau, B. & Voegelé, J. (1979) Mise en évidence d'une nouvelle espèce de Trichogramme [Hym.: Trichogrammatidae]. Phénomène d'intersexualité, étude enzymatique. *Annales de la Societé Entomologique de France* 15, 635–642.

Kfir, R. (1982) Reproduction characteristics of *Trichogramma brasiliensis* and *T. lutea*, parasitising eggs of *Heliothis armiger*. *Entomologia Experimentalis et Applicata* 32, 249–255.

Knutson, L. (1980) Symbiosis of biosystematics and biological control. In: Papavizas, G.C. (ed.) *Biological Control in Crop Production. BARC Symposium* No. 5, pp. 61–78.

Kostadinov, D.N. (1988) Description of *Nuniella bistrae* gen. n., sp. n. (Hymenoptera, Trichogrammatidae) from Bulgaria. *Acta Zoologica Bulgarica* 36, 49–51.

Kostadinov, D.N. & Pintureau, B. (1991) A possibility to discriminate females of three closely related species of *Trichogramma* (Hymenoptera: Trichogrammatidae) with special purpose analysis of the type of *Trichogramma evanescens* Westwood. *Annales de la Societé Entomologique de France* 27, 393–400.

Kryger, J.P. (1918) The European Trichogramminae. *Entomologiske Meddelelser* 12, 257–354.

Lin, K. (1981) Genera of the Trichogrammatidae (Hymenoptera: Chalcidoidea) of Taiwan with descriptions of new taxa. *Journal of Agricultural Research, China* 30, 426–443.

Lin, N. (1991) Descriptions of five new species of *Megaphragma* and *Paramegaphragma* gen. nov. (Hymenoptera: Trichogrammatidae) from China. *Entomotaxonomia* 14, 129–138.

Marchal, P. (1936) Recherches sur la biologie et développement des Hyménoptères parasites. Les Trichogrammes. *Annales Épiphyties et de Phytogénétique* 2, 447–550.

Mayr, E. (1963) *Animal Species and Evolution.* Harvard University Press, Cambridge, Massachusetts, 797 pp.

Miller, D.D. & Westphal, N.J. (1967) Further evidence on sexual isolation within *Drosophila athabasca. Evolution* 21, 479–492.

Nagaraja, H. (1973) On some new species of Indian *Trichogramma* (Hymenoptera: Trichogrammatidae). *Oriental Insects* 7, 275–290.

Nagaraja, H. (1978a) Studies on *Trichogrammatoidea* (Hymenoptera: Trichogrammatidae). *Oriental Insects* 12, 489–530.

Nagaraja, H. (1978b) Experimental hybridization between some species of *Trichogrammatoidea* Girault (Hymenoptera: Trichogrammatidae). *Journal of Entomological Research* 2, 192–198.

Nagaraja, H. (1987) Recent advances in biosystematics of *Trichogramma* and *Trichogrammatoidea* (Hymenoptera, Trichogrammatidae). *Proceedings of the Indian Academy of Sciences* 96, 469–477.

Nagaraja, H. & Nagarkatti, S. (1973) A key to some New World species of *Trichogramma* (Hymenoptera: Trichogrammatidae), with descriptions of four new species. *Proceedings of the Entomological Society of Washington* 75, 288–297.

Nagarkatti, S. (1970) The production of a thelytokous hybrid in an interspecific cross between two species of *Trichogramma*. *Current Science* 39, 76–78.

Nagarkatti, S. (1975) Two new species of *Trichogramma* (Hym.: Trichogrammatidae) from the USA. *Entomophaga* 20, 245–248.

Nagarkatti, S. & Fazaluddin, M. (1973) Biosystematic studies on *Trichogramma* species (Hymenoptera: Trichogrammatidae). II. Experimental hybridization between some *Trichogramma* spp. from the New World. *Systematic Zoology* 22, 103–117.

Nagarkatti, S. & Nagaraja, H. (1968) Biosystematic studies on *Trichogramma* species: experimental hybridization between *Trichogramma australicum* Girault, *T. evanescens* Westwood and *T. minutum* Riley. *CIBC Technical Bulletin* 10, 81–96.

Nagarkatti, S. & Nagaraja, H. (1971) Redescriptions of some known species of *Trichogramma* (Hymenoptera: Trichogrammatidae) showing the importance of the male genitalia as a diagnostic character. *Bulletin of Entomological Research* 61, 13–31.

Nagarkatti, S. & Nagaraja, H. (1977) Biosystematics of *Trichogramma* and *Trichogrammatoidea* species. *Annual Review of Entomology* 22, 157–176.

Novicky, S. (1946) Weitere Beschreibungen von Trichogrammiden. *Zentralblatt für das Gesamtgebiet der Entomologie* 1, 44–50.

Noyes, J.S. & Valentine, E.W. (1989) Chalcidoidea (Insecta: Hymenoptera). Introduction, and review of genera in smaller families. *Fauna of New Zealand* No. 18, 91 pp.

Oatman, E.R. & Pinto, J.D. (1987) A taxonomic review of *Trichogramma* (*Trichogrammanza*) Carver (Hymenoptera: Trichogrammatidae), with descriptions of two new species from Australia. *Journal of the Australian Entomological Society* 26, 193–201.

Oatman, E.R. & Platner, G.R. (1973) Biosystematic studies of *Trichogramma* species: 1. Populations from California and Missouri. *Annals of the Entomological Society of America* 66, 1099–1102.

Oatman, E.R. & Platner, G.R. (1985) Biological control of two avocado pests. *California Agriculture* (Nov–Dec), 21–23.

Oatman, E.R., Platner, G.R. & Greany, P.D. (1968) Parasitization of imported cabbageworm and cabbage looper eggs on cabbage in southern California, with notes on the colonization of *Trichogramma evanescens*. *Journal of Economic Entomology* 61, 724–730.

Oatman, E.R., Platner, G.R. & Gonzalez, D. (1970) Reproductive differentiation of *Trichogramma pretiosum*, *T. semifumatum*, *T. minutum* and *T. evanescens*, with notes of the geographic distribution of *T. pretiosum* in the southwestern United States and in Mexico (Hymenoptera: Trichogrammatidae). *Annals of the Entomological Society of America* 63, 633–635.

Oatman, E.R., Pinto, J.D. & Platner, G.R. (1982) *Trichogramma* (Hymenoptera: Trichogrammatidae) of Hawaii. *Pacific Insects* 24, 1–24.

Oldroyd, H. & Ribbands, C.R. (1936) On the validity of *Trichogramma* trichiation as a systematic character. *Proceedings of the Royal Entomological Society of London* 5, 148–152.

Pang, X.-F. & Chen, T.-L. (1974) *Trichogramma* of China (Hymenoptera: Trichogrammatidae). *Acta Entomologica Sinica* 17, 441–454.

Parker, F.D. (1970) Seasonal mortality and survival of *Pieris rapae* (Lepidoptera:

Pieridae) in Missouri and the effects of introducing an egg parasite, *Trichogramma evanescens*. *Annals of the Entomological Society of America* 63, 985–994.

Peck, O., Boucek, Z. & Hoffer, A. (1964) Keys to the Chalcidoidea of Czechoslovakia (Insecta: Hymenoptera). *Memoirs of the Entomological Society of Canada* 34, 1–120.

Pinto, J.D. (1990) The genus *Xiphogramma*, its occurrence in North America, and remarks on closely related genera (Hymenoptera: Trichogrammatidae). *Proceedings of the Entomological Society of Washington* 92, 538–543.

Pinto, J.D. (1992) Novel taxa of *Trichogramma* from the New World tropics and Australia (Hymenoptera: Trichogrammatidae). *Journal of the New York Entomological Society* 100, 621–633.

Pinto, J.D. (1994) The family Trichogrammatidae. In: Hanson, P. & Gauld, I. (eds) *Hymenoptera of Costa Rica*. Oxford University Press, Oxford (in press).

Pinto, J.D. & Oatman, E.R. (1985) Additions to Nearctic *Trichogramma* (Hymenoptera: Trichogrammatidae). *Proceedings of the Entomological Society of Washington* 87, 176–186.

Pinto, J.D. & Oatman, E.R. (1988) *Trichogramma* species in a chaparral community of southern California, with a description of a new species (Hymenoptera: Trichogrammatidae). *The Pan-Pacific Entomologist* 64, 391–402.

Pinto, J.D. & Viggiani, G. (1987) Two new Trichogrammatidae (Hymenoptera) from North America: *Ittysella lagunera* Pinto and Viggiani (n. gen, n. sp.) and *Epoligosita mexicana* Viggiani (n. sp.). *The Pan-Pacific Entomologist* 63, 371–376.

Pinto, J.D. & Viggiani, G. (1991) A taxonomic study of the genus *Ceratogramma* (Hymenoptera: Trichogrammatidae). *Proceedings of the Entomological Society of Washington* 93, 719–732.

Pinto, J.D., Platner, G.R. & Oatman, E.R. (1978) Clarification of the identity of several common species of North American *Trichogramma* (Hymenoptera: Trichogrammatidae). *Annals of the Entomological Society of America* 71, 169–180.

Pinto, J.D., Oatman, E.R. & Platner, G.R. (1986) *Trichogramma pretiosum* and a new cryptic species occurring sympatrically in southwestern North America (Hymenoptera: Trichogrammatidae). *Annals of the Entomological Society of America* 79, 1019–1028.

Pinto, J.D., Velten, R.K., Platner, G.R. & Oatman, E.R. (1989) Phenotypic plasticity and taxonomic characters in *Trichogramma* (Hymenoptera: Trichogrammatidae). *Annals of the Entomological Society of America* 82, 414–425.

Pinto, J.D., Stouthamer, R., Platner, G.R. & Oatman, E.R. (1991) Variation in reproductive compatibility in *Trichogramma* and its taxonomic significance (Hymenoptera: Trichogrammatidae). *Annals of the Entomological Society of America* 84, 37–46.

Pinto, J.D., Kazmer, D.J., Platner, G.R. & Sassaman, C.A. (1992) Taxonomy of the *Trichogramma minutum* complex (Hymenoptera: Trichogrammatidae): allozymic variation and its relationship to reproductive and geographic data. *Annals of the Entomological Society of America* 85, 413–422.

Pinto, J.D., Platner, G.R. & Sassaman, C.A. (1993) An electrophoretic study of two closely related species of North American *Trichogramma*, *T. pretiosum* and *T. deion* (Hymenoptera: Trichogrammatidae). *Annals of the Entomological Society of America* 86, 702–709.

Pintureau, B. (1987) Systématique évolutive du genre *Trichogramma* Westwood en

Europe. Thesis, Université Paris VII, 311 pp.

Pintureau, B. (1991) Indices d'isolement reproductif entre espèces proches de Trichogrammes (Hym.: Trichogrammatidae). *Annales de la Societé Entomologique de France* 27, 379–392.

Pintureau, B. (1993) Enzyme polymorphism in some African, American and Asiatic *Trichogramma* and *Trichogrammatoidea* species (Hymenoptera: Trichogrammatidae). *Biochemical Systematics and Ecology* 21, 557–573.

Pintureau, B. & Babault, M. (1981) Caractérisation enzymatique de *Trichogramma evanescens* et *T. maidis*, étude des hybrides. *Entomophaga* 26, 11–22.

Pintureau, B. & Babault, M. (1982) Comparaison des enzymes chez 10 souches de *Trichogramma* (Hym. Trichogrammatidae). In: Voegelé, J. (ed.) *Les Trichogrammes. Les Colloques de l'INRA* 9, 31–44.

Pintureau, B. & Babault, M. (1988) Systématique des espèces africaines des genres *Trichogramma* Westwood et *Trichogrammatoidea* Girault (Hym. Trichogrammatidae). In: Voegelé, J., Waage, J.K. & van Lenteren, J.C. (eds) *Trichogramma and Other Egg Parasites – Les Trichogrammes et autres parasitoïdes oophages. Les Colloques de l'INRA* 43, 97–120.

Pintureau, B. & Keita, F.B. (1989) Nouvelles données sur les estérases des Trichogrammes (Hym. Trichogrammatidae). *Biochemical Systematics and Ecology* 17, 603–608.

Pintureau, B. & Voegelé, J. (1980) Une nouvelle espèce proche de *Trichogramma evanescens*: *T. maidis* (Hym.: Trichogrammatidae). *Entomophaga* 25, 431–440.

Pintureau, B., Voegelé, J. & Pizzol, J. (1982) A propos du statut de *Trichogramma maidis* [Hym. Trichogrammatidae]. *Bulletin de la Societé Entomologique de France* 87, 319–321.

Quednau, W. (1960) Uber die Identitat der *Trichogramma* arten und einiger ihrer Okotypen. *Mitteilungen aus der Biologischen Bundesanstalt für Land und Fortwirtschaft, Berlin-Dahlem* 100, 11–50.

Raven, P.N. (1986) Modern aspects of the biological species in plants. In: Iwatsuki, P., Raven, P. & Bock, W. (eds) *Modern Aspects of Plants*. University of Tokyo Press, Tokyo, pp. 11–29.

Richardson, B.J., Baverstock, P.R. & Adams, M. (1986) *Allozyme Electrophoresis*. Academic Press, New York, 410 pp.

Richardson, P.M., Holmes, W.P. & Saul, G.B. (1987) The effect of tetracycline on nonreciprocal cross incompatibility in *Nasonia vitripennis*. *Journal of Invertebrate Pathology* 50, 176–183.

Rousset, F., Bouchon, D., Pintureau, B., Juchault, P. & Solignac, M. (1992) *Wolbachia* endosymbionts responsible for various alterations of sexuality in arthropods. *Proceedings of the Royal Society of London* 250, 91–98.

Ruiz E.R. & Korytkowski, C.A. (1979) Contribucion al conocimiento de los Trichogrammatidae (Hymenoptera: Chalcidoidea) del Peru. *Revista Peruana de Entomologia* 22, 1–8.

Russo, J. & Pintureau, B. (1981) Etude biométrique de quatre espèces de *Trichogramma* Westwood [Hym. Trichogrammatidae]. *Annales de la Societé Entomologique de France* 17, 241–258.

Salt, G. (1937) The egg-parasite of *Sialis lutaria*: a study of the influence of the host upon a dimorphic parasite. *Parasitology* 29, 539–553.

Smith, S.M., Carrow, J.R. & Laing, J.E. (eds) (1990) Inundative release of the egg

parasitoid, *Trichogramma minutum* (Hymenoptera: Trichogrammatidae), against forest insect pests such as the spruce budworm, *Choristoneura fumiferana* (Lepidoptera: Tortricidae). *The Ontario Project 1982–1986. Memoirs of the Entomological Society of Canada* No. 153, 87 pp.

Sorokina, A.P. (1984) New species of the genus *Trichogramma* Westw. (Hymenoptera, Trichogrammatidae) from the USSR. *Entomological Review* 63, 139–152

Sorokina, A.P. (1987a) Variability of morphological characters in species of the genus *Trichogramma* (Hymenoptera, Trichogrammatidae) associated with parasitization in eggs of different hosts. In: Sugonyaev, E.S. (ed.) *Taxonomy and Biology of Parasitic Hymenoptera – Chalcidoid and Proctotrupoid Wasps. Proceedings of the Zoological Institute, Leningrad* 191, 79–89.

Sorokina, A.P. (1987b) Biological and morphological substantiation of the specific distinctness of *Trichogramma telengai* sp. n. *Entomological Review* 1987, 20–34.

Sorokina, A.P. (1992) New data on species of the genus *Trichogramma* Westw. (Hymenoptera, Trichogrammatidae) in the USSR with remarks on synonymy. *Entomological Review* 70, 31–43.

Sorokina, A.P. (1993) *Key to Species of the Genus* Trichogramma *Westw. (Hymenoptera, Trichogrammatidae) of the World Fauna.* Kolos Publishing House, Moscow, 77 pp.

Stouthamer, R. (1989) Causes of thelytoky and crossing incompatibility in several *Trichogramma* species. PhD Thesis. University of California, Riverside, 112 pp.

Stouthamer, R. (1990) Evidence for microbe-mediated parthenogenesis in Hymenoptera. *Proceedings and Abstracts, Vth International Colloquium on Invertebrate Pathology and Microbial Control, Adelaide, Australia*, pp. 417–421.

Stouthamer, R. (1991) Effectiveness of several antibiotics in reverting thelytoky to arrhenotoky in *Trichogramma* spp. In: Wajnberg, E. & Vinson, S.B. (eds) Trichogramma *and Other Egg Parasitoids. 3rd International Symposium. Les Colloques de l'INRA* 56, 119–122.

Stouthamer, R. & Werren, J.H. (1993) Microorganisms associated with pathogens in wasps of the genus *Trichogramma. Journal of Invertebrate Pathology* 61, 6–9.

Stouthamer, R., Luck, R.F. & Hamilton, W.D. (1990a) Antibiotics cause parthenogenetic *Trichogramma* to revert to sex. *Proceedings of the National Academy of Sciences of the USA* 87, 2424–2427.

Stouthamer, R., Pinto, J.D., Platner, G.R., Luck, R.F. (1990b) Taxonomic status of thelytokous forms of *Trichogramma. Annals of the Entomological Society of America* 83, 475–581.

Stouthamer, R., Breeuwer, J.J., Luck, R.F. & Werren, J.H. (1993) Molecular identification of microorganisms associated with parthenogenesis. *Nature* 361, 66–68.

Sugonyaev, E.S. (1986) New systematic data on the genus *Trichogramma* (Hymenoptera, Trichogrammatidae). *Entomological Review* 1986, 143–148.

Sugonyaev, E.S. (1988) On some actual problems of *Trichogramma* taxonomy. In: Voegelé, J., Waage, J.K. & van Lenteren, J.C. (eds) Trichogramma *and Other Egg Parasites – Les Trichogrammes et autres parasitoïdes oophages. Les Colloques de l'INRA* 43, 123–124.

Templeton, A.R. (1989) The meaning of species and speciation: a genetic perspective. In: Otte, D. & Endler, J.A. (eds) *Speciation and Its Consequences.* Sinauer Assoc, Sunderland, Massachusetts, pp. 3–27.

Thorpe, K.W. (1984) Seasonal distribution of *Trichogramma* (Hymenoptera: Trichogrammatidae) species associated with a Maryland soybean field. *Environmental*

Entomology 13, 127–132.

Vanlerberghe, F. (1991) Mitochondrial DNA polymorphism and relevance for *Trichogramma* population genetics. In: Wajnberg, E. & Vinson, S.B. (eds) *Trichogramma and Other Egg Parasitoids. 3rd International Symposium. Les Colloques de l'INRA* 56, 123–126.

Vargas, P. & Cabello, T. (1985) A new species of *Trichogramma* [*T. cordubensis* n. sp] [Hym.: Trichogrammatidae]. Parasitoid of *Heliothis* eggs in cotton crops in the SW of Spain. *Entomophaga* 30, 225–230.

Velten, R.K. & Pinto, J.D. (1990) *Soikiella* Novicky (Hymenoptera: Trichogrammatidae): occurrence in North America, description of a new species, and association of the male. *The Pan-Pacific Entomologist* 66, 246–250.

Verrell, P.A. & Arnold, S.J. (1989) Behavioral observations of sexual isolation among allopatric populations of the mountain dusky salamander, *Desmognathus ochrophaeus*. *Evolution* 43, 745–755.

Viggiani, G. (1971) Ricerche sugli Hymenoptera Chalcidoidea XXVIII. Studio morfologico comparativo dell'armatura genitale esterna maschile dei Trichogrammatidae. *Bollettino del Laboratorio di Entomologia Agraria 'Filippo Silvestri' di Portici* 29, 181–222.

Viggiani, G. (1981) Note su alcune specie di *Oligosita* Walker (Hym. Trichogrammatidae) e descrizione di quattro nuove specie. *Bollettino del Laboratorio di Entomologia Agraria 'Filippo Silvestri' di Portici* 38, 125–132.

Viggiani, G. (1982) Description of *Hayatia*, n. gen., n. sp. (Hymenoptera Trichogrammatidae) from India. *Bolletino del Laboratorio di Entomologia Agraria 'Filippo Silvestri' di Portici* 39, 27–29.

Viggiani, G. (1989) Description of *Pintoa nearctica*, gen. nov., sp. nov. (Hymenoptera: Trichogrammatidae). *Bolletino del Laboratorio di Entomologia Agraria 'Filippo Silvestri' di Portici* 45, 23–29.

Viggiani, G. & Laudonia, S. (1989) Le specie italiane di *Trichogramma* Westwood (Hymenoptera: Trichogrammatidae), con un commento sullo stato della tassonomia del genere. *Bolletino del Laboratorio di Entomologia Agraria 'Filippo Silvestri'* 46, 107–124.

Voegelé, J. & Bergé, J.B. (1976) Les Trichogrammes (Insectes Hymenop. Chalcidiens, Trichogrammatidae), caractéristiques isoestérasiques de deux espèces: *Trichogramma evanescens* Westw. et *T. acheae* Nagaraja, Nagarkatti. *Comptes Rendus des Séances de l'Académie des Sciences, Paris* 283, 1501–1503.

Voegelé, J. & Pintureau, B. (1982) Caractérisation morphologique des groupes et espèces du genre *Trichogramma* Westwood. In: Voegelé, J. (ed.) *Les Trichogrammes. Les Colloques de l'INRA* 9, 45–75.

Voegelé, J. & Pointel, J.G. (1979) *Trichogramma oleae*, n. sp., espèce jumelle de *Trichogramma evanescens*. *Annales de la Société Entomologique de France* 15, 643–648.

Voegelé, J. & Russo, J. Jr. (1981) Les Trichogrammes. Vc. Découverte en Alsace de deux espèces nouvelles de Trichogrammes *Trichogramma schuberti* et *T. rhenana* (Hym. Trichogrammatidae) sur pontes d'*Ostrinia nubilalis* Hubn. (Lepid. Pyralidae). *Annales de la Société Entomologique de France* 17, 535–541.

Voegelé J., Cals-Usciati, J., Pihan, J.P. & Daumal, J. (1975) Structure de l'antenne femelle des Trichogrammes. *Entomophaga* 20, 161–169.

Wang, F. & Zhang, S. (1988) Studies on *Trichogramma pintoi*, its deuterotokous

reproduction, artificial propagation and field releases. *Chinese Journal of Biological Control* 4, 149–151.

Westwood, J.O. (1833) Descriptions of several new British forms amongst the parasitic Hymenopterous insects. *The London, Edinburgh and Dublin Philosophical Magazine and Journal of Science* 2, 443–445.

White, M.J.D. (1978) *Modes of Speciation.* W.H. Freeman & Co., San Francisco, 455 pp.

Woolley, J.B. & Browning, H.W. (1987) Morphometric analysis of uniparental *Aphytis* reared from chaff scale, *Parlatoria pergandii* Comstock, on Texas citrus (Hymenoptera: Aphelinidae, Homoptera: Diaspididae). *Proceedings of the Entomological Society of Washington* 89, 77–94.

Wright, C.A. (1980) Chemosystematics: perspectives, problems and prospects for the zoologist. In: Bisby, F.A., Vaughan, J.G. & Wright, C.A. (eds) *Chemosystematics: Principles and Practice.* Academic Press, New York, pp. 29–38.

Yousuf, M. & Shafee, S.A. (1986a) Catalogue of genus-group names of world Trichogrammatidae (Hymenoptera). *Indian Journal of Systematic Entomology* 3, 13–27.

Yousuf, M. & Shafee, S.A. (1986b) Checklist of species and bibliography of the world Trichogrammatidae (Hymenoptera). *Indian Journal of Systematic Entomology* 3, 29–82.

Yousuf, M. & Shafee, S.A. (1987) Taxonomy of Indian Trichogrammatidae (Hymenoptera: Chalcidoidea). *Indian Journal of Systematic Entomology* 4, 55–200.

Zerova, M.D. & Fursov, V.N. (1989) A catalogue of species of the genus *Trichogramma* Westwood of the world (Hymenoptera, Trichogrammatidae). *Institute of Zoology of the Science Academy of the Ukrainian SSR. Publ* 89(4), 52 pp.

Zchori-Fein, E., Roush, R.T. & Hunter, M. (1992) Male production induced by antibiotic treatment in *Encarsia formosa*, an asexual species. *Experientia* 48, 102–105.

Zucchi, R.A. (1988) New species of *Trichogramma* (Hym., Trichogrammatidae) associated with the sugar cane borer *Diatraea saccharalis* F. (Lep. Pyralidae) in Brazil. In: Voegelé, J., Waage, J.K. & van Lenteren, J.C. (eds) Trichogramma *and Other Egg Parasites – Les Trichogrammes et autres parasitoïdes oophages. Les Colloques de l'INRA* 43, 133–140.

Worldwide Use of Trichogramma for Biological Control on Different Crops: A Survey

Li-Ying Li
Guangdong Entomological Institute, 106 Xingang Road West,
Guangzhou 510260, China

Abstract

In recent years, a total area of over 32 million ha of agriculture and forestry in the world has been treated annually with *Trichogramma* for controlling insect pests. In the worldwide use of *Trichogramma*, the former USSR ranked first, followed by China and Mexico. Extensive utilization of this parasitoid was developed on corn, rice, sugar-cane, cotton, vegetables and pines. The most important pests controlled by *Trichogramma* were corn borers, sugar-cane borers and cotton bollworm. The selection of suitable species, quality of mass-reared parasitoids, reasonable release rate per hectare, release methods (inundative or inoculative), methods of conducting release (manual or mechanical), climate during releases, timing of release, and integration with other methods of pest control (particularly with chemical control) are the main factors that determine the efficiency of using *Trichogramma*. Some biological, economical, ecological and theoretical criteria are used in evaluating the effectiveness of the commercial use of *Trichogramma*.

Introduction

The history of the use of *Trichogramma* for controlling insect pests has been recorded for a long time. But only since 1926, when Flanders developed the first mass-production system with *Sitotroga cerealella* eggs (Flanders, 1929; Olkowski & Zhang, 1990), has the utilization of *Trichogramma* been realized in many countries. The last 20 years have seen considerable use of this parasitoid on a particularly large scale on corn, sugar-cane, cotton, fruit trees and vegetables in more than 30 countries. Progress in utilizing *Trichogramma*

resulted from factors such as its persistent economic efficiency demonstrated for many years in several countries (e.g. the former USSR, China, etc.), the more elaborate construction and successful utilizations of biofactories for mass-production of *Trichogramma* and the accumulation of research and practical data concerning technology of commercial use and methods for evaluating wasp efficiency.

In this chapter, we give a survey of the scale of worldwide *Trichogramma* use in different countries on a variety of crops as well as their insect pests, the ways of utilization, effectiveness of use and the methods of its evaluation.

Scale of Worldwide Use of *Trichogramma*

Trichogramma has been introduced to countries where the government and public opinion support biological control for ecological and economical reasons, and where the mechanization of mass-rearing *Trichogramma* has been developed for a long time, or where the labour cost is lower, as in developing countries with large populations. The former USSR claimed to have treated about 27.6 million ha with *Trichogramma* in 1990, the largest area of *Trichogramma* use in the world (Filippov, 1990). China (including mainland and Taiwan, about 2.1 million ha; Cheng, 1991; Tseng, 1991; Li, 1992) and Mexico (2 million ha; Filippov, 1990) ranked second and third, respectively. The USA, western Europe and other countries lagged behind due largely to the lack of adequate research investment in biocontrol and to the widespread use of insecticides highly toxic to *Trichogramma* (Olkowski & Zhang, 1990). According to the data collected by Hassan (1988, 1990) and others (Filippov, 1990; Li, 1992) over 32 million ha of agriculture and forestry were likely treated annually with *Trichogramma*. In Table 2.1, a comparative survey of the estimated scale of *Trichogramma* use for biological control of insect pests is given.

Table 2.1. Estimated area of *Trichogramma* use in different countries (data before 1992).

Area (ha)	Country
2–27.6 million	Former USSR, China (including Taiwan), Mexico
140,000–350,000	USA, Philippines, Colombia
10,000–35,000	Iran, Bulgaria, Egypt, India, France
Less than 6000	Germany, Switzerland, Australia, Honduras, Thailand, Bolivia, Canada, Portugal, Italy, Uruguay, Nicaragua, Tunisia, The Netherlands, etc.

Crops and Their Insect Pests Treated with *Trichogramma*

The worldwide use of *Trichogramma* is concentrated on forage crops (corn, rice, wheat, sorghum), crops of primary industries (sugar-cane, sugarbeet, cotton, soyabean), vegetables, and fruit and forest trees (apple, plum, citrus, avocado, vineyards, pines, spruce). Because of their environmental and food uses, non-toxicity of these crops is extremely important to man. Certainly, the use of an insecticide for controlling a pest is restricted, and effective pest control with *Trichogramma* alone or integrated with other control measures is greatly advantageous to man. Table 2.2 shows 22 types of crop and tree on which *Trichogramma* was used to control their main insect, mostly lepidopterous, pests. Among them, the most extensively treated crops were corn, sugar-cane and cotton. In most of Asia, Europe and America, where corn is a very important crop, corn borers (*Ostrinia furnacalis* in southeastern Asia, and *O. nubilalis* in the other continents) were the main target insect pests controlled by *Trichogramma*.

In southern and southeastern Asia, Africa and South America, where sugar-cane is extensively cultivated, sugar-cane borers (*Chilo* spp., *Diatraea* spp., etc.) are the most important pests controlled by *Trichogramma*. Cotton is also economically important in many Asian and South American countries, the USA and certain republics of the former USSR, so *Trichogramma* was widely used in cotton integrated pest management to control cotton bollworm (*Heliothis* spp.). Since numerous species of insect pests other than cotton bollworm occurred with overlapping generations over the years, sometimes inundative releases of *Trichogramma* could not give an effective and economic guarantee of high harvest. In particular, when the outbreak of cotton weevil occurred, it was impossible to avoid the use of chemicals which in turn interfered with the activity of *Trichogramma* and its population establishment.

In more and more countries, development of *Trichogramma* for use in fields of vegetables (cabbage, tomato, beans, beet, etc.), vineyards, fruit orchards and forests is taking place. China has the greatest diversity of crops and trees using *Trichogramma* (Li, 1992). Before political changes, the former USSR also claimed to use *Trichogramma* to control a variety of insect pests of multiple crops and trees (Filippov, 1990). It is encouraging that a wide use of *Trichogramma* on various plants was found in South American countries, such as Mexico and Colombia (Gusev 1978; Hassan, 1988, 1990).

Although species of the genus *Trichogramma* attack more than 400 species in 203 genera, 44 families and 7 orders (Bao & Chen, 1989), the worldwide commercial use of *Trichogramma* is still limited to a small number of them. This shows that there is a great potential for the development of *Trichogramma* applications.

Table 2.2. Various crops and trees on which *Trichogramma* spp. were used for controlling insect pests in different countries (data before 1991).

Crop/tree	Country
Corn	Former USSR, China (including Taiwan), Mexico, Philippines, Colombia, Bulgaria, France, Germany, Switzerland, USA, Italy, Austria, former Czechoslovakia, Romania
Sugar-cane	China (including Taiwan), Philippines, Colombia, Iran, Egypt, Cuba, India, Uruguay, Mexico
Cotton	Former USSR, USA, Colombia, Mexico, China, Iran
Tomato	Former USSR, China, Mexico, Colombia, USA
Cabbage	Former USSR, China, Bulgaria, The Netherlands, former Czechoslovakia
Apple	Former USSR, Bulgaria, China, Germany, Poland
Beet	Former USSR, Bulgaria, China
Rice	China, Iran, India
Soyabean	Colombia, USA, China
Sorghum	Mexico, Colombia, China
Pine	China, Bulgaria
Vine	Former USSR, Bulgaria
Forage grass	Former USSR
Cayenne pepper	China
Tobacco	Bulgaria
Wheat	Former USSR
Citrus	China
Avocado	USA
Spruce	Canada
Olive	Tunisia
Plum	Bulgaria
Stored products	USA

Trichogramma Commercially Used in Different Countries

More than 70 species of *Trichogramma* have been used around the world (Li, 1984). However, until now, only about 20 species were mass-reared for field use (Table 2.2). Among them *T. dendrolimi*, *T. evanescens*, *T. chilonis* (= *T. confusum*), *T. japonicum*, *T. pretiosum*, *T. maidis* (= *T. brassicae*) and *T. ostriniae* were used on large areas of agricultural, horticultural and forest crops worldwide.

For an effective commercial use, it is essential to select suitable species (either introduced or indigenous) to control the target insect pests. In China, for instance, *Pieris* spp. were controlled quite well by *T. evanescens* in the Inner Mongolia Autonomous Region, Shanxi and Shaanxi Provinces and Shanghai suburbs. However, no species of *Trichogramma* has been found naturally occurring in Guangdong Province. In the laboratory, besides *T. evanescens* introduced from Shanxi or *T. nagarkatti* from France, no *Trichogramma* species could parasitize the eggs of *Pieris* spp. collected from the Guangzhou suburbs (Li, 1984). The cause still remains to be clarified. In the paddy-fields of China, *T. japonicum* is the dominant species of *Trichogramma*, and shows a greater competition ability in parasitizing the eggs of lepidopterous pests than *T. confusum* (= *T. chilonis*) and *T. dendrolimi*. It was found that *T. ostriniae* was the dominant species in the corn fields of Jiling Province, Beijing suburb, but *T. dendrolimi* was dominant in Helongjiang and Liaoning Provinces. Thus, dominant species of *Trichogramma* were recommended to be mass-reared and utilized in each province respectively.

In the Philippines, India and Taiwan, *T. chilonis* (= *T. confusum*) is a successful biological control agent against sugar-cane borers (Hassan, 1988; Alba, 1990; Cheng, 1991). In China, besides *T. confusum*, *T. dendrolimi* is also widely used to control sugar-cane borers, because it is easy to mass-rear in the laboratory and it is more plastic. At the same time, in South America, sugar-cane borers were controlled by *T. pretiosum* (Hassan, 1988). The European corn borer was controlled by *T. maidis* (= *T. brassicae*) in Switzerland (Bigler, 1986), France, Italy, Austria, etc. (Hassan, 1988). But in the former USSR, Germany and Bulgaria *T. evanescens* was used to control the corn borer (Hassan, 1988; see Table 2.3). The results of Hassan & Guo (1991) show that only the following three strains, from a total of 20 tested, satisfactorily parasitized the eggs of the European corn borer (in order of effectiveness): *T. ostriniae* from China, *T. evanescens* from Moldavia and *T. evanescens* from Germany. Based on research on intraspecific differentiation, four types of *T. evanescens* were selected in the former USSR especially for controlling noctuids, corn borer, cabbage butterfly and codling moth (Beglyarov & Smitnik, 1977). In India, a type of *T. australicum*, tolerant to high temperature (38°C) and drought (10% RH), was selected by directional hybridization through 33 generations (Abraham, 1976).

Some species, such as *T. evanescens*, *T. pretiosum* and *T. dendrolimi*, are quite commonly used in many countries against a number of insect pests. This indicates that they are polyphagous, and it also shows the plasticity of their response to the diversity of habitats. There are also some *Trichogramma* species that are found in only one country, and their utilization is restricted.

Generally, the selection for suitable species for commercial use is made empirically in a small area by experiments or laboratory trials. Since the

Table 2.3. Main insect pests, controlled with *Trichogramma* spp. in different countries (data collected from Pak & Oatman, 1982; Burbutis & Goldstein, 1983; Hassan, 1988; Hassan *et al.*, 1988).

Country	*Trichogramma* spp.	Target pest
China, Germany, former USSR	*T. dendrolimi*	*Adoxophyes* spp., *Agrotis* spp., *Cydia pomonella*, *Dendrolimus* spp., *Heliothis armigera*, *Pieris* spp., *Leguminivora glycinivorella*, *Ostrinia furnacalis*, *Pandemis heparana*, *Rhyacinonia buoliana*
India, China (including Taiwan), Thailand, Philippines	*T. confusum* (= *T. chilonis*)	*Agrotis* spp., *Chilo* spp., *Argyroploce schistaceana*, *Cnapholocrocis medinalis*, *Diatraea* spp., *Heliothis armigera*, *Leguminivora glycinivorella*, *Spodoptera exigum*
Colombia, Mexico, USA, Nicaragua, Uruguay	*T. pretiosum*	*Alabama argillacea*, *Anticarsia* spp., *Cadra cautella*, *Diatraea* spp., *Heliothis* spp., *Trichoplusia ni*, *Kaiferio lycopersicolla*, *Plodia interpunctella*, *Scrobipalpula absoluta*
Bulgaria, China, Germany, former Czechoslovakia, Egypt, Philippines, The Netherlands, former USSR	*T. evanescens*	*Agrotis* spp., *Clysia ambiguella*, *Lobesis* spp., *Loxostege sticticalis*, *Mamestra brassicae*, *Pieris* spp., *Ostrinia nubilalis*
Bulgaria, Poland, former USSR	*T. cacoeciae*	*Clysia ambiguella*, *Lobesia* spp., *Cydia pomonella*, *Laspeyresia* spp., *Orgyia antiqua*
Austria, France, Iran, Switzerland, The Netherlands, Romania, Italy	*T. maidis* (= *T. brassicae*)	*Mamestra brassicae*, *Pieris* spp., *Ostrinia nubilalis*

Former USSR	*T. pintoi*	*Heliothis armigera, Ostrinia nubilalis*
Philippines, Thailand	*T. chilotraea*	*Chilo* spp., *Diatraea* spp.
Philippines	*T. nana*	*Chilo* spp., *Diatraea* spp.
China, India	*T. japonicum*	*Chilo suppressalis, Cnaphalocrocis medinalis*
India	*T. brasiliensis*	*Heliothis armigera*
Australia	*T. carverae*	*Heliothis armigera*
Bulgaria, former USSR	*T. embryophagum*	*Cydia pomonella*
Canada	*T. minutum*	*Choristoneura fumiferana*
USA	*T. nubilale*	*Ostrinia nubilalis*
Tunisia	*T. oleae*	*Prays oleae*
China (including Taiwan)	*T. ostriniae*	*Ostrinia furnacalis*
USA	*T. platneri*	*Amorbia cuneana*
Iran	*T. rhenana*	*Heliothis armigera*

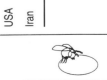

taxonomy of *Trichogramma* has developed rapidly in recent years, more and more entomologists understand the importance of identification and distribution of natural populations of *Trichogramma* which are well-adapted to the given climate conditions. Information regarding the pest one wishes to control and the crops on which the pest feeds also appears to be necessary. Usually, the utilization of an indigenous dominant species is preferred. However, sometimes, introduction of exotic species is also necessary, if available. In this case, there is a risk of eliminating, by competition, the original species. For example, it is known that *T. dendrolimi* eliminates *T. evanescens*, *T. evanescens* eliminates *T. buesi*, and *T. euproctidis* eliminates *T. evanescens* (Voegelé, 1988). Thus, research on the biology, ecology and behaviour of indigenous, as well as introduced, species of *Trichogramma* before their utilization, always takes precedence.

Methods of Utilization

There are two methods of commercial use of *Trichogramma*: 'inundative' release and 'inoculative' release (or seasonal colonization).

Inundative release

The field augmentation of naturally occurring introduced *Trichogramma* species is most commonly achieved via 'inundative' release of *Trichogramma*, wherein the crop is 'flooded' with insectary-reared parasitoids to raise the field parasitism rate sufficiently to prevent economic injury to the crop (Olkowski & Zhang, 1990). At the beginning of the season, when adult hosts start to oviposit, *Trichogramma* are released at a lower density than when host egg density is high or starts to decrease. The total number of released parasitoids should be large enough to control the population of target insect pests at the permissible economic level, which varies depending on the type and growth of crops, climate, species and density of pests, biocenose, etc. Generally, the release rate of *Trichogramma* is defined by practical experiments, long-term field experiences and theoretical calculations. Table 2.4 shows the number of *Trichogramma* released per hectare for some important target insect pests in China (Li, 1984): for corn borer, 45,000–345,000; for rice leafroller, 60,000–750,000; for cotton bollworm, 215,000–645,000. In the former USSR when the target insect pest, such as the cabbage army worm, laid egg clusters the number of parasitoids released was defined as the density ratio between parasitoid and host egg = 1 : 20. But, when the eggs of the target insect pest were laid separately, such as the eggs of cutworm, this ratio was 1 : 10 (Schepetilnikova, 1979). In the former USSR, the release number of *Trichogramma* per hectare for different crops was said to be as follows: for beet,

Table 2.4. Effectiveness of utilizing *Trichogramma* spp. to control insect pests in China (1982–1992).

Target pest	*Trichogramma* spp.	Released per hectare per generation (×1000)	Parasitism (%)	Reduction of pest population (%)	Decrease of damage (%)
Adoxophyes spp., *Lampronedata cristata, Fentonia osypete, Phalera assimilis*	*T. dendrolimi*	375–600	80.8–98.7		47.0–72.0
Agrotis ypsilon	*T. dendrolimi*	150–225	77.4–98.9	89.2–94.8	73.5–98.6
Argyroploce schistaceana, Chilo infuscatellas, Chilo sacchariphagous	*T. confusum*	45–150	60.5–93.7	72.3–85.7	84.9–94.5
Barathra brassicae	*T. leucaniae* *T. evanescens*	395–450	80.4–88.8	77.8–83.3	
Chilo suppressalis	*T. japonicum*	450	89.1		80.0
Clanis bilineata	*T. closterae*	285	77.2		
Cnaphalocrocis medinalis	*T. japonicum*	60–750	70.0–95.0	72.8–89.5	63.0–80.7
Dendrolimus spp.	*T. dendrolimi*	1500–4500	51.8–96.2	68.0–72.0	
Heliothis armigera	*T. confusum*	215–645	60.0–90.0	80.0–89.4	63.5–78.6
Leguminivora glycinivorella	*T. confusum*	300	6.5–66.7		52.2–67.2
Ostrinia furnacalis, Spodoptera exigum	*T. dendrolimi* *T. ostriniae* *T. confusum*	40–300 345 225	71.0–90.3 94.1 75.0–85.4	50.0–92.0 88.1	56.1–64.0 85.3 61.3–77.3

15,000–20,000; for wheat, 25,000–40,000; for cabbages, 60,000–100,000; for corn, 25,000–40,000; and for cotton, 80,000–100,000 (Varenik & Padzievskii, 1976).

Some models of optimizing the release number of *Trichogramma* have been designed. Kanour & Burbutis (1984) presented a hypothetical model for controlling European corn borer:

Daily female *Trichogramma nubilale* release rate ha^{-1}

$$= \frac{(cm^2 \ plant^{-1}) \ (plants \ ha^{-1})}{2800} \times 0.052$$

In order to calculate the release rate of *T. nubilale* females per day, one needs to measure the average leaf surface area ($cm^2 \ plant^{-1}$) and determine the number of plants per hectare. Need & Burbutis (1979) defined 2800 cm^2 of the leaf surface as equivalent to 1 searching unit (SU) per plant for *T. nubilale*. There was a relatively constant density level of 5.2 females per 100 SU ha^{-1}. It was calculated that a release rate of ca. 2400 and 12,000 females $day^{-1} \ ha^{-1}$. would be needed to achieve 80% parasitism of first and second brood respectively.

Li *et al*. (1984) suggested a model for predicting the optimum number (X) of *Trichogramma japonicum* per mu (= 1/15 ha) that should be released in the paddy-fields of Guangdong Province to get a definite controlling effect (Y) depending on the density of the rice leafroller moth (X_1). After field verification, the model appears to be reliable:

At the beginning of rice tiller:

$$X = \frac{Y - 40.9499 - 0.362903 \ X_1}{1.72275}$$

At the end of rice tiller:

$$X = \frac{Y - 30.5207 + (2.448e^{-2}) \ X_1}{4.9985}$$

Inoculative release

Inundative release of naturally occurring *Trichogramma* requires the avail-ability of a large quantity of parasitoids at precise times in the life cycle of the target insect pest. The close coordination between rearing facilities and field staff, as well as timing of releases, are critical to success. Sometimes the outbreak of a pest population cannot be suppressed by inundative augmenta-tion of a large number of *Trichogramma*, even though the parasitism rate would be very high. Inoculative release, on the other hand, requires relatively fewer parasitoids introduced early in the season to preclude the pest populations from reaching the injury level. Wang *et al.* (1988) introduced an alternative control tactic against major cotton pests to the conventional 'inundative' release, using inoculative release of *T. dendrolimi*, based on their field experiments from 1977 to 1984. It was found that the populations of naturally occurring cutworm (*Agrotis ypsilon*) and other vegetable pests supported the build-up of the *T. dendrolimi* population which emigrated into cotton fields and in turn exerted effective control of the cotton pests in adjacent fields. In practice, the ratio between the area of *Trichogramma*-released vegetable gardens and the adjacent cotton fields in Shanxi Province of China was 1 : 14 to 1 : 50. During late March to mid-April, *Trichogramma* were released 2–3 times at a rate of 75,000 ha^{-1} for each release. In late summer the population of *Trichogramma* became 46–70-fold of that released in the beginning of spring, with reference to the normal rate of inundative release (375,000–495,000 ha^{-1}) practised by the local farmers to control cotton pests. The number of *Trichogramma* multiplying from those released in 1 ha of a vegetable garden in spring equalled the number released in 24 ha of cotton in late summer, so there were enough *Trichogramma* to control pests in the adjacent cotton fields, in comparison with inundative releases. There are three distinctive advantages of this method of inoculative release: (i) the number of released parasitoids necessary for effective control is substantially reduced thus reducing greatly the cost of biological control; (ii) the pest populations of both vegetables and cotton are effectively suppressed with the same release; (iii) parasitoids, naturally reared in the vegetable garden are of better quality than those reared in laboratories, making them more effective.

Inoculative releases of *T. dendrolimi* for controlling corn borer and rice leafroller were also carried out on 50,000 ha in Henan Province of China (Shen *et al.*, 1988). For the corn borer, the release rate was 7000–150,000 ha^{-1} at the end of May.

There is high farmer acceptance of the inoculative method of pest control in China where production of *Trichogramma* naturally in the fields fits easily into the multi-crop farming methods that have been practised for hundreds of years.

Methods of Releasing *Trichogramma*

Releases of *Trichogramma* are conducted either manually or mechanically (with tractor-sprayer or airplane) (see Chapter 6). Manual release is very common in most countries where *Trichogramma* is mass-used as a biological control agent. Parasitoid pupae (just before emergence or at the beginning of emergence) are released with the container (made of bamboo, cardboard, paper or gelatine capsule) set on plants. It can also be done by hanging parasitized egg-cards on plants or rolling them in the leaves of corn or sugar-

cane. Protection from predators and from unfavourable weather is important. Manual release requires much labour, which is more available in most of the developing countries.

Mechanization of releasing *Trichogramma* was studied and intensively implemented in the former USSR, involving release by machines or airplane (Shepetilnikova, 1979; Sochta & Hakimov, 1984; Barabash & Pasiko, 1990; Kiku *et al.*, 1991; Krakovetzkii *et al.*, 1991; Pushkarev, 1991). In 1985, an area of 15,000 ha in the former USSR was treated with *Trichogramma*, released by tractor-sprayer, which treated 140–150 ha in 8 h. An area of 700,000 ha was treated with *Trichogramma* by airplane which covered 250 ha h^{-1} (Filippov, 1986, 1989). According to data of 1977–1984, aerial application of *Trichogramma* in the former USSR was said to increase the productivity by 100–150 times than the manual release (Barabash & Pasiko, 1990). After the aerial application of *T. evanescens*, the average egg parasitism on corn and beet was 58–72% and 60–67% for cotton (Barabash & Pasiko, 1990). Later, it was reported that improved equipment 'WART-1', positioned at the sprayer on helicopter 'KA-26' was constructed and used for aerial delivery of pure *Trichogramma* in *Sitotroga* eggs (Barabash & Pasiko, 1990). The release rate was 0.4–0.5 g (of parasitized *Sitotroga* eggs) ha^{-1}. With such a device, 180 ha was said to be treated per hour. *Trichogramma* was released at 5 m above the land and the flying speed was 100 km h^{-1}.

In the central Asian republics of the former USSR, *Trichogramma* was released in a water–air stream by tractor-sprayer (30 g of parasitoid pupae in *Sitotroga* eggs with 45 l water) to prevent dryness. The survival of the parasitoid pupae in water (78.2%) was similar to that of controls (81.5%) after 3 h. The release rate was 1.5–2.5 g of parasitized *Sitotroga* eggs ha^{-1} (while it was 1.7 g ha^{-1} for the manual release). Parasitism was 48.2–54.9%, while it was 58.3% in the area treated manually. The mechanical release increased the productivity tenfold in comparison with manual releases (Sochta & Hakimov, 1984). Mostly, *Trichogramma* is delivered in the uniform stage of late pupae. For this purpose, in some countries, such as America, the parasitized *Sitotroga* eggs on the 8th day of development at 27°C were placed at 16.7°C for 6 days, then they were transferred to 27°C again for another 4–5 days before being released. The reduced temperature induced a high level (73%) of adult emergence within 4 h after field releases (Morrison *et al.*, 1978).

Evaluation of Effectiveness of Using *Trichogramma*

Besides direct control, *Trichogramma* releases may also have a preventative effect. It provides benefits not only to the current period of utilization, but also to successive years of crop cultivation, even if the direct control effect is sometimes not so high as chemical treatment. It is difficult to analyse all the benefits which result after the utilization of *Trichogramma*. However, there are still some criteria and methods popularly used for evaluating the success of releasing *Trichogramma* in many countries. They are: increase of egg parasitism of target insect pests, decrease of pest density and damage, increasing or decreasing trends of pest populations, decrease of production costs, increase of the density of natural enemies, improvement of crop quality and yield, reduction of toxic chemical residues, etc. Table 2.4 shows the effectiveness of releasing *Trichogramma* spp. to control insect pests in China. At Jiangang Farm in Jiangsu Province, *Trichogramma* spp. have been used for controlling cotton bollworm for 10 years on an area of 3546 ha every year from 1975 to 1984 (Zhou, Li-tzu, 1988). It is indicated that, even if the egg-parasitism of cotton bollworm was expressed at a moderate level (28.2–68%) within ten years, the pest density decreased significantly in comparison with areas where chemicals were used. Augmentative releases of *Trichogramma* prevented the use of wide-spectrum chemicals, and thus protected natural enemies, which suppressed the pest continuously. According to the practices of three counties in Guangdong Province, China, where *Trichogramma* spp. have been mass-released on an area of about 11,000 ha for more than 20 years to control sugar-cane borers, there have been more evident economic and ecological effects than for the chemical control.

1. The parasitism of borers' eggs was raised by 1–3 times and the average parasitism reached 80%.
2. The percentage of dead heart caused by borers dropped to a maintained level of about 2%, a decrease of 70–80% relative to the control, so that the growth of principle effective canes was protected.
3. The percentage of damaged sugar-cane joints decreased by 50–80% relative to the control and remained consistently below the allowed level (10%).
4. The percentage of dead shoots was between 5–10%, similar to the level obtained when chemicals were used.
5. 2.6–7.5 tonnes of sugar-cane and 24.7–43.5 kg of sugar were added per hectare.
6. The farmers actually gained an US$200 ha^{-1} extra because of reduction of chemical application, labour and additional gain in salvage from the retrieved production.
7. The sugar mills have made approximately US$250,000 extra each year because of the increased production of sugar and sugar-cane stalks.
8. The Biological Control Stations made a total income of US$33,000 every year.

As to the multi-level economic benefits of the above aspects, the economic profits which were made by the three counties during 1981 to 1985 reached US$9,030,000 with the application of *Trichogramma* to control sugar-cane borers, for an investment of only US$340,000 (i.e. an increase of income over

expenditure of 25 times). The application of 112 kg ha^{-1} of chemicals was eliminated by releasing *Trichogramma*. The population of natural enemies in the released area increased by 2–5 times as compared to the chemically controlled area, so pest density and their damage decreased year by year. It was determined by residual toxicity analysis in 1983 that the organic chloride content in sugar-cane stalks and in the soil in the released area were respectively 34.84–63.41% and 51.16–61.74% lower than in the chemically controlled area (Zhou, Shen-zheng, 1988).

In Switzerland, Bigler summarized the efficiency of *T. maidis* (= *T. brassicae*) to control *Ostrinia nubilalis* from 1978 to 1985 (Bigler, 1986). The parasitization rate varied from 75% to 93% in these years. The reduction of attacked plants was 70.2–82.1%, except for the low effectiveness in 1980 because of the deteriorated quality of *Trichogramma*. Biological control of corn borers by inundative releases of *T. maidis* has been used in Switzerland for 15 years and has become an important tool in integrated control against this pest.

Ground releases of 12 million *T. minutum* per hectare in 12- to 20-year-old white spruce, *Picea alba*, in Canada resulted in 87% parasitism of spruce budworm, *Choristoneura funiferana*, egg masses, while the natural parasitism level was less than 4%. As a consequence, the overwintering second instar budworm population reduced by 80% (Smith *et al.*, 1987; Olkowski & Zhang, 1990).

In the former USSR, *Trichogramma* was applied seasonally against *Mamestra brassicae*, *Heliothis armigera*, *Scotia segetum*, *Ostrinia nubilalis*, *Lespeyresia pomonella* and tortricids on vegetables, cereals, cotton, maize and other crops in orchards and vineyards. Scientifically based and timely application of high-quality parasitoid material in regions favourable for its activity yielded 60–80% parasitizations of noctuid pests eggs, 40–60% of eggs of corn borer (Filippov, 1989) and 24–75% of beet leafminer eggs (Pastuch, 1991). According to the data, evaluated by the Atakskaya Regional Biological Laboratory of Moldavia from 1970–1984, utilization of *Trichogramma* to control *Mamestra brassicae* was said to give biological effectiveness of 76.8–92.0%, and increased the yield by 2400 kg ha^{-1}. The income from each rouble spent on releasing parasitoids was 8 roubles. At the same time, in controlling corn borer, the biological effectiveness of utilizing *Trichogramma* was 84.4% with an increase of yield of 200–300 kg ha^{-1}. The ratio of cost to income was 1 : 3 (Zilberg, 1985). In regions favourable for *Trichogramma* activities, average yield increase of wheat was 170–200 kg ha^{-1}; sugarbeet, 200–350 kg ha^{-1}; and cabbages, 2000–3000 kg ha^{-1} (Tzibulskaya *et al.*, 1988).

Evaluation of the effectiveness of natural enemies can be conducted theoretically by means of a life table and the Morris-Watt mathematical analysis, based on the data of several continuous years. In the integrated rice pest controlled area of Guangdong Province, China (about 120,000 ha), where insecticide use was limited to a very low level after the extensive use of *Trichogramma* in controlling rice leafroller, about 97.9% of the rice leafrollers, 78.9% of the plant hoppers, 46.0% of the yellow stem borers and 85.0–99.0% of the rice gallmidges were destroyed by a variety of species of conserved natural enemies. Life table experiments indicated that, eliminating the function of parasitoids, the index of population trend of the next generation of rice leafrollers would be increased by 5.09 times. Theoretically, in the absence of all parasites, predators and diseases, the index of population trend would have increased to 258.69 times against the original (Pang *et al.*, 1984). After the implementation of integrated control with *Trichogramma*, the expenditure of insecticides decreased by 50% and the co-relation ratio between pests and predators was maintained below 1 during the rice growth season (Li, 1982).

References

Abraham, C. (1976) Studies on developing races of *Trichogramma australicum* Girault suitable for high temperature, low humidity conditions. *Madras Agriculture Journal* 63, 550–556.

Alba, M.C. (1990) Use of natural enemies to control sugarcane pests in the Philippines. In: *The Use of Parasites and Predators to Control Agricultural Pests. Proceedings of the International Seminar 1989, October 2–7, Tsukuba.* FFTC Book Series No. 40. Manila, Philippines, pp. 124–133.

Bao, Jian-zhong & Chen, Xiu-hao (1989) *Research and Application of* Trichogramma *in China.* Academia Books and Periodicals Science Press, Beijing, 220 pp.

Barabash, A.V. & Pasiko, A.K. (1990) For aerial release of *Trichogramma. Zaschita Rastenii* 3, 27.

Beglyarov, G.A. & Smitnik, A.I. (1977) Seasonal colonization of entomophages in the USSR. In: Ridway, R.L. & Vinson, S.B. (eds) *Biological Control by Augmentation of Natural Enemies.* Plenum Press, New York, pp. 283–329.

Bigler, F. (1986) Mass production of *Trichogramma maidis* Pint. & Voeg. and its field application against *Ostrinia nubilalis* Hbn. in Switzerland. *Sonswesruck aus Bd* 1, 23–29.

Burbutis, P.P. & Goldstein, L.F. (1983) Mass rearing *Trichogramma nubilalis* on European corn borer, its natural host. *Protection Ecology* 5, 269–275.

Cheng, Wen-yi (1991) Research and development of sugarcane insects in Taiwan. *Chinese Journal of Entomology* 7, 167–181.

Filippov, N.A. (1986) For developing biomethod. *Zaschita Rastenii* 1, 18–21.

Filippov, N.A. (1989) The present status and future outlook of biological control in the USSR. *Acta Entomologia Fennica* 53, 11–18.

Filippov, N.A. (1990) Problem of biomethod. *Zaschita Rastenii* 10, 3–5.

Flanders, S.E. (1929) The mass production of *Trichogramma minutum* Riley and observations on the natural and artificial parasitism of the codling moth egg. *Proceedings of the 4th International Congress of Entomology* 2, 110–130.

Gusev, G.V. (1978) Biological control in Mexico. *Zaschita Rastenii* 6, 60–61.

Hassan, S.A. (ed.) (1988) Trichogramma *News,* Vol. 4. Federal Biological Research

Li-Ying Li

Centre for Agriculture and Forestry, Braunschweig, pp. 12–14.

Hassan, S.A. (ed.) (1990) Trichogramma *News*, Vol. 5. Federal Biological Research Centre for Agriculture and Forestry, Braunschweig, pp. 12, 13, 28.

Hassan, S.A. & Guo, Ming-fang (1991) Selection of effective strains of egg parasites of the genus *Trichogramma* (Hymenoptera: Trichogrammatidae) to control the European corn borer, *Ostrinia nubilalis* Hbn. (Lepidoptera: Pyralidae). *Journal of Applied Entomology* 111, 335–341.

Hassan, S.A., Kohler, E. & Rost, W.M. (1988) Mass production and utilization of *Trichogramma*: 10. Control of the codling moth, *Cydia pomonella* and the summer fruit tortrix moth, *Adoxophyes orana* (Lepidoptera: Tortricidae). *Entomophaga* 33, 413–420.

Kanour, W.G.W. Jr & Burbutis, P.P. (1984) *Trichogramma nubilale* field release in corn and a hypothetical model for control of European corn borer. *Journal of Economic Entomology* 77, 103–107.

Kiku, B.B., Malitzkii, E.A., Jigen, A.I. & Abashkin, A.S. (1991) Manufacture of apparatus for distribution of *Trichogramma* on vineyard. In: *Proceedings of the 3rd All-Union Symposium on* Trichogramma, *1991, March 11-16, Kishinev.* Moscow, p. 113.

Krakovetzkii, N.N., Alenchkova, T.F. & Biriuchenskii, P.S. (1991) Releasing *Trichogramma* with apparatus PRA-35. In: *Proceedings of the 3rd All-Union Symposium on* Trichogramma, *1991, March 11–16, Kishinev.* Moscow, p. 118.

Li, Li-Ying (1982) Integrated rice insect pest control in the Guangdong Province of China. *Entomophaga* 27, 81–88.

Li, Li-Ying (1984) Research and utilization of *Trichogramma* in China. In: *Proceedings of the Chinese Academy of Sciences/United States National Academy of Sciences Joint Symposium on Biological Control of Insects, September 25–28, 1982, Beijing.* Science Press, Beijing, pp. 204–223.

Li, Li-Ying (1992) Recent status of biological control of insect pests in China. In: Hirose, Y. (ed.) *Biological Control in South and East Asia.* Kyushu University Press, Kyushu, pp. 1–10.

Li, Li-Ying, Liu, Wen-hui, Li, Kai-huang & Guo, Ming-fang (1984) The behaviour, ecology, resistance to pesticides and *in vitro* rearing of *Trichogramma japonicum* Ashmead. In: *Proceedings of the 17th International Congress of Entomology, 1984, August 20–26, Hamburg.* Science Press, Beijing, p. 789.

Morrison, R.K., Jones, S.L. & Lopez, J.D. (1978) A unified system for the production and preparation of *Trichogramma pretiosum* for field release. *Southwestern Entomologist* 3, 62–68.

Need, J.T. & Burbutis, P.P. (1979) Searching efficiency of *Trichogramma nubilale*. *Environmental Entomology* 8, 224–247.

Olkowski, W. & Zhang, Anghe (1990) *Trichogramma* – A modern day frontier in biological control. *IPM Practitioner* 12, 1–28.

Pak, G.A. & Oatman, E.R. (1982) Comparative life table, behaviour and competition studies of *Trichogramma brevicapilium* and *T. pretiosum*. *Entomologia Experimentalis et Applicata* 32, 68–79.

Pang, Xiong-fei, Liang, Guan-wen & Zeng, Ling (1984) Evaluation of the effectiveness of natural enemies. *Acta Entomologica Sinica* 4, 1–11.

Pastuch, S.K. (1991) Role of *Trichogramma semblidis* in controlling pest population in Ukraine. In: *Proceedings of the 3rd All-Union Symposium on* Trichogramma, *1991,*

March 11–16, Kishinev. Moscow, pp. 135–136.

Pushkarev, B.V. (1991) On possibility of using ultra-light aerial apparatus for plant protection. In: *Proceedings of the 3rd All-Union Symposium on* Trichogramma, *1991, March 11–16, Kishinev.* Moscow, p. 137.

Schepetilnikova, V.A. (1979) *Trichogramma* in USSR. *Zaschita Rastenii* 1, 26–27.

Shen, Xiao-cheng, Wang, Ke-zen & Meng, Guang (1988) The inoculative release of *Trichogramma dendrolimi* for controlling corn borer and rice leafroller. In: Voegelé, J., Waage, J. & van Lenteren, J.C. (eds) Trichogramma *and Other Egg Parasites. 2nd International Symposium. Les Colloques de l'INRA* 43, 575–583.

Smith, S.M., Hubbes, M. & Carrow, J.R. (1987) Ground releases of *Trichogramma minutum* Riley (Hymenoptera: Trichogrammatidae) against the spruce bud worm, *Choristoneura fumiferana* (Lepidoptera: Tortricidae). *Canadian Entomologist* 119, 251–263.

Sochta, A.A. & Hakimov, A.H. (1984) Mechanization of releasing *Trichogramma* on corn. *Zaschita Rastenii* 4, 34.

Tseng, Ching-tien (1991) Research and development on the control methods for upland crop in insect pests. *Chinese Journal of Entomology* 7, 183–202.

Tzibulskaya, G.N., Gegtyarev, B.G., Yanishevskaya, L.V., Konverskaya, V.P. & Palii, M.V. (1988) Strategy and tactics of utilizing *Trichogramma* in Ukraine. In: Grinberg, A.M. (ed.) Trichogramma *in Plant Protection.* Agropromizdat Press, Moscow, pp. 120–136.

Varenik, I.A. & Padzievskii, L.L. (1976) What is given by biological plant protection? *Zaschita Rastenii* 8, 18–19.

Voegelé, J. (1988) Reflections upon the last ten years of research concerning *Trichogramma.* In: Voegelé, J., Waage, J. & van Lenteren, J.C. (eds) Trichogramma *and Other Egg Parasites. 2nd International Symposium. Les Colloques de l'INRA* 43, 17–29.

Wang, Fuchen, Zhang, Shuyi & Hou, Shourang (1988) Inoculative release of *Trichogramma dendrolimi* in vegetable gardens to regulate populations of cotton pests. In: Voegelé, J., Waage, J. & van Lenteren, J.C. (eds) Trichogramma *and Other Egg Parasites. 2nd International Symposium. Les Colloques de l'INRA* 43, 613–619.

Zhou, Li-tzu (1988) Study on parasitizing efficiency of *Trichogramma confusum* Viggiani in controlling *Heliothis armigera* and its modelling. In: Voegelé, J., Waage, J. & van Lenteren, J.C. (eds) Trichogramma *and Other Egg Parasites. 2nd International Symposium. Les Colloques de l'INRA* 43, 641–644.

Zhou, Shen-zheng (1988) Advance in extension of *Trichogramma* utilization in Guangdong Province of China. In: Voegelé, J., Waage, J. & van Lenteren, J.C. (eds) Trichogramma *and Other Egg Parasites. 2nd International Symposium. Les Colloques de l'INRA* 43, 633–639.

Zilberg, L.P. (1985) Experience of Arakskaya Biolaboratory in production and utilization of *Trichogramma.* In: *Proceedings of the 2nd All-Union Symposium on* Trichogramma, *1985, Oct. 14–18, Kishinev.* Ataki, Moldavia, pp. 71–72.

Strategies to Select Trichogramma Species for Use in Biological Control

3

SHERIF A. HASSAN
*Institute for Biological Pest Control, BBA, Heinrichstr.
243, D-6100 Darmstadt, Germany*

Abstract

The *Trichogramma* genus comprises more than 100 species. As evidence is accumulating that considerable variation between species exists, especially with respect to host preference, searching capacity and tolerance to weather conditions, more attention is being given to the selection of suitable candidates for use in biological control.

Approaches used to select strains for use in biological control are discussed and standard procedures to compare candidate *Trichogramma* strains are suggested. These procedures are based on laboratory (host preference and suitability), semi-field (searching capacity) and field (efficacy) experiments. Results of experiments to select effective parasite species to control several lepidopterous pests are given.

The use of internationally approved standard methods to select effective species would make the comparison of results from different laboratories in different parts of the world possible. A catalogue of the attributes of different parasite species should be produced.

Introduction

Species of the hymenopterous genus *Trichogramma* occur worldwide on a broad range of crops and hosts. Research on the use of this egg parasite in biological control started at the turn of the century and a method of mass-producing *Trichogramma* on eggs of the Angoumois grain moth *Sitotroga cerealella* was developed in 1930 in the USA. Later, similar mass-production methods using the eggs of this species or other lepidopterous stored product

pests gave rise to a worldwide use of this parasite as a biological control agent (Flanders, 1930; Lebedev, 1970; Dysart, 1973; Voegelé *et al.*, 1975; Morrison *et al.*, 1976; Hassan *et al.*, 1978; Hassan, 1981; see Chapter 2 for a recent survey).

The genus *Trichogramma* includes more than 100 species (Voegelé *et al.*, 1988; see also Chapter 1) which predominantly attack eggs of Lepidoptera. *Trichogramma* spp. have traditionally been considered as polyphagous (Mokrzecki & Bragina, 1916; Hase, 1925; Salt, 1935; Sweetman, 1958). However, evidence is accumulating that considerable inter- and intraspecific variation in host preference exists (Kot & Plewka, 1974; Stschepetilnikova, 1974; Dijken *et al.*, 1986; Pak, 1988; Hassan, 1989). *Trichogramma* wasps are used more than any other entomophagous species for biological control of insect pests (Stinner, 1977; King *et al.*, 1985). Control with *Trichogramma* is mostly attempted through mass releases against at least 28 different phytophagous pest species on some 20 different crops (Hassan, 1988; Voegelé *et al.*, 1988). However, because of considerable variability in success of releases and little evidence of consistently successful application of *Trichogramma*, the usefulness of these parasitoids is currently being debated.

Trichogramma species vary greatly in their searching behaviour, host preference and response to environmental conditions. Consequently, they also vary in their suitability for use in biological control. Failures of egg parasites to control agricultural pests could be due to the use of less suitable *Trichogramma* strains (Hassan, 1989). Several authors have shown variations between strains of the same *Trichogramma* species and stressed the importance of choosing an appropriate strain for use in biological control. Others have indicated particular attributes that would increase the biological abilities of the parasite and its effectiveness in the field (Lenteren, 1986; Pak, 1988; Bigler, 1989; Pak *et al.*, 1991; Pavlik, 1993; see Chapters 5 and 12). In recent years there has been an increasing awareness of this problem among *Trichogramma* research workers.

When a suitable host egg is encountered, the female *Trichogramma* examines the egg by antennal drumming, drills into it with her ovipositor and lays one or more eggs within the host egg, depending on its size. When a *Trichogramma* female finds 'preferred' host eggs, it will usually stay on or near them for a longer period of time until all or most of them are parasitized. Less preferable host eggs may be totally rejected or the parasite might lay a few eggs before leaving the location to search for more suitable hosts. Diversity in host preference of *Trichogramma* species has now been fully recognized as an important factor with regard to biological control.

For the parasite to be successful in the field it must locate food, host eggs and shelter. The presence and distribution of food differ from crop to crop and from year to year. Adult parasites that do not rapidly find food and moisture have a much shorter life span and less chance of locating hosts. If food is scarce and adults spend more time and energy finding it, the choice of parasites with high searching capacity and energy reserve is crucial. The abundance and distribution of the host is equally important. If an adult has to fly from one tree to another in search for a host, it would need high flying and walking abilities to find a host before exhaustion. These attributes are important both for the survival of the parasite and for its usefulness as a biological control agent.

Judging by the numerous attempts at biological control with *Trichogramma* over the past 80 years, it is remarkable how little basic research has been conducted on these insects. The entomophages' searching behaviour in the field following its release is a key process in the successful parasitization of hosts. Unfortunately, this is an area on which there is little information available (Noldus, 1989). Salt (1940, 1958), one of the first to realize the importance of the study of behaviour of parasitoids for their utilization as biological control agents, made a detailed study of behavioural and physiological aspects of parasitism by *Trichogramma*. His work has been continued in recent years by, among others, Klomp *et al.* (1980), Pak (1988) and Schmidt & Smith (1989) (see Chapter 9). Their studies have focused on the processes that occur after a wasp has contacted a host, namely host acceptance and host suitability. Searching behaviour preceding host contact has received less detailed attention. *Trichogramma* exhibits a strong preference for Lepidoptera species, including a large number of important agricultural pests. Adult *Trichogramma* require both water and food (honey) for maximum fecundity. They spend most of their time seeking this food, host eggs, or looking for a mate. The searching activity of the adult parasite is generally believed to be stimulated by chemical substances (i.e. kairomones) produced and left by its hosts.

A common practice among the users of *Trichogramma* in many parts of the world is to release a strain that was collected from one pest to control another. Evaluation in commercial releasing areas is seldom undertaken. For practical reasons, attempts to use an available, but less effective species, to control a particular pest is often knowingly undertaken. The purity of the parasite in mass-rearing is also important to ensure effectiveness. Ways to select appropriate strains and secure methods of rearing them are needed to ensure success of the augmentative approach.

Potential of Egg Parasites as Agents for Biological Control

A survey of the natural abundance of egg parasites and their potential for use in biological control was reported by Hassan (1992). Results from 20 different countries showed that 11 genera of egg parasites, including 54 species, occurred regularly in fields on 34 crops involving 69 pests. The advantage of using egg parasites over larval parasites in biological control is that the former prevent hatching. Moreover, egg parasites can be mass-reared more easily in

large numbers. Success or failure in the use of egg parasites in biological control will depend, among other factors, on the choice of species.

The survey showed the wide distribution of egg parasite species throughout the world, especially those of the genus *Trichogramma* and indicated their importance as natural enemies of pests. This natural potential and the possibility of using it in biological control should not be left unexplored. The attributes of the different egg parasite species that occur on the large number of crops in the diverse types of agroecosystems throughout the world should be assessed. Due to variation in climatic conditions and other factors, the adaptation process of the parasite is likely to differ between locations.

Strategy for Selection of Strains

It is generally known that most *Trichogramma* species exhibit a strong preference for certain hosts, crops and climatic conditions. Before field releases are undertaken, a suitable *Trichogramma* strain of known qualities should be chosen. The effectiveness of *Trichogramma* in the field largely depends on its searching behaviour (habitat location, host location), host preference (recognition, acceptance, suitability) and tolerance to environmental conditions. Different methods were tried by several authors to test some of these factors (Quednau, 1955, 1956; Kochetova, 1969; Schieferdecker, 1969; Gonzalez *et al.*, 1970; Need & Burbutis, 1979; Lenteren *et al.*, 1982; Dijken *et al.*, 1986; Wäckers *et al.*, 1987). While searching behaviour should be tested in the presence of plants, in semi-field or field experiments, host preference and host suitability can be successfully tested in the laboratory (Hassan, 1989).

Until recent years, little attention was given to the selection of *Trichogramma* species for use in biological control. Consequently, a strategy for selection was not often discussed. In practice, *Trichogramma* is often collected in the field, mass-reared and released to control pests on the same crop where the parasite was found but often also on other crops. Biological control projects are often poorly funded and do not allow pre-introductory evaluation. Little attention is given to the study of the literature in relation to strain selection. Sometimes parasites for mass-rearing are obtained from a research laboratory or from a commercial insectary with little information as to origin, host insect and crop range. Until recently, pre-introductory evaluation or screening of relevant strains was seldom carried out.

It is a valid practice to choose an indigenous *Trichogramma* strain to mass-produce and release on the same crop. A local strain is preferable to an imported one if efficacy is adequate. Although local strains are likely to be more adapted to the environment, if the efficacy of the stain is not satisfactory, efforts to explore for better parasites should be made.

Use of Imported Strains

Strains from different agroecosystems may differ in their potential. Therefore, a useful strategy would be to compare all relevant species according to literature studies. Indigenous and imported strains of these species from different localities may also be compared. The strains should first be screened in laboratory experiments, with more effective ones further tested in semi-field and field experiments. Furthermore, the efficacy of commercial parasite releases should be regularly assessed (see Chapter 5).

As mentioned above, a local *Trichogramma* strain is often chosen for use in biological control because the parasite is expected to be well adapted to the local environmental conditions. However, a local strain might not be the most effective one for the particular purpose. More effective parasites with better adaptation to the crop and/or the host may be found elsewhere in areas where the crop has been grown for a longer time. Evolution or adaptation of the parasite in different ecosystems are unlikely to be identical. In the course of generations, the genetic structure of certain populations and the host searching behaviour on a particular type of plant might have evolved to make the parasite more effective than in other similar agroecosystems. For biological control to compete effectively with chemical pesticides, increased efficacy is important. Efforts to acquire more effective strains is therefore justified.

An FAO Code of Conduct has recently been drafted to ensure the safe use of biological control agents and prevent the accidental introduction of unwanted organisms into importing countries. In particular, the code aims to harmonize cooperation between research scientists and authorization officers.

Simultaneous Release of More Than One Species

To increase efficacy, a combination of more than one *Trichogramma* species may be used to control one pest. The choice for such a combination should be based on the results of laboratory, semi-field and field experiments. One advantage of such practice would be that the two species may be complementary to each others in aspects such as time of emergence, searching behaviour, area of activity (upper or lower part of the plant), longevity and tolerance to extreme weather conditions.

Field experiments to optimize the use of *Trichogramma* to control the codling moth (*Cydia pomonella*) and the summer fruit tortrix moth (*Adoxophyes orana*) were conducted between 1988 and 1990 (Hassan *et al.*, 1988). In three field experiments, a combination of two egg parasites, *Trichogramma dendrolimi* and *T. embryophagum*, was found to increase effectiveness by about 10% compared with the release of the same total number of *T. dendrolimi*

alone. Three or four releases, each at 2.5 million parasites ha^{-1}, were conducted in each case. With the mixture of species, the reduction in codling moth damage was 52.9, 84.3 and 74.1% compared with 42.0, 78.0 and 66.7%, respectively, for the release of *T. dendrolimi* alone. The damage of the summer fruit tortrix moth in the same experiment was reduced by 39.2 and 85.4% when the mixture was used, compared with 23.5 and 70.7% when only *T. dendrolimi* was released. In further trials, a combination of *T. dendrolimi* and *T. cacoeciae* was used to control the codling moth. The effectiveness increased by 7.0 or 11.1% compared with each of the two strains alone. The increase with summer fruit tortrix moth in the same experiments was 9.1 or 14.1% in comparison to the other strains alone (Hassan & Rost, 1993). The commercial use of this combination of species has commenced in Germany in 1992.

Tolerance to Weather Conditions

Tolerance of prospective *Trichogramma* strains to extreme weather conditions in the relevant area is important. However, with an augmentative release strategy, only the weather conditions during the growing season are important. Because inundatory releases are carried out for only a few months during the growing season, differences in off-season climatic factors between the exploration area and the treatment location are usually of secondary importance.

Tolerance to extreme weather conditions during the growing season should be carefully considered and tested. Although most *Trichogramma* strains are susceptible to low humidity and drought, natural selection has produced stains of *Trichogramma* in many parts of the world tolerant to extreme high and low temperatures. To ensure survival in the release area, a test for resistance to maximum and minimum temperatures likely to be experienced in the relevant location should be conducted.

Lack of success in biological control programmes is doubtless often caused by high mortality of natural enemies due to climatic extremes. Although diapause has been studied extensively, surprisingly relatively little attention has been given to hot, dry or cold resistance of egg parasites. Chapter 11 summarizes aspects related to cold resistance, quiescence and diapause.

Off-season conditions are important when the parasite is to be established in a new ecosystem. The outstanding success of the use of *Trichogramma evanescens* to control the Asian corn borer *Ostrinia furnacalis* (Guenée) and the establishment of the species in the Island of Mindanaw, Philippines, could partly be attributed to the suitable weather conditions in the area for the parasite (high humidity and moderate temperature). The almost continuous cropping of corn all year round also helped the establishment of the species (Hassan, 1984; Tran & Hassan, 1986; Tran *et al.*, 1986).

Host Preference as an Attribute for Selection

Comparing the efficacy of strains by field testing is very time consuming and rational screening procedures are therefore needed. Pre-introductory evaluation should involve, among other factors, the criteria of host preference and host suitability. Laboratory and semi-field experiments are recommended, not as alternatives to field testing but as a first screening. Several research workers have shown that these characters can easily be tested in simple laboratory experiments (Quednau, 1955, 1956; Kochetova, 1969; Schieferdecker, 1969; Gonzalez *et al.*, 1970; Lenteren *et al.*, 1982). Two different methods used to assess the host preference of *Trichogramma* species will be discussed in more detail. In both methods, the behaviour of a single parasite female released on different types of host eggs is observed. In one method, observations are carried out at defined time intervals after parasite release with parasitism of the host eggs assessed at the end of the 5 day experiment (Hassan, 1989; Hassan & Guo, 1991; Wührer & Hassan, 1993). The other is based on the continuous and direct observation of the parasite and the recording of all its activities (Dijken *et al.*, 1986).

The contact and parasitism method

The host preference of *Trichogramma* was tested by offering a single parasite female the choice between eggs of one target pest and eggs of a standard factitious host (Hassan, 1989; Hassan & Guo, 1991; Wührer & Hassan, 1993). A single *Trichogramma* female (about 24 h old) is released in a glass tube (100 mm long and 26 mm in diameter) together with 80 eggs (two patches each of 40 eggs) of the target pest and 80 eggs (two patches each of 40 eggs) of the standard mass-rearing host *Sitotroga cerealella*. Host eggs were glued near the four corners of a piece of paper (2 × 2 cm) and a drop of honey/agar was added in the centre. Monitoring was carried out by: (i) checking all the tubes eight times during the first 7 h of the experiment and recording the location of the parasite (on target pest eggs, on *S. cerealella* eggs, or elsewhere); and (ii) assessing parasitism 5 days later.

Observing the location of the parasite in the first monitoring reflected the preference of *Trichogramma* to search for, contact and remain for some time on the 'preferred' eggs, compared to the standard rearing host eggs. Counting the number of *Trichogramma* developing in the host eggs (parasitism) in the second monitoring showed the preference of the parasite for laying eggs and indicated the ability of the parasite to develop in these eggs (i.e. host suitability). This method was based on the repeated field observation: When a parasite female finds 'preferred' host eggs, it will stay on or near them for a longer period of time and often totally parasitizes the eggs. In the test, this was reflected in higher numbers of contacts of the female with the 'preferred'

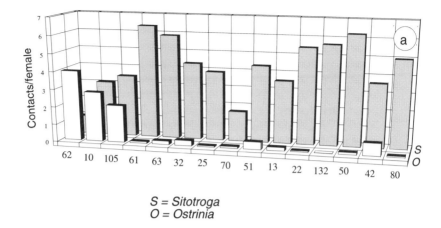

S = *Sitotroga*
O = *Ostrinia*

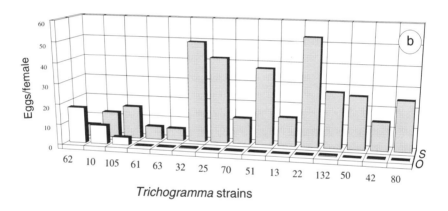

Trichogramma strains

Fig. 3.1. Results of the experiments done to compare the host preference of 15 *Trichogramma* strains between the natural host (*Ostrinia nubilalis*) and a factitious host (*Sitotroga cerealella*). (a) Average number of contacts per female for each host. (b) Average number of eggs laid per host for each host.

eggs and higher parasitism. *Trichogramma brassicae* (= *T. maidis*) females, from Moldavia, were frequently seen on *Ostrinia nubilalis* egg masses in releasing fields in Germany. No contacts were observed in plots where other *Trichogramma* species were released on this crop, the parasite left the eggs to search for a more 'preferred' host.

The results of the experiments to compare the host preference of 15 candidate *Trichogramma* strains to contact and parasitize *O. nubilalis* eggs are given in Fig. 3.1. Figure 3.1(a) shows the number of observed parasite contacts with host eggs, Fig. 3.1(b) the number of eggs laid per *Trichogramma* female (average of at least 30 replicates).

Results show that when the parasite was offered the choice between eggs of the natural host *O. nubilalis* and eggs of the mass-rearing host *S. cerealella*, *T. ostrinae* from China (strain no. 62) showed strong preference for the eggs of the European corn borer *O. nubilalis*. The data clearly indicate that *T. ostrinae* had the strongest preference for corn borer eggs compared with all the other *Trichogramma* strains tested. The *Trichogramma* adult females had many more contacts with and parasitized higher numbers of *O. nubilalis* eggs than *S. cerealella* eggs. It was unexpected to find that *T. ostrinae* from Taiwan (61) exhibited no preference for the pest eggs and had comparatively low egg-laying capacity. This strain has apparently lost its preference for the natural host, possibly due to the continuous laboratory rearing on alternative hosts. All the other *Trichogramma* strains tested preferred to contact and parasitize *S. cerealella* over *O. nubilalis*. Among the strains tested, *T. dendrolimi* (strain 20, not shown on Fig. 3.1) and *T. chilonis* (32) were the most fecund, parasitizing the highest number of *Sitotroga* eggs.

Zhang (1986) showed that *T. ostriniae* was the dominant species of parasitoid attacking corn borer eggs of all generations and is a key factor affecting populations of the Asian corn borer in the Beijing suburbs. It accounted for over 95% of the *Trichogramma* recorded, whereas *T. dendrolimi*, *T. chilonis*, *T. evanescens* and *T. closterae* together account for less than 5%. He showed, in laboratory and field experiments, that *T. ostriniae* was much more effective than *T. dendrolimi* which, according to Wang (1986), is being annually released in practice on an area of about 5,000,000 mu (about 340,000 ha) in five provinces of China.

The preference of *T. brassicae* from Moldavia (10) and *T. evanescens* from Germany (105) for corn borer eggs was significantly lower than *T. ostriniae* (62) but higher than all the other strains. The differences between these two strains was also significant. The level of significance among these three strains was 5% for contacts (Fig. 3.1) and at 1% for number of eggs laid per female (Fig. 3.1b). The results of this test agree with those of a field experiment that was carried out by Hassan (1985) near Gross-Gerau, Germany, to compare the effectiveness of different *Trichogramma* strains. Single field releases at 150,000 parasites ha^{-1} resulted in 86.1% egg parasitism with *T. brassicae* Moldavia (10) and 81.3% reduction in the number of *Ostrinia* larvae compared with 64.5% parasitism and 65.4% larval reduction for *T. evanescens* Germany (105). *T. brassicae* (10) in Moldavia and *T. ostrinia* (62) in China may have had better conditions to adapt to the corn borer than the strain *T. evanescens* (105) in Germany. In Germany, this species occurs mainly in cabbage fields, not in corn. These experiments showed that two imported *Trichogramma* strains were more effective than the local one. The fact that these two species are known in practice to be effective against the corn borer indicates the usefulness of the contact and parasitism method. *T. ostriniae* has been successfully used in Asia for a long period of time and *T. brassicae* in Europe since 1974.

Results confirm that the choice for *T. brassicae* to control the European corn borer in western Europe seems justified. However, the better performance of *T. ostrinia* in the host preference test suggests its possible value as an alternative. About 27,000 ha of corn are being treated annually with the Moldavia strain of *T. brassicae* or with another very similar species in France, Switzerland and Germany. In Germany, the reduction in the field number of *O. nubilalis* larvae obtained by releases, compared with untreated control plots, varied between 70 and 93%. Two treatments each at the rate of 75,000 ha^{-1} are being carried out starting at the beginning of the *Ostrinia* adult flight, as indicated by light traps (Hassan, 1985; Hassan *et al.*, 1978, 1986).

The contact and parasitism method was also used by Hassan (1989) to select suitable *Trichogramma* strains to control the codling moth *Cydia pomonella*, the two summer fruit tortrix moths *Adoxophyes orana* and *Pandemis heparana*. Seventeen strains were screened for their suitability and *Trichogramma dendrolimi* from China as well as a local strain of *Trichogramma embryophagum* were chosen for further testing in the field. The results of the inundative release experiments between 1984 and 1986 (Hassan, 1988) showed that *T. dendrolimi* reduced the damage of *C. pomonella* and *A. orana* by 72.85 and 61.35%, respectively. *T. embryophagum* reduced the damage of *C. pomonella* and *A. orana* by 50.80 and 50.07% respectively.

Wührer & Hassan (1993) used the same method to test 47 *Trichogramma* strains and two *Trichogrammatoidea* strains against the diamondback moth *Plutella xylostella*. All strains accepted eggs of the diamondback moth, but varied greatly in their preference for the pest. *Trichogrammatoidea bactrae*, *Trichogramma chilonis* and *T. pintoi* had a high egg laying capacity and showed a strong preference for diamondback moth eggs. The ratios of parasitism of *Plutella:Sitotroga* eggs were 34.4 : 6.1, 32.8 : 12.1 and 32.0 : 9.1 respectively. The ratios of contacts of *Plutella:Sitotroga* eggs were 2.6 : 0.2, 1.5 : 0.7 and 3.5 : 0.3 respectively. *T. ostriniae* had a slightly lower egg laying capacity than the other three species but had a strong preference for *P. xylostella* (parasitism, 19.6 : 7.3 and contacts, 2.8 : 0.2). *T. pretiosum* had a high egg laying capacity, but showed only preference in contacts (2.6 : 1.0) and not in parasitism (28.2 : 25.5). *T. cacoeciae* had a low egg laying capacity, but a clear preference for diamondback moth eggs (parasitism, 8.3 : 1.0 and contacts, 2.0 : 0.0).

With these experiments, the suitability of the method for the selection of species was shown once again. The species selected by the method were reported by other authors to be active against the diamondback moth. *Trichogramma chilonis* (Nguyen & Nguyen, 1982; Iga, 1985; Okada, 1989; Hirashima *et al.*, 1990), *Trichogramma confusum* and *Trichogrammatoidea bactrae* (Keinmeesuke *et al.*, 1990) were recorded as important natural enemies of *P. xylostella*.

The continuous observation method

The method developed by Dijken *et al.* (1986) involves the continuous observation of a single *Trichogramma* female released on host eggs of several pest species. Eight eggs of each species were arranged alternately in a rectangular grid. The distance between two eggs was 2 mm. This distance was based on *Trichogramma*'s limit of sight (Glas *et al.*, 1981). The eggs were offered to a single *Trichogramma* female which was observed continuously through a dissecting microscope during 90 min. Time started at first antennal contact with a host. Observation ceased when all eggs of one host species had been parasitized or when the female flew away. The following parameters were recorded: (i) various behavioural components and their duration, as they have been described by Glas *et al.* (1981); (ii) the eggs that were parasitized (identified by coordinates); (iii) the number of eggs laid in the host (through observation of oviposition movements); and (iv) the primary sex ratio (see Suzuki *et al.*, 1984). As a measure for host preference, the acceptance : contacted (a : c) ratio (i.e. the total number of accepted eggs of one host species, divided by the total number of contacts made with this host species before acceptance occurs) was also recorded.

Wäckers *et al.* (1987) compared the continuous observation method with the contact and parasitism method. The continuous observation test was found to be more time-consuming and may be conducted where time and necessary equipment are available. The contact and parasitism method was time-saving and easy to perform. The results of the two methods corresponded, but their degree of discrimination of strains were different.

Facilities and Requirements Needed for the Test

Certain facilities and requirements are needed to conduct laboratory experiments to select *Trichogramma* strains suitable for use in biological control:

1. A continuous mass-rearing of a suitable factitious host such as the Angoumois grain moth *Sitotroga cerealella* (Oliv.), the flour moth *Ephestia kuehniella* or the rice moth *Corcyra cephalonica*. The factitious host in the experimental laboratory should be free of *Trichogramma* at all times. To prevent contamination, the rearing of the factitious host should be located away from any *Trichogramma* rearing and should be operated early in the day before any *Trichogramma* rearing has been dealt with.
2. Rearing of the candidate parasite stains. Experience has shown that different *Trichogramma* strains could easily be reared fairly close to each other without the risk of being mixed. Rearings of a large number of stains in an environmental cabinet or in a small room were established in many laboratories in several countries without major problems.

About 50 species or strains were kept at the institute in Darmstadt, Germany, in small glass tubes for many years without contamination. Each strain was kept in several glass tubes (145 mm long and 26 mm in diameter) with a cloth cover. The tubes were confined in a plastic container with one side darkened by black paper to keep the photopositive parasites away from the cover of the tubes. The tubes were kept in environmental chambers at 18°C, 80% RH and light : dark (LD) 8 : 16. Adult *Trichogramma* emerging in the rearing tubes are supplied with host eggs. To prevent escape, the tube is knocked on a soft object just before opening it. The parasites fall to the bottom of the tube. After providing the adults with host eggs, the tubes are transferred into other climatic chambers kept at 26°C, 60% RH and LD 8 : 16. At this temperature, and without food, the adults die within 2 or 3 days, and the parasitized host eggs turn black during the fourth or fifth day. The egg-card with the parasitized host eggs are taken out, reduced to 20% and transferred to a new tube. The newly established rearing tubes with the *Trichogramma* pupae are kept in the cooler chamber until emergence. To ensure high level of parasitism, it is important to control all the tubes in this chamber and provide the parasites with host eggs shortly after the beginning of emergence.

Semi-field Test to Assess the Searching Capacity

The searching capacity of *Trichogramma* strains was tested by Hassan & Guo (1991) by releasing adult parasites in cages (90 × 60 × 70 cm) with cloth-covered walls and roofs. Each cage included four potted corn plants that were previously grown in a greenhouse. *Ostrinia nubilalis* eggs laid on small pieces of paper were placed on the plant leaves evenly distributed in the cage. The egg cards were collected and replaced three times at intervals of 2 to 3 days. The number of egg patches that included parasitized eggs as well as the total number of parasitized eggs per cage were recorded. The results of these experiments showed that the two species *T. ostrinia* (62) and *T. brassicae* (10) were equally effective against the European corn borer. The number of corn borer eggs parasitized by the two species was not significantly different in all the experiments conducted.

The searching capacity of six *Trichogramma* species on cabbage plants with naturally laid *Plutella xylostella* eggs was tested by Wührer & Hassan (1993). The release of *Trichogrammatoidea bactrae*, *Trichogramma ostriniae*, *T. chilonis*, *T. pretiosum*, *T. pintoi* and *T. cacoeciae* in a ratio of one female to 20 eggs, lead to a parasitism of 55.8, 40.3, 28.7, 16.7, 4.9 and 4.2% and a larval reduction of 66.7, 54.1, 42.6, 20.1, 14.2 and 20.8%, respectively. When releasing *T. ostriniae*, *Trichogrammatoidea bactrae* and *Trichogramma chilonis* in a ratio of 1 : 1 : 1, larval reduction was 89.3, 84.6 and 74.7%, respectively.

Both the corn and cabbage experiments confirmed the results of the laboratory trails. Species that were effective in the laboratory also had the highest efficacy in the semi-field test.

Field Test to Assess the Efficacy

The release of the parasite in the field should be done as close as possible to methods used in practice. These aspects include the releasing device, the number of parasites per unit, the distance between releasing points and the number of releases. In field crops, the plot should be square in shape and include at least 25 releasing points, monitoring should be restricted to the centre part. The distance between plots should be as large as possible (i.e. a minimum of about 30 m in a corn field). The size of the plot in fruit orchards will depend on the type of trees, about 25 larger trees in a square shaped plot will allow monitoring of the nine central ones.

Both the increase in egg parasitism and the reduction in larval infestation should be monitored. Voegelé *et al.* (1975) developed a method to assess parasitism of the European corn borer eggs in *Trichogramma* releasing experiments. This method was used by Hassan (1981). Eggs laid on 21 selected corn plants surrounding releasing points are examined at regular time intervals and the number of parasitized *O. nubilalis* eggs as well as emerged pest larvae are counted. The advantage of this method is that the eggs are observed at intervals directly in the field until they turn black or the host larvae hatch. With this, they are exposed to the parasite during all its development time in the field.

The reduction in larval infestation in corn fields was estimated by Hassan *et al.* (1978) by dissecting about 8×25 plants per plot. Half of these plants were taken at the releasing point, the other half between releasing points. In fruit orchards, the number of damaged fruits related to the total number of fruit per certain number of trees is counted.

The Need for International Cooperation

Due to the importance of the choices of species for the success of biological control with *Trichogramma*, experiments to compare relevant species and strains should be conducted. Due to the amount of work involved and the complexity of such projects, international cooperation to divide the tasks could be helpful. If standard methods to test the effectiveness of *Trichogramma* strains as biological control agents could be developed, finding an effective strain would be easier and the egg parasites could be used more successfully worldwide. The use of standard methods to test the effectiveness of *Trichogramma* strains in biological control would make it possible to compare results carried out in different laboratories in different parts of the world, saving time

and effort. A catalogue showing the effectiveness of the different parasite species and strains to control each pest under different climate conditions could then be presented. Large amounts of data could be accumulated from research workers, making it possible for biological control experts to find these vital pieces of information. Besides the supply of suitable strains for use in biological control, the exchange of information on the methodology to test the purity of the parasite in mass-rearing as an aspect of quality control should also be encouraged.

References

Bigler, F. (1989) Quality assessment and control in entomophagous insects used for biological control. *Journal of Applied Entomology* 108, 390–400.

Dijken, M.J. van, Kole, M., van Lenteren, J.C. & Brand, A.M. (1986) Host preference studies with *Trichogramma evanescens* Westwood (Hym., Trichogrammatidae) for *Mamestra brassicae, Pieris brassicae* and *Pieris rapae. Zeitschrift für Angewandte Entomologie* 101, 64–85.

Dysart, R.J. (1973) The use of *Trichogramma* in the USSR. *Proceedings of the Tall Timbers Conference on Ecological Animal Control Habitat Management.* Tallahassee, Florida, pp. 165–173.

Flanders, S.E. (1930) Mass production of egg parasites of the genus *Trichogramma. Hilgardia* 4, 465–501.

Glas, P.C., Smits, P.H., Vlaming, P.H. & van Lenteren, J.C. (1981) Biological control of lepidopteran pests in cabbage crops by means of inundative releases of *Trichogramma* species (*T. evanescens* Westwood and *T. cacoeciae* March.) a combination of field and laboratory experiments. *Mededelingen van de faculteit Landbouwwetenschappen Rijksuniversiteit Gent* 46, 487–497.

Gonzalez, D., Orphanides, G., van den Bosch, R. & Leigh, T.F. (1970) Field cage assessment of *Trichogramma* as parasites of *Heliothis zea*: development of methods. *Journal of Economical Entomology* 63, 1292–1296.

Hase, A. (1925) Beiträge zur Lebensgeschichte der Schlupfwespe *Trichogramma evanescens* Westwood. *Arbeiten aus der Biologischen Reichsanstalt für Land und Forstwirtschaft, Berlin-Dahlem* 14, 171–224.

Hassan, S.A. (1981) Mass-production and utilization of *Trichogramma.* 2. Four years successful biological control of the European corn borer. *Mededelingen van de faculteit Landbouwwetenschappen Rijksuniversiteit Gent* 46, 417–427.

Hassan, S.A. (1984) Bekämpfung des Maiszünslers mit *Trichogramma* in der Bundesrepublik Deutschland und in der Republik Philippinen. *Nachrichtenblatt des Deutschen Pflanzenschutzdienstes Braunschweig* 36, 124–125.

Hassan, S.A. (1985) Massenproduktion und Anwendung von *Trichogramma*: 7. Siebenjährige Erfahrungen bei der Bekämpfung des Maiszünslers *Ostrinia nubilalis* Hübner. *Gesunde Pflanzen* 37, 197–202.

Hassan, S.A. (ed.) (1988) Trichogramma *News*, Vol. 4. Federal Biological Research Centre for Agriculture and Forestry, Braunschweig, 32 pp.

Hassan, S.A. (1989) Selection of suitable *Trichogramma* strains to control the codling moth *Cydia pomonella* and the summer fruit tortrix moth *Adoxophyes orana*, *Pandemis heparana* (Lep., Tortricidae). *Entomophaga* 34, 19–27.

Hassan, S.A. (ed.) (1992) Trichogramma *News*, vol. 6. Federal Biological Research Centre for Agriculture and Forestry, Braunschweig, 46 pp.

Hassan, S.A. & Guo, M.F. (1991) Selection of effective strains of egg parasites of the genus *Trichogramma* (Hym., Trichogrammatidae) to control the European corn borer *Ostrinia nubilalis* Hb. (Lep., Pyralidae). *Zeitschrift für Angewandte Entomologie* 111, 335–341.

Hassan, S.A. & Rost, W.M. (1993) Massenzucht und Anwendung von *Trichogramma*: 13. Optimierung des Einsatzes zur Bekämpfung des Apfelwickler *Cydia pomonella* L. und des Apfelschalenwicklers *Adoxophyes orana* F.R. *Gesunde Pflanzen* 45, 296–300.

Hassan, S.A., Langenbruch, G.A. & Neuffer, G. (1978) Der Einfluß des Wirtes in der Massenzucht auf die Qualität des Eiparasiten *Trichogramma evanescens* bei der Bekämpfung des Maiszünslers, *Ostrinia nubilalis*. *Entomophaga* 23, 321–329.

Hassan, S.A., Stein, E., Dannemann, K. & Reichel, W. (1986) Massenproduktion und Anwendung von *Trichogramma*: 8. Optimierung des Einsatzes zur Bekämpfung des Maiszünslers *Ostrinia nubilalis* Hbn. *Zeitschrift Angewandte Entomologie* 101, 508–515.

Hassan, S.A., Kohler, E. & Rost, W.M. (1988) Erprobung verschiedener *Trichogramma*-Arten zur Bekämpfung des Apfelwicklers *Cydia pomonella* L. und des Apfelschalenwicklers *Adoxophyes orana* F.R. (Lep., Tortricidae). *Nachrichtenblatt des Deutschen Pflanzenschutzdienstes Braunschweig* 40, 71–75.

Hirashima, Y., Nohara, K. & Miura, T. (1990) Studies on the biological control of the diamondback moth *Plutella xylostella* (Linnaeus). 1. Insect natural enemies and their utilization. *Science Bulletin of the Faculty of Agriculture, Kyushu University* 3, 65–70.

Iga, M. (1985) The seasonal incidence and life-tables of the diamondback moth, *Plutella xylostella* (L.) (Lepidoptera: Yponomeutidae). *Japanese Journal of Applied Entomology and Zoology* 29, 119–125.

Keinmeesuke, P., Vattanatangum, A., Sarnthoy, O., Sayampol, B., Miyata, T., Saito, T. & Nakasuji, F. (1990) Life table of diamondback moth and its egg parasite, *Trichogrammatoidea bactrae* in Thailand. *Abstracts of the Second International Workshop, Tainan, Taiwan. 10–14 December 1990*, pp. 34–35.

King, E.G., Bull, D.L., Bouse, L.F. & Phillips, J.R. (eds) (1985) Biological control of bollworm and tobacco budworm in cotton by augmentative releases of *Trichogramma*. *Southwestern Entomologist* 8, 1–198.

Klomp, H., Teerink, B.J. & Ma, W.C. (1980) Discrimination between parasitized and unparasitized hosts in the egg parasite *Trichogramma embryophagum* (Hymenoptera: Trichogrammatidae): a matter of learning and forgetting. *Netherlands Journal of Zoology* 30, 254–277.

Kochetova, H.I. (1969) Adoption of parasitism by several egg parasites of the genus *Trichogramma* (Hym., Trichogrammatidae). *Russian Zoologiceskij Zhurnal* 48, 1816–1823.

Kot, J. & Plewka, T. (1974) Biology and ecology of *Trichogramma* spp. In: Shumakow, E.M., Gusew, G.V. & Fedorinchik, N.S. (eds) *Biological Agents for Plant Protection*. Kolos Publishing House, Moscow, pp. 183–200.

Lebedev, G.I. (1970) Utilisation des méthodes biologiques de lutte contre les insectes

nuisibles et les mauvaises herbes en Union Soviétique. *Annales de Zoologie, Ecologie Animale* 3, 17–23.

Lenteren, J.C. van (1986) Evaluation, mass production, quality control and release of entomophagous insects. In: Franz, J.M. (ed.) *Biological Plant and Health Protection.* Fischer, Stuttgart, pp. 31–56.

Lenteren, J.C. van, Glas, P.C.G. & Smith, P.H. (1982) Evaluation of control capabilities of *Trichogramma* and results of laboratory and field research on *Trichogramma* in the Netherlands. In: Voegelé, J. (ed.) *Les Trichogrammes. Les Colloques de l'INRA* 9, 257–268.

Mokrzecki, S.A. & Bragina, A.P. (1916) The rearing of *Trichogramma semblidis* Aur. and *T. fasciatum* P. in the laboratory and temperature experiments on them. *Applied Entomology* 5, 155–156.

Morrison, R.K., Stinner, R.E. & Ridgway, R.L. (1976) Mass production of *Trichogramma pretiosum* on eggs of the Angoumois grain moth. *Southwestern Entomology* 1, 74–80.

Need, J.T. & Burbutis, P.P. (1979) Searching efficiency of *Trichogramma nubilale.* *Environmental Entomology* 8, 224–227.

Nguyen, I.T. & Nguyen, S.T. (1982) The use of *Trichogramma* in Vietnam. *Zaschifa Rastenij* 1, 52.

Noldus, L.P.J.J. (1989) Semiochemicals, foraging behaviour and quality of entomophagous insects for biological control. *Journal of Applied Entomology* 108, 425–451.

Okada, T. (1989) Parasitoids of the diamondback moth in Malaysia. In: Talekar, N.S. & Griggs, T.D. (eds) *Diamondback Moth Management: Proceedings of the First International Workshop,* Tainan, Taiwan, pp. 25–34.

Pak, G.A. (1988) Selection of *Trichogramma* for inundative biological control. PhD Thesis, Wageningen Agricultural University, The Netherlands, 224 pp.

Pak, G.A., Berkhout, H. & Klapwijk, J. (1991) Do *Trichogramma* look for hosts? In: Wajnberg, E. & Vinson, S.B. (eds) Trichogramma *and Other Egg Parasitoids. Third International Symposium. Les Colloques de l'INRA* 56, 77–80.

Pavlik, J. (1993) Variability in the host acceptance of European corn borer, *Ostrinia nubilalis* Hbn. (Lep., Pyralidae) in strains of the egg parasitoid *Trichogramma* spp. (Hym., Trichogrammatidae). *Journal of Applied Entomology* 115, 77–84.

Quednau, W. (1955) Über einige neue *Trichogramma*-Wirte und ihre Stellung im Wirt-Parasit-Verhätnis. Ein Beitrag zur Analyse des Parasitismus bei Schlupfwespen. *Nachrichtenblatt Deutschen Pflanzenschutzdienstes Braunschweig* 7, 145–148.

Quednau, W. (1956) Die biologischen Kriterien zur Unterscheidung von *Trichogramma*-Arten. *Zeitschrift für Plantzenkrankheiten und Planzenschutz* 63, 333–344.

Salt, G. (1935) Experimental studies in insect parasitism. III. Host selection. *Proceedings of the Royal Society* 117, 413–435.

Salt, G. (1940) Experimental studies in insect parasitism. VII. The effects of different hosts on the parasite *Trichogramma evanescens* Westw. (Hym. Chalcidoidea). *Proceedings of the Royal Entomological Society of London* 15, 81–95.

Salt, G. (1958) Parasite behaviour and the control of insect pests. *Endeavour* 17, 145–148.

Schieferdecker, H. (1969) Zur Eignung von Lepidoptereneiern als Wirte der Eiparasiten *Trichogramma cacoeciae* Marchal und *Trichogramma evanescens* Westwood

(Hym., Trichogrammatidae). *Wanderversammlung deutscher Entomologen, Dresden* 3, 495–511.

Schmidt, J.M. & Smith, J.J.B. (1989) Host examination walk and oviposition site selection of *Trichogramma minutum*: studies on spherical hosts. *Jounal of Insect Behavior* 2, 143–171.

Stinner, R.E. (1977) Efficacy of inundative releases. *Annual Review of Entomology* 22, 513–531.

Stschepetilnikova, V.A. (1974) Use of *Trichogramma* in the U.S.S.R. In: Shumakov, E.M., Gusev, G.V. & Fedorinchik, N.S. (eds) *Biological Agents for Plant Protection.* Kolos Publishing House, Moscow, pp. 160–182.

Suzuki, Y., Tsuji, H. & Sasakawa, M. (1984) Sex allocation and effects of super-parasitism on secondary sex ratios in the gregarious parasitoid *Trichogramma chilonis* (Hymenoptera: Trichogrammatidae). *Animal Behaviour* 32, 478–484.

Sweetman, H.L. (1958) *The Principles of Biological Control.* Brown, Dubuque, 560 pp.

Tran, L.C. & Hassan, S.A. (1986) Preliminary results on the utilization of *Trichogramma evanescens* Westw. to control the Asian corn borer *Ostrinia furnacalis* (Guenée) in the Philippines. *Journal of Applied Entomology* 11, 18–23.

Tran, L.C., Bustamente, R. & Hassan, S.A. (1986) Release and recovery of *Trichogramma evanescens* Westw. in corn fields in the Philippines. In: Voegelé, J., Waage, J. & van Lenteren, J.C. (eds) Trichogramma *and Other Egg Parasites. 2nd International Symposium. Les Colloques de l'INRA* 43, 597–607.

Voegelé, J., Stengel, M., Schubert, G., Daumal, J. & Pizzol, J. (1975) Les Trichog-rammes. Va. Premiers résultats sur l'introduction en Alsace sous forme de lâchers saisonniers de l'écotype moldave de *Trichogramma evanescens* Westw. contre la pyrale du maïs, *Ostrinia nubilalis* Hubn. *Annales de Zoologie, Ecologie Animale* 7, 535–551.

Voegelé, J., Waage, J.K. & van Lenteren, J.C. (eds) (1988) Trichogramma *and Other Egg Parasites. 2nd International Symposium. Les Colloques de l'INRA* 43, Paris, 644 pp.

Wäckers, F.L., de Groot, I.J.M., Noldus, L.P.J.J. & Hassan, S.A. (1987) Measuring host preference of *Trichogramma* egg parasites: an evaluation of direct and indirect methods. *Mededelingen van de Faculteit Landbouwwetenschappen Rijksuniversiteit Gent* 52, 339–348.

Wang, C.-L. (1986) Biological control of *Ostrinia furnacalis* with *Trichogramma* sp. In: Voegelé, J., Waage, J. & van Lenteren, J.C. (eds) Trichogramma *and Other Egg Parasites. 2nd International Symposium. Les Colloques de l'INRA* 43, 609–612.

Wührer, B. & Hassan, S.A. (1993) Selection of effective species/strains of *Trichogramma* (Hym., Trichogrammatidae) to control the diamondback moth *Plutella xylostella* L. (Lep., Plutellidae). *Zeitschrift Angewandte Entomologie* 116, 80–89.

Zhang, Z.-L. (1986) *Trichogramma* sp. parasiting the eggs of Asian corn borer *Ostrinia furnacalis* and its efficacy in Beijing suburb. In: Voegelé, J., Waage, J. & van Lenteren, J.C. (eds) Trichogramma *and Other Egg Parasites. 2nd International Symposium. Les Colloques de l'INRA* 43, 629–633.

Rearing of *Trichogramma* and Other Egg Parasitoids on Artificial Diets

4

Simon Grenier

Laboratoire de Biologie Appliquée, Bât. 406. INSA, LA INRA 227,
20 avenue Albert Einstein, 69621 Villeurbanne Cedex, France

Abstract

Utilization of egg parasitoids to control many insect pests all over the world
has increased year after year. Rearing on artificial diets may facilitate the
multiplication of these entomophagous insects and allow progress in research
on their physiology and behaviour. In order to succeed in such rearing, it is
necessary to consider different parameters and cues, such as nutritional
needs, leading to a definition of a suitable artificial diet. Two main kinds of diet
are used, with or without insect haemolymph. The high respiration rate of
developing larvae and pupae is to be considered in order to find an acceptable
artificial egg shell. Among egg parasitoids, about 18 species (14 Trichogrammatidae, one Eupelmidae, one Eulophidae, one Scelionidae and one
Encyrtidae) were reared from egg to adult *in vitro*. Some *Trichogramma* species
produced in artificial conditions were released in the fields and successfully
controlled pests.

Introduction

Insect parasitoids and especially oophagous species are being used increasingly in biological control programmes. However, such programmes are
currently limited by the availability of a large number of insects, especially for
inundative releases. In order to simplify the production of biocontrol agents,
studies have been conducted in several laboratories in order to develop
techniques allowing mass culture in artificial media. For mass-production,
media are suitable if they are inexpensive relative to the use of conventional
hosts, and if they support normal growth of entomophagous insects.

Chemically defined media as well as those containing raw materials derived from insects have been used.

Artificial rearing can be conducted for two purposes. The availability of large numbers of insects to be released is the applied aim. The artificial rearing technique also provides tools for basic research on the biology, physiology and behaviour of egg parasitoids. For the latter purpose, it is advisable to use a quite well-defined diet in order to get accurate results (Grenier *et al.*, 1986). Indeed, the use of *in vitro* techniques allows a much greater degree of control over the parasitoid environment. Such techniques can be very useful, for example to investigate the role of teratocytes or ovipositional behaviour. Thanks to artificial medium, Xin & Li (1989) investigated the ovipositional behaviour of *Anastatus japonicus* and defined precisely the different sequences of the egg laying process.

A film (1986), directed by Z.L. Pu, with K.J. Dai and W.H. Liu as technical advisors, entitled 'The cradle – artificial ova and *Trichogramma*' was made by the Shanghai Scientific and Educational Film Studio, and described the research conducted on artificial rearing of *Trichogramma*. Thanks to artificial media, very good scenes, especially of the egg laying, were filmed.

Cultures in artificial media will also allow the testing of insecticides and eventually the selection of resistant strains, because traditional methods using pulverization are very difficult and risky to conduct with such tiny insects, and lead to frequent breakdown of the population along the selective process, as observed with *Trichogramma japonicum* and *A. japonicus* (Xu *et al.*, 1986; Zhang *et al.*, 1989).

Artificial rearing might permit introduction of exogenous material into *Trichogramma* eggs by biomolecular techniques, thanks to the possibility of collecting eggs easily, and further development in controlled conditions after treatment. Artificial media also open possibilities for studying the preimaginal conditioning of parasitoids (Greany *et al.*, 1984).

Artificial rearing of egg parasitoids concerns five hymenopterous families only: Trichogrammatidae, Eupelmidae, Scelionidae, Eulophidae and Encyrtidae.

For endoparasitoids, the diet is not only the food source, but also their environment in which they are bathed for all their larval life. So, besides the nutritional needs, the medium must provide all requirements for essential physiological functions and protection from desiccation (Grenier *et al.*, 1986). Nevertheless, many egg parasitoids can develop in killed host eggs revealing the low probabilities of strict host–parasitoid interactions, especially at hormonal level. No studies have been conducted on the possible influence of hormones *in vitro*, but generally no hormones were added to artificial media for egg parasitoids.

For this overview we will first consider the main physiological functions implicated in artificial rearing: nutrition and respiration. Then, we will analyse the role on parasitoid development of pH, osmotic pressure and of some preservatives added in the diet. Artificial host eggs fabrication and egg laying stimulation will also be considered. Finally, some examples of successful *in vitro* rearing of egg parasitoids will be given.

Nutrition

Of the various physiological needs of egg parasitoids, one of the most important to be satisfied is the supply of nutrients needed for the development and the growth of the larvae. Unlike the species evolving at the expense of growing stages, egg parasitoids develop in closed systems, independently of external nutritional support, having an exceptional physiological dynamism and a short larval life (Mellini, 1986). Consequently they need very rich and concentrated food.

The basic qualitative nutritional requirements of parasitoids are similar to those of free-living insects. Few true nutritional studies of the needs in precise components have been conducted on oophagous parasitoids.

Definition of medium composition

Parasitoids are carnivorous species, so they need a protein-rich diet, sometimes with some special requirements in aromatic amino acids for cuticle tanning. Concentrated energy sources coming from carbohydrates or lipids are also needed. All needs have to be supplied taking into account osmotic pressure and dietary balances (between and within nutrient classes).

Nutritional composition of the diet can be defined from food analyses (host egg). For *Trichogramma* rearing, analyses of the host eggs of *Antherea pernyi* and *Dendrolimus punctatus* (and also *A. pernyi* pupae), in proteins, amino acids, lipids (phospholipids, cholesterol) and carbohydrates were performed by Hubei Coop. Research Group. (1979, 1987) and Cao *et al.* (1988). The media used by Grenier & Bonnot (1988) were based on analyses of *Ephestia kuehniella* eggs. The needs for *Trichogramma* in amino acids were evaluated by analyses of some lepidopterous eggs used as alternative hosts and compared with egg parasitoid wasp content (Barrett & Schmidt, 1991). Female secretions into hosts and their role in modifications of host content were also studied (I.G. Yazlovetsky, unpubl.).

Whole parasitoid body analyses allowed control of the biochemical content of the parasitoids obtained *in vitro* and led to an adjustment in the medium composition and thus an improvement of yields and quality of insect produced (S. Grenier *et al.*, unpubl.).

Studies of the feeding sequences and development of *Trichogramma brassicae* in alternative host eggs of *E. kuehniella*, gave some useful indications in order to improve artificial rearing (Hawlitzky & Boulay, 1982). Anatomical

studies of the alimentary canal in *Anastatus* sp. were also fruitful to look at the possibility of food consumption by the larvae, for example concerning the size of eatable particles (Lu & Yang, 1988).

After the first definition and tests of a medium for one species, it was often tried for another species.

Types of media

Artificial media can be divided into two main categories according to the presence or the absence of insect components (host egg content, haemolymph or holotissue extract).

Egg content of the host is generally used as artificial medium, or as a part of it, to test whether the method of artificial culture is suitable for the growth of a precise species of egg parasitoid (Guan *et al.*, 1978; Volkoff *et al.*, 1992). Adjunction of egg homogenate increased the yield in *Trichogramma pretiosum* (Xie *et al.*, 1986a). Basically, for *Trichogramma*, artificial media with insect components contain haemolymph or holotissue extract from lepidopterous species (20–43%), hen's egg yolk (15–34%), cow milk (15–30%) and Neisenheimer's salt solution (5–14%). When pupal tissue fluid is used, for example from *Antherea pernyi*, it is necessary to eliminate manually or mechanically undigested food dregs to obtain normal development of *Trichogramma* (Zhong & Zhang, 1992).

Artificial media devoid of insect components and containing yeast hydrolysate (20%), calf fetus serum (20%), Grace tissue medium (20%), chicken embryo extract (10%), cow's milk (15%) and hen's egg yolk (15%), allow egg laying and development to adult of *Trichogramma dendrolimi*, *Trichogramma confusum* and *T. pretiosum*, in wax eggs and hanging drops, with 17–36% of pupation and 1–2% of emergence (Liu & Wu, 1982). With similar media, containing a basic solution of amino acids, sugars, vitamins, salts, organic acids (20%), and calf serum (10–18%), chicken embryo extract (20%), wheat germ (8–10%), yeast extract (10–13%), egg yolk (20–30%) and Grace tissue medium (0–20%), the pupation rate of *T. dendrolimi* was 36–53% and the adult emergence rate was 8–16%. Omission of basic solution or egg yolk reduced the pupation rate and no adults were produced. A high level of amino acids and presence of yolk are necessary for the successful culture of *T. dendrolimi*. There is some coincidence in the composition of the media which induces oviposition of the adults and the one that provides nutritional requirements of the developing larvae (Wu *et al.*, 1982; Qin & Wu, 1988).

Precise nutritional parasitoid requirements are difficult to define because most of the tested media contained raw materials, which are characterized by their complex composition.

Nitrogen sources (proteins, peptides and amino acids)

Many parasitoids required the ten essential amino acids. The most commonly used proteins (hydrolysed or not) are casein, lactalbumin, serum albumin and soyabean extract. Extracts and hydrolysates of different yeasts are also commonly used (Bratti, 1990). Hydrolysates are very active on osmotic pressure; enzymatic hydrolysates are less active than acid ones. Such considerations are very important to take into account because osmotic pressure is a key factor, as explained below. High yields in the different stages were obtained with casein.

Lipids

Dietary sterols are required by a great number of parasitoids. Cholesterol is added, especially in artificial media devoid of hen's egg yolk. Fatty acids may be presented as free fatty acids or triglycerides, but in any case, in the absence of egg yolk, it is necessary to use emulsifying agents (Tween 20 or 80, lauryl sulfate, lecithin) to obtain a good dispersion of lipids in the diet.

Hen's egg yolk, used in many artificial media, at a rate of 10–30%, is difficult to replace by a mixture of fatty acids, cholesterol and phospholipids. Egg yolk used as the only source of lipids in defined artificial media for *T. brassicae* was necessary to obtain high percentages of larvae and adults (Wu *et al.*, 1982; Grenier & Bonnot, 1988).

Carbohydrates

The energy supply may come from either carbohydrates or lipids. Carbohydrates may improve growth. One of them could be necessary for pupation of *T. pretiosum* (Irie *et al.*, 1987). Glucose, fructose (monosaccharides), sucrose (oligosaccharides) are the most common in artificial diets; glycogen or other polysaccharides are more rare. Trehalose, the main carbohydrate of insect haemolymphs was quite rarely used, for example in media tested for some *Trichogramma* species and *Telenomus heliothidis* (Strand & Vinson, 1985; Grenier & Bonnot, 1988; Strand *et al.*, 1988). The choice is directed by the tolerance to high osmotic pressure of the species to be reared.

Cow's milk is also a very common component in artificial media at a rate of 15–20%; it mainly contains lipids, proteins, carbohydrates, mineral salts and vitamins. Sometimes it has been replaced by malt solution. The needs in vitamins are very difficult to study (Grenier *et al.*, 1986).

Host factors

Xie *et al.* (1986a) and Nettles (1990) revealed the need of host factors for *Trichogramma pretiosum*. Addition of *Manduca sexta* egg homogenate to a diet composed of haemolymph, egg yolk and bovine milk produced a yield of 76% of adult *T. pretiosum*, with an increase of 7–13-fold compared to a diet without egg homogenate (Xie *et al.*, 1986a).

Haemolymph is a very complex component, containing mineral salts, trehalose, free amino acids, peptides, proteins, lipids (cholesterol), organic acids, amines and others. Haemolymph from different lepidopterous species has been used in artificial media to rear *Trichogramma*: *Antherea pernyi*, *Philosamia cynthia*, *Bombyx mori*, *Heliothis zea*, *Manduca sexta* and *Spodoptera littoralis*. Heating haemolymph at 98°C for 30 min prevented adult development of *Trichogramma*, probably by denaturation of essential components, but no undesirable effects were observed at 67–70°C for 1 min (Hubei Coop. Research Group, 1979). In most experiments, haemolymph is heated at 60°C for 5–6 min; this treatment is generally sufficient to inhibit enzymes responsible for melanization.

The Encyrtid *Ooencyrtus nezarae*, an egg parasitoid of the green bug *Nezara viridula*, develops up to the last larval instar in an artificial medium containing 50% *B. mori* haemolymph, 25% hen's egg yolk, and 25% cow milk. Many factors, not only nutritional ones, may be involved in the failure of pupation (Takasu & Yagi, 1992).

Centrifuged haemolymph of *B. mori* is not suitable for the development of *T. brassicae*, because an essential factor may be eliminated by high-speed centrifugation (Grenier & Bonnot, 1988).

If not replaced by other components, haemolymph has to be used in high percentages generally between 20 and 50%; pure haemolymph rarely allows a normal development (Hubei Coop. Research Group, 1979; Liu *et al.*, 1979). Hoffman *et al.* (1975) used haemolymph from *Heliothis zea* to rear *T. pretiosum* on filter paper with some success. Haemolymph from *M. sexta* (20%) was necessary to obtain pupation of *T. pretiosum* in artificial diets with free amino acids, intralipids (soyabean oil, glycerol, phospholipids), cholesterol, carbohydrates, inorganic salts, vitamins, organic acids and bovine serum albumin (Strand & Vinson, 1985). It seemed that at least two and possibly several polar low-molecular-weight chemicals present in *M. sexta* haemolymph were responsible for growth and development of *T. pretiosum* to pupal stage. Only partial isolation and purification of these factors was possible, but one of them could be a carbohydrate (Irie *et al.*, 1987).

Female factors

Media with host egg content from parasitized *Nezara viridula* eggs, allowed better development of *Trissolcus basalis* (Scelionidae) than egg content from unparasitized eggs, indicating that female wasps may inject some factors within the host which improve the nutritional quality of the eggs for development of *T. basalis*. Media containing haemolymph of the lepidopterous *Manduca sexta* and hen's egg yolk allow only the larval development of the scelionid (Volkoff *et al.*, 1992).

Respiration

The integument wettability is important to consider for larval respiration in artificial medium. Parasitoid larvae with wettable integument sink into the medium and may have some difficulties in respiring. Respiration takes place through the cuticle or through the spiracles. Early first instar larvae of hymenopterous species respire mainly by cutaneous diffusion, so liquid medium may be used. For egg parasitoids the problems of gas exchange are very acute because their respiratory rate is very high. Importance of ventilation for normal development of *Trichogramma* was reported by Hubei Coop. Research Group (1979).

Respiratory rate varies greatly according to the developmental stage. The consumption of oxygen by *Trichogramma dendrolimi* Matsumura is quite low during the egg stage and at the beginning of the larval stage (1 to 4 μl h^{-1} for 100 insects). It rises at the end of this stage, reaching 7 μl h^{-1} at the prepupal stage and decreases thereafter. At the end of the pupal stage, the rate increases again (5 μl h^{-1}) (Dai *et al.*, 1987). Therefore, to culture *Trichogramma* species *in vitro*, it will be of prime importance to make sure that a normal respiration of prepupae and pupae is possible.

Artificial rearing is easier for some species, like *Ooencyrtus pityocampae* (Encyrtidae), which have a special respiratory anatomical adaptation, i.e. a stalk protruding from the host egg or plastic film which allows them to respire through the host cuticle or through an artificial membrane. However, some difficulties arise before pupation when mature larvae lose their respiratory stalk (Battisti *et al.*, 1990) and, in artificial conditions, a high larval mortality was observed at this time (Masutti *et al.*, 1992).

Physico-chemical Factors

pH

Hydrogen ion concentration in insect haemolymph ranges between 6.0 and 8.2, but is generally restricted between 6.4 and 6.8. Most of the artificial

media for parasitoids show a pH close to these latter values. Despite the fact that pH is important for success in artificial rearing, few studies are dealing with the effect of this factor on the development of parasitoids.

With a diet containing holotissue of *A. pernyi* pupae, egg yolk, milk and water, optimal pH for parasitization, pupation and adult rates was found to be between 6.7 and 6.95 for *T. confusum* (Zhong & Zhang, 1989). High percentages of adults of *Trichogramma* were obtained when pH ranged between 6.6 and 6.8 (Hubei Coop. Research Group, 1979; Strand & Vinson, 1985). For *T. pretiosum*, a diet with a pH between 6.5 and 9.0 was suitable, with an optimum at 7.0 (Nettles *et al.*, 1983).

Osmotic pressure

For oophagous parasitoids of the genus *Trichogramma*, an osmolarity higher than 450 mOsm kg^{-1} generally does not allow egg hatching and normal development (Grenier *et al.*, 1986). Most of the artificial media have an osmotic pressure around 300 mOsm, so for *T. dendrolimi* and *T. brassicae*, 320 mOsm seems to be an optimal value (Grenier & Bonnot, 1988). For *Telenomus heliothidis*, the osmotic pressure used was 390 mOsm (Strand *et al.*, 1988), but, for *T. pretiosum*, the osmotic pressure used was surprisingly higher: 460 mOsm (Strand & Vinson, 1985).

The actual osmotic pressure undergone by parasitoids is often higher than the osmotic pressure measured on the fresh medium, because of desiccation occurring when small quantities of medium are distributed. The degree of tolerance to high osmotic pressure may depend on both the characteristics of the egg chorion and the larval integument. Unwettable integument avoids direct contact with the medium. Many egg parasitoid hymenopterous may lack such integumentary characteristics and suffer osmotic shock. In order to maintain the osmotic pressure within acceptable values, part of the free amino acids may be supplied as proteins or protein hydrolysates, while carbohydrates are supplied as a mixture of oligosaccharides (mainly glucose and sucrose) and polysaccharides (mainly glycogen).

Role of Teratocytes

Among egg parasitoids, teratocytes are reported in Scelionidae and Trichogrammatidae, and in some species, the role of teratocytes may be important for normal development (Dahlman, 1990). They are necessary *in vitro* to obtain high percentages of pupae and adults of *Telenomus heliothidis*; survival of egg and larvae were similar in the presence or absence of teratocytes, but in their absence, pupation and adult emergence were reduced from 64 to 42%, and 44 to 16% respectively (Strand *et al.*, 1988). Teratocytes may have trophic functions and play a hydrolytic rather than a direct nutritional role, probably by secretion of digestive enzymes (Strand *et al.*, 1988; Volkoff & Colazza, 1992).

Use of Preservatives

Because artificial media are very rich, contaminations by microorganisms can develop easily. Preservatives, like antibiotics and antifungals, are necessary for mass-production. However they can present some deleterious effects.

Different molecules were used as antibiotics, mainly penicillin, streptomycin and gentamycin, but also Yan 1 in China. In *T. pretiosum*, Xie *et al.* (1986b) showed negative effects on egg hatching when 0.5–2% sorbic acid or 0.1–0.4% methyl hydroxybenzoate were added to the medium, but development was not affected when 0.5% kanamycin sulfate, 0.15% gentamycin, 0.5% streptomycin sulfate, or 30 µg ml^{-1} amphotericin B were added. Unfortunately, the latter levels are too low to allow efficient control of contaminants.

It is quite easy to control bacteria, but it is more difficult for fungi. Fungi can quickly spread from an initial contamination point thanks to their mycelia and spores, so antifungal agents are often necessary. Unfortunately many preservatives are noxious for insects and especially for parasitoids, which are generally more sensitive than free-living insects. Before adding preservatives to the medium, special studies are necessary to test the effects on the parasitoids. Species are more or less sensitive, but most of the classical fungicides are toxic to oophagous *Trichogramma* at effective levels against fungi (sodium benzoate, sodium propionate, potassium sorbate). Antifungals used in tissue culture media, such as amphotericin B, geneticin and nystatin incorporated at the proper level (50–100 mg, 10–30 mg and 500–800 × 10^3 IU (100 g)$^{-1}$ respectively) can control moulds efficiently without reducing significantly the yields in parasitoids (Grenier & Liu, 1990, 1991).

The Artificial Host Egg (Container as Egg Shell)

The components used in the artificial medium, its presentation and the containers used must be well suited to the physiological requirements of each species. The diet must reflect some special needs of oophagous species such as the necessity to limit the quantity of food for each individual. It is also desirable that both the artificial egg shell and the medium induce egg laying.

Oophagous species need a limited quantity of food, because they cannot regulate the intake of food, leading to developmental abortion, mainly at pupation, or abnormalities in the adults produced. Moreover, excess of medium or dew produces moisture, in the presence of which egg parasitoids

usually fail to pupate. Thus, with many species of *Trichogramma* cultured *in vitro* under unsuitable conditions, bloated larvae and adults with enlarged abdomen and/or unexpanded wings are obtained (Guan *et al.*, 1978; Liu *et al.*, 1979; Nettles *et al.*, 1985; Strand & Vinson, 1985; Grenier & Bonnot, 1988). If possible, the food has to be divided into small quantities necessary and sufficient for the complete development of one individual, especially for solitary species (for example *Telenomus heliothidis* or *Ooencyrtus pityocampae*). For non-cannibalistic species (gregarious species mainly, for example many *Trichogramma* spp.), it is also possible to try to control the number of parasitoid eggs laid to suit the total amount of food available.

Different methods have been used to culture egg parasitoids *in vitro* giving a limited quantity of food: hanging drop, wax eggs, paraxylylene eggs, artificial egg cards.

Hanging drop technique

This technique was used to test the artificial medium itself. It needs a previous oviposition in natural or alternative host eggs or in wax eggs. The wasp eggs are collected by dissection and transferred on glass slides (Guan *et al.*, 1978; Liu *et al.*, 1979; Grenier & Bonnot, 1988). Microtitre (multiwell) plates used for tissue culture and Petri dishes could be used for rearing *T. pretiosum* on a large scale, but the eggs have to be first collected from wax spheres and then transferred into the plates or dishes (Xie *et al.*, 1986b). To create a real artificial egg, it is necessary to have an egg shell.

Wax eggs

Wax eggs are obtained with a capillary tube dipped into a warmed diet solution covered by a paraffin–vaseline mixture and touched on a wax paper (Rajendram & Hagen, 1974; Guan *et al.*, 1978). The stickiness and the hardness of the wax may be factors preventing penetration of the ovipositor (Rajendram, 1978a). Nevertheless, most of the *Trichogramma* species tested laid numerous eggs in proper wax eggs except *T. japonicum*.

There are also problems if the medium container interacts with respiration. The artificial host egg shell may constitute a barrier against gaseous exchange. In a completely closed system such as a spherical egg shell, it is necessary to have an artificial membrane which is highly permeable to oxygen and carbon dioxide, but less permeable to vapour, in order to avoid desiccation. Thus, it appears that, for complete development of egg parasitoids, a wax membrane is not suitable.

Morphological studies of the structure of the egg shell of *A. pernyi*, and of the ovipositor of different *Trichogramma* species to be reared, gave indications about the optimal thickness of the artificial membrane (Hsia & Wang, 1979). This thickness varies from 15 μm for small species (like *T. japonicum*) to 64 μm for *T. dendrolimi*.

Paraxylylene egg shells

Artificial host eggs can be made by encapsulating frozen droplets of medium of suitable size with a vapour deposition of paraxylylene. In these artificial host eggs, containing different diets even without insect component, *T. dendrolimi*, *T. brassicae* and *Trichogramma ostriniae* laid numerous eggs. However in most cases, the development stopped at the larval stage, and prepupae were obtained only with a diet containing haemolymph. In the latter diet, the development probably failed because of the difficulties for gaseous exchange; in fact the right thickness of the paraxylylene membrane is not easy to obtain (Grenier & Bonnot, 1988).

Artificial egg cards

Two polyethylene sheets, sealed by heating, may be used as egg shell, but successful rearing necessitates the piercing of some holes in one sheet to facilitate aeration (Hubei Coop. Research Group, 1979). A good system consists of a semi-opened container with plastic film (PVC, polyethylene or polypropylene) as described by many authors. Adapted artificial egg plates made of plastic film with an air reserve, were defined for *Trichogramma* (Liu *et al.*, 1983; Dai *et al.*, 1988; Li *et al.*, 1988). The thickness of the film (generally between 20 and 50 μm) must be adapted to the size of the ovipositor of the species to be reared. For *T. japonicum*, egg laying strongly depends on the quality of the artificial egg shell: only a few eggs were obtained in wax capsules, but a lot of eggs were laid in plastic cupules (Liu *et al.*, 1983). The volume of the cupules is of prime importance and must be adapted to the size of the species.

For *Ooencyrtus pityocampae*, egg shape and distribution of the eggs on the egg card modify the egg laying: tight rows are better than isolated or clustered eggs (Masutti *et al.*, 1991). For *Trichogramma australicum*, there is no correlation between the thickness of the membrane (cellophane, silicon film, polypropylene or polyethylene), within the range of 20–89 μm, and the number of the eggs laid.

Artificial host eggs made of a plastic capsule and medium containing host material allow the development of many species of *Trichogramma* (see below). Parasitizing efficiency varied according to oviposition behaviour and length and width of the ovipositor (Li *et al.*, 1989).

For oophagous species, it soon appears that direct oviposition of the

female into the artificial medium is necessary for mass-rearing. Oviposition is controlled by a series of very complex physiological and chemical cues (Heath et al., 1990; see also Chapter 9). Stimulation appears necessary to obtain regularly egg laying in vitro (Grenier et al., 1993).

Stimulation of Egg Laying

The number of eggs laid in artificial host eggs is often lower than those obtained in natural or alternative host eggs. Stimulation of egg laying, by components present on the artificial egg shell or in the medium, allows normal parasitization rates to be restored.

Surface treatment of the shell

The parasitization rate by T. dendrolimi on artificial host eggs was increased when they were treated with haemolymph from A. pernyi pupae or with sodium carbonate (Na_2CO_3) (Dai et al., 1988). Application of $NaHCO_3$ or $MgSO_4$ has very little effect on egg laying into wax capsules by T. japonicum (Liu et al., 1983). Lepidopterous scale extracts or a blend of saturated hydrocarbons deposited on artificial eggs acted as a kairomone, stimulating the oviposition by T. brassicae. The total number of eggs laid was six to 15 times higher than in the control. Organic solvent alone improved the egg laying by modification of the surface of the membrane (Grenier et al., 1993).

Components of the media

Oviposition by T. californicum was influenced by the concentration of the salt solution used (Rajendram, 1978b). Nettles et al. (1982, 1983) showed that some cations (Ca^{2+}) may inhibit oviposition and that a diluted solution of KCl and $MgSO_4$ encapsulated inside wax spheres strongly improved oviposition by T. pretiosum and T. minutum, allowing collection of numerous eggs that could be used for mass-production. $MgSO_4$ alone was not active, but a synergistic effect occurred when $MgSO_4$ and KCl were combined. Successful oviposition of T. pretiosum was obtained through a silicone polycarbonate membrane into a diluted solution of KCl and $MgSO_4$ (Morrison et al., 1983). Addition to this mixture of α-aminobutyric acid and leucine + isoleucine + ornithine solution, improved the egg laying in Trichogramma neustadt (Xie et al., 1991b). Nevertheless, the KCl–$MgSO_4$ solution cannot support larval development and is detrimental for Trichogramma eggs when they remain too long inside. So, it is necessary to transfer the eggs under other conditions (Xie et al., 1986a, 1988).

With *T. dendrolimi*, the mixture of leucine, phenylalanine and histidine had better effect on oviposition in wax eggs compared to the haemolymph of the natural host. Mineral salts, sugars, vitamins and organic acids had little or no effect (Wu & Qin, 1982; Qin & Wu, 1988). According to Liu & Wu (1982) yeast hydrolysate improved the oviposition in *T. dendrolimi*, but according to Nettles *et al.* (1985) casein hydrolysate, glucose solution or free amino acids had little or no effect on the oviposition of *T. pretiosum*. Protein hydrolysate at 1% was a good medium for oviposition (Lu *et al.*, 1979; Lu & Long, 1979).

These contradictory results could be explained by the difficulties of testing some components of the stimulation of the oviposition, taking into consideration all the parameters that could act or interfere (basic medium composition, pH and osmotic pressure, concentration of the components to be tested, interaction between components). There are also interactions with both surface components and diet components. *Tetrastichus schoenobii* laid its eggs into various solutions encapsulated in artificial eggs which were coated with hairs from the host egg mass. Oviposition occurred (in decreasing order regarding the number of eggs laid), in TC10 tissue culture medium from Gardiner & Stockdale, 2% enzymatic hydrolysate of casein + Neisenheimer's salt solution and casein hydrolysate alone. No eggs were laid in distilled water and Neisenheimer's solution (Ding *et al.*, 1980b). In this species, oviposition is elicited by kairomones from the host (the yellow paddy borer, *Tryporyza incertulus*) located in the hairs of the anal tuft of the female moth, deposited on the egg mass (Ding *et al.*, 1981). Glutamine, alanine and glycine act as ovipositional stimulants for *Anastatus japonicus* (Xie *et al.*, 1991a).

To be sure that the whole development was entirely obtained in an artificial medium, the only way consists in obtaining eggs laid by the female directly into the diet or by dissection of the female to collect caesarean (ovarian) eggs. Transfer from one medium to another does not alter the rigour of the experiment but makes the experiment more difficult.

Some Examples of Success in Artificial Rearing and Production

Up to now, about 18 species of egg parasitoids were reared in different artificial media, if we consider successes as development from egg to adult (Grenier *et al.*, 1986; Bratti, 1990).

The major steps for *in vitro* rearing of oophagous parasitoids were achieved with the repeated rearing in reliable conditions of *T. dendrolimi* in a medium containing insect components (in 1978–1979), then its development in a medium devoid of insect material and finally the rearing of other species in similar media (in 1980–1982).

Trichogramma dendrolimi was the first oophagous species reliably reared *in*

vitro in a medium with insect material (Guan *et al.*, 1978; Hubei Coop. Research Group, 1979; Liu *et al.*, 1979) where 35 generations were continuously cultured by Gao *et al.* (1982). It can also develop in artificial media devoid of insect material (Wu *et al.*, 1980; Wu & Qin, 1982).

Liu *et al.* (1985) reported the successful use of *T. confusum* reared on artificial host eggs against sugar-cane borers (*Proceras venosetus*, *Argyroploce shistaceana*, *Chilo infuscatellus*), with a parasitization rate of 81–95% in a 6.7 ha experimental field. Shen *et al.* (1991) released *in vitro*-reared *T. dendrolimi* to control the Asian corn borer, *Ostrinia furnacalis*, with a good success. *T. japonicum* was reared on artificial host eggs for three generations with better viability than in the natural eggs of *A. pernyi* (Xin & Li, 1989).

Recently, in China, large-scale production of *Trichogramma* in artificial host egg cards was developed to control different insect pests: *Dendrolimus punctatus*, *Philosamia cynthia*, *Heliothis armigera*, sugar-cane borers, *Maruca testulais*, *Diaphania indica* and *Heliothis assulta*. The daily production could reach 30–43 million of *T. dendrolimi* or 17–23 million of *Trichogramma chilonis*, for a cost 50% less than on oak silkworm eggs. The results for *T. dendrolimi* and *T. chilonis* were respectively: 95–90% parasitization, 90–85% adult emergence, 90–85% female, and 80–70% normal wasps. The treated area was more than 1300 ha with a parasitization rate usually more than 80% (Dai *et al.*, 1991).

T. ostriniae was successfully cultured in artificial eggs with medium mainly composed of haemolymph or holotissues of *A. pernyi*, egg yolk and milk. Parasitization rate was 90% and adult production 56–63% (Zhong, 1987).

One of the best examples of the success of artificial rearing was the use of *Trichogramma* pupae produced in artificial host eggs, as a prey to rear the predacious mite *Amblyseius nicholsi*. The mite, reared for two generations and released in citrus orchards, significantly controlled the citrus red mite *Panonychus citri* (Zhang & Cao, 1993).

At least 14 species of *Trichogramma* successfully developed in artificial media containing pupal haemolymph from silkworm (or holotissue extracts), hen's egg yolk, cow's milk or malt solution and inorganic salts, encapsulated by plastic film like polyethylene: *T. brassicae* (= *T. maidis*), *T. cacoeciae*, *T. chilonis* (= *T. confusum*), *T. chilotraeae*, *T. cordubensis*, *T. dendrolimi*, *T. embryophagum*, *T. evanescens*, *T. japonicum*, *T. nagarkattii*, *T. nubilalae*, *T. ostriniae*, *T. pretiosum*, *T. trjapitzini* (Hubei Coop. Research Group, 1979; Liu *et al.*, 1983; Li, 1986; Li *et al.*, 1989; Dai *et al.*, 1991).

Outside the *Trichogramma* genus, a few species of oophagous were successfully reared in artificial media. *Tetrastichus shoenobii* (Tetrastichidae) was reared from caesarean eggs, obtained by dissection of females, to normal adults in an artificial diet with a pupation rate of 70–80% and emergence rate of 78–83%. Fecundity of adults was normal and their body size and longevity were higher than those for natural hosts (Ding *et al.*, 1980a). *Anastatus japonicum* and *Anastatus* sp. were reared on an artificial diet containing silkworm haemolymph, hen's egg yolk, cow's milk and inorganic salts, with 93% pupation and 66–89% adult emergence. The egg laying took place directly into the plastic artificial host eggs. *Anastatus* spp. were released on a large scale in litchi orchards to control the litchi stink bug *Tessaratoma papillosa*, resulting in 89–90% parasitization of pest eggs (Liu *et al.*, 1986, 1988; Han *et al.*, 1988; Li *et al.*, 1988). Eggs of *Telenomus heliothidis* (Scelionidae), collected from parasitized eggs of *Heliothis virescens*, developed in artificial media containing *M. sexta* haemolymph, hen's egg yolk, cow's milk and trehalose solution with a survival rate of 42% for pupae and 16% for adults (Strand *et al.*, 1988). *Ooencyrtus pityocampae* (Encyrtidae) was reared in artificial host eggs (cupules in PVC film) after direct egg laying into the cupules containing a diet free from insect material, with hen's egg yolk, casein and yeast hydrolysates, trehalose, Wesson salts, RNA solution and vitamins. But only a very low percentage of adults was obtained (Masutti *et al.*, 1992).

Conclusions

The possibility of continuous artificial mass-rearing of parasitoids and predators is currently limited. A better understanding of their physiology, behaviour and genetics is needed. But, in order to have good quality parasitoids produced, it seems inadvisable to maintain the parasitoids for many generations *in vitro*, as recommended now for mass-production, even on alternative hosts.

Up to now, mainly parasitoids of lepidopterous species have developed in artificial conditions, but tests are in progress with some egg parasitoids of Heteroptera. Best results were obtained with hymenopteran oophagous compared to other parasitoids since they kill their host quickly and do not display a great dependence on living host physiology (Campadelli & Dindo, 1987). In fact, many egg parasitoids such as *Trichogramma* spp. can develop in alternative host eggs killed by UV radiation, and *T. nubilalae* developed in irradiated frozen eggs (Nagarkatti *et al.*, 1991). Utilization of scelionid wasps should be emphasized, as suggested by Orr (1988), eventually using artificial rearing.

Even if continuous culture of some species is possible for some generations, it is not desirable to maintain parasitoids under artificial conditions for too long because problems of loss of vigour and performance could arise, possibly implicating genetic factors (Mackauer, 1976). Unfortunately, few investigations have been done in this field; some studies, however, have been done on laboratory colonies on unnatural hosts (Thompson, 1986). In tachinids, a long rearing in the laboratory on an alternative host induced the modification of the capability to develop in artificial medium (Grenier & Pintureau, 1991).

References

Barrett, M. & Schmidt, J.M. (1991) A comparison between the amino acid composition of an egg parasitoid wasp and some of its hosts. *Entomologia Experimentalis et Applicata* 59, 29–41.

Battisti, A., Ianne, P., Milani, N. & Zanata, M. (1990) Preliminary accounts on the rearing of *Ooencyrtus pityocampae* (Mercet) (Hym., Encyrtidae). *Journal of Applied Entomology* 110, 121–127.

Bratti, A. (1990) Tecniche di allevamento *in vitro* per gli stadi larvali di insetti entomofagi parassitoidi. *Bolletino dell' Istituto di Entomologia della Università degli studi di Bologna* 44, 169–220.

Campadelli, G. & Dindo, M.L. (1987) Recenti progressi nello studio delle diete artificiali per l'allevamento di insetti entomofagi parassiti. *Bolletino dell' Istituto di Entomologia della Università degli studi di Bologna* 42, 101–118.

Cao, A.H., Li, S.W., Dai, K.J. & Gao, Y.G. (1988) Study on the artificial host egg for Telenominae. 1. Biochemical analysis in the compositions of *Dendrolimus punctatus* (Walker) eggs. *Natural Enemies of Insects* 10, 125–128.

Dahlman, D.L. (1990) Evaluation of teratocyte functions: an overview. *Archives of Insect Biochemistry and Physiology* 13, 159–166.

Dai, K.J., Cao, A.H. & Li, X.Y. (1987) Oxygen consumption in different stages of *Trichogramma dendrolimi* during its development. *Journal of Wuhan University* 4, 169–177.

Dai, K.J., Zhang, L.W., Ma, Z.J., Zhong, L.S., Zhan, Q.X., Cao, A.H., Xu, K.J., Li, Q. & Gao, Y.G. (1988) Research and utilization of artificial host egg for propagation of parasitoid *Trichogramma*. In: Voegelé, J., Waage, J.K. & van Lenteren, J.C. (eds) Trichogramma *and Other Egg Parasites. Les Colloques de l'INRA* 43, 311–318.

Dai, K.J., Ma, Z.J., Zhang, L.W., Cao, A.H., Zhan, Q.X., Xu, K.J., Pan, D.S. & Zhang, J.L. (1991) Research on technology of industrial production of the artificial host egg of *Trichogramma*. In: Wajnberg, E. & Vinson, S.B. (eds) Trichogramma *and Other Egg Parasitoids. 3rd International Symposium. Les Colloques de l'INRA* 56, 137–139.

Ding, D.C., Qiu, H.G. & Hwang, C.B. (1980a) *In vitro* rearing of an egg-parasitoid *Tetrastichus schoenobii* Ferriere (Hymenoptera: Tetrastichidae). *Contributions Shanghai Institute of Entomology* 1, 55–58.

Ding, D.C., Zhang, T.P. & Zhong, Y.K. (1980b) Studies on *Tetrastichus schoenobii* Ferriere (Hymenoptera: Tetrastichidae): oviposition into artificial media. *Contributions Shanghai Institute of Entomology* 1, 59–61.

Ding, D.C., Qiu, H.G., Du, J.W., Fu, W.J. & He, L.F. (1981) Studies on the kairomone influencing oviposition behavior of *Tetrastichus schoenobii* Ferriere: source and extraction. *Acta Entomologica Sinica* 24, 262–268.

Gao, Y.G., Dai, K.J. & Shong, L.S. (1982) *Trichogramma* sp. and their utilization in People's Republic of China. In: Voegelé, J. (ed.) *Les Trichogrammes. Les Colloques de l'INRA* 9, 181.

Greany, P.D., Vinson, S.B. & Lewis, W.J. (1984) Insect parasitoids: finding new opportunities for biological control. *BioScience* 34, 690–696.

Grenier, S. & Bonnot, G. (1988) Development of *Trichogramma dendrolimi* and *T. maidis* (Hymenoptera, Trichogrammatidae) in artificial media and artificial host eggs.

In: Voegelé, J., Waage, J.K. & van Lenteren, J.C. (eds) Trichogramma *and Other Egg Parasites. Les Colloques de l'INRA* 43, 319–326.

Grenier, S. & Liu, W.H. (1990) Antifungals: mold control and safe levels in artificial media for *Trichogramma* (Hymenoptera, Trichogrammatidae). *Entomophaga* 35, 283–291.

Grenier, S. & Liu, W.H. (1991) Mold control and safe level of antifungals in artificial media for egg parasitoids (Hymenoptera). In: Wajnberg, E & Vinson, S.B. (eds) Trichogramma *and Other Egg Parasitoids. 3rd International Symposium. Les Colloques de l'INRA* 56, 141–144.

Grenier, S. & Pintureau, B. (1991) Essai d'élevage en milieux artificiels de *Lixophaga diatraeae* (Dipt. Tachinidae), parasitoïde des Lépidoptères foreurs de la canne à sucre. Importance de la souche. In: Pavis, C. and Kermarrec, A. (eds) *Rencontres caraïbes en lutte biologique. Les Colloques de l'INRA* 58, 451–458.

Grenier, S., Delobel, B. & Bonnot, G. (1986) Physiological interactions between endoparasitic insects and their hosts: physiological considerations of importance to the success of *in vitro* culture: an overview. *Journal of Insect Physiology* 32, 403–408.

Grenier, S., Veith, V. & Renou, M. (1993) Some factors stimulating oviposition by the oophagous parasitoid *Trichogramma brassicae* Bezd. (Hym., Trichogrammatidae) in artificial host eggs. *Journal of Applied Entomology* 115, 66–76.

Guan, X.C., Wu, Z.X., Wu, T.N. & Feng, H. (1978) Studies on rearing *Trichogramma dendrolimi* Matsumura *in vitro. Acta Entomologica Sinica* 21, 122–126.

Han, S.C., Chen, Q.X., Xu, X., Zhang, M.L., Zhu, D.F. & Liu, W.H. (1988) *In vitro* rearing *Anastatus japonicus* Ashmead (Hym.: Eupelmidae) for controlling litchi stink bug, *Tessaratoma papillosa* Drury (Hem.: Pentatomidae). *Natural Enemies of Insects* 10, 170–173.

Hawlitzky, N. & Boulay, C. (1982) Régimes alimentaires et développement chez *Trichogramma maidis* Pintureau et Voegelé (Hym. Trichogrammatidae) dans l'oeuf d'*Anagasta kuehniella* Zeller (Lep. Pyralidae). In: Voegelé, J. (ed.) *Les Trichogrammes. Les Colloques de l'INRA* 9, 101–106.

Heath, R.R., Ferkovich, S.M., Greany, P.D., Eller, F.J., Dueben, B.D. & Tilden, R.L. (1990) Progress in the isolation and characterization of a host hemolymph ovipositional kairomone for the endoparasitoid *Microplitis croceipes. Archives of Insect Biochemistry and Physiology* 13, 255–265.

Hoffman, J.D., Ignoffo, C.M. & Dickerson, W.A. (1975) *In vitro* rearing of the endoparasitic wasp, *Trichogramma pretiosum. Annals of the Entomological Society of America* 68, 335–336.

Hsia, P.Y. & Wang, M.H. (1979) The structure of oak silkworm egg shell and its relation to Trichogrammatid parasitism. *Acta Entomologica Sinica* 22, 30–33.

Hubei Coop. Research Group (1979) Studies on the artificial host egg of the endoparasitoid wasp *Trichogramma. Acta Entomologica Sinica* 22, 301–309.

Hubei Coop. Research Group (1987) A preliminary analysis of the ingredients of *Antheraea pernyi* egg – host of *Trichogramma. Journal of Wuhan University* 131–138.

Irie, K., Xie, Z.N., Nettles, W.C., Morrison, R.K., Chen, A.C., Holman, G.M. & Vinson, S.B. (1987) The partial purification of a *Trichogramma pretiosum* pupation factor from hemolymph of *Manduca sexta. Insect Biochemistry* 17, 269–275.

Li, L.Y. (1986) Mass production of natural enemies (parasites and predators) of insect

pests. *Natural Enemies of Insects* 8, 52–62.

Li, L.Y., Liu, W.H., Chen, C.S., Han, S.C., Shin, J.C., Du, H.S. & Feng, S.Y. (1988) *In vitro* rearing of *Trichogramma* spp. and *Anastatus* sp. in artificial 'eggs' and the methods of mass production. In: Voegelé, J., Waage, J.K. & van Lenteren, J.C. (eds) Trichogramma *and Other Egg Parasites. Les Colloques de l'INRA* 43, 339–352.

Li, L.Y., Chen, Q.X. & Liu, W.H. (1989) Oviposition behavior of twelve species of *Trichogramma* and its influence on the efficiency of rearing them *in vitro*. *Natural Enemies of Insects* 11, 31–35.

Liu, W.H. & Wu, Z.X. (1982) Recent results in rearing *Trichogramma in vitro* with artificial media devoid of insectan additives. *Acta Entomologica Sinica* 25, 160–163.

Liu, W.H., Xie, Z.N., Xiao, G.F., Zhou, Y.F., Ou Yang, D.H. & Li, L.Y. (1979) Rearing of the *Trichogramma dendrolimi* in artificial diets. *Acta Phytophylacica Sinica* 6, 17–25.

Liu, W.H., Zhou, Y.F., Chen, C.S., Zhang, Y.H. & Han, S.C. (1983) *In vitro* rearing of *Trichogramma japonicum* and *T. nubilalae*. *Natural Enemies of Insects* 5, 166–170.

Liu, Z.C., Sun, Y.R., Wang, Z.Y., Liu, J.F., Zhang, L.W., Dai, K.J. & Gao, Y.G. (1985) Field release of *Trichogramma confusum* reared on artificial host eggs against sugarcane borers. *Natural Enemies of Insects* 3, 1–5.

Liu, Z.C., Wang, Z.Y., Sun, Y.R., Liu, J.F. & Yang, W.H. (1986) Mass propagation of *Anastatus* sp., a parasitoid of litchi stink bug, with artificial host eggs. *Chinese Journal of Biological Control* 2, 54–58.

Liu, Z.C., Wang, Z.Y., Sun, Y.R., Liu, J.F. & Yang, W.H. (1988) Studies on culturing *Anastatus* sp. a parasitoid of litchi stink bug, with artificial host eggs. In: Voegelé, J., Waage, J.K. & van Lenteren, J.C. (eds) Trichogramma *and Other Egg Parasites. Les Colloques de l'INRA* 43, 353–360.

Lu, A.P. & Yang, Q.Y. (1988) A study of alimentary canal in *Anastatus* sp. (Hymen.: Eupelmidae). *Natural Enemies of Insects* 10, 83–87.

Lu, W.Q. & Lang, S. (1979) Studies on behaviour of parasitic animals. V. Oviposition behaviour of *Trichogramma australicum* (Trichogrammatidae, Hymenoptera) through flat surface of artificial membranes. *Natural Enemies of Insects* 1, 16–24.

Lu, W.Q., Lang, S., Xie, Z.N. & Chang, Y.H. (1979) Oviposition behaviour of Trichogrammatid wasps. *Acta Entomologica Sinica* 22, 361–363.

Mackauer, M. (1976) Genetic problems in the production of biological control agents. *Annual Review of Entomology* 21, 369–385.

Masutti, L., Slavazza, A. & Battisti, A. (1991) Oviposition of *Ooencyrtus pityocampae* (Mercet) into artificial eggs (Hym., Encyrtidae). *Redia* 74, 457–462.

Masutti, L., Battisti, A., Milani, N. & Zanata, M. (1992) First success in the *in vitro* rearing of *Ooencyrtus pityocampae* (Mercet) (Hym. Encyrtidae). Preliminary note. *Redia* 75, 227–232.

Mellini, E. (1986) Importanza dell'età dell'uovo, al momento della parassitizzazione, per la biologia degli Imenotteri oofagi. *Bolletino dell' Istituto di Entomologia dell' Università degli studi di Bologna* 41, 1–21.

Morrison, R.K., Nettles, W.C., Ball, D. & Vinson, S.B. (1983) Successful oviposition by *Trichogramma pretiosum* through a synthetic membrane. *Southwestern Entomologist* 8, 248–251.

Nagarkatti, S., Giroux, K.J. & Keeley, T.P. (1991) Rearing *Trichogramma nubilale* (Hymenoptera, Trichogrammatidae) on eggs of the Tobacco Hornworm, *Manduca sexta* (Lepidoptera, Sphingidae). *Entomophaga* 36, 443–446.

Nettles, W.C. (1990) *In vitro* rearing of parasitoids – role of host factors in nutrition. *Archives of Insect Biochemistry and Physiology* 13, 167–175.

Nettles, W.C., Morrison, R.K., Xie, Z.N., Ball, D., Shenkir, C.A. & Vinson, S.B. (1982) Synergistic action of potassium chloride and magnesium sulfate on parasitoid wasp oviposition. *Science* 218, 164–166.

Nettles, W.C., Morrison, R.K., Xie, Z.N., Ball, D., Shenkir, C.A. & Vinson, S.B. (1983) Effect of cations, anions and salt concentrations on oviposition by *Trichogramma pretiosum* in wax eggs. *Entomologia Experimentalis et Applicata* 33, 283–289.

Nettles, W.C., Morrison, R.K., Xie, Z.N., Ball, D., Shenkir, C.A. & Vinson, S.B. (1985) Effect of artificial diet media, glucose, protein hydrolyzates, and other factors on oviposition in wax eggs by *Trichogramma pretiosum*. *Entomologia Experimentalis et Applicata* 38, 121–129.

Orr, D.B. (1988) Scelionid wasps as biological control agents: a review. *Florida Entomologist* 71, 506–528.

Qin, J.D. & Wu, Z.X. (1988) Studies on cultivation of *Trichogramma in vitro*: ovipositional behaviour and larval nutritional requirements of *T. dendrolimi*. In: Voegelé, J., Waage, J.K. & van Lenteren, J.C. (eds) Trichogramma *and Other Egg Parasites*. *Les Colloques de l'INRA* 43, 379–387.

Rajendram, G.F. (1978a) Oviposition behavior of *Trichogramma californicum* on artificial substrates. *Annals of the Entomological Society of America* 71, 92–94.

Rajendram, G.F. (1978b) Some factors affecting oviposition of *Trichogramma californicum* (Hymenoptera: Trichogrammatidae) in artificial media. *Canadian Entomologist* 110, 345–352.

Rajendram, G.F. & Hagen, K.S. (1974) *Trichogramma* oviposition into artificial substrates. *Environmental Entomology* 3, 399–401.

Shen, X.C., Wang, W.X., Kong, J., Zhang, G.F. & Dai, K.J. (1991) Inoculative release of *Trichogramma dendrolimi* reared with artificial host eggs for controlling corn borer *Ostrinia furnacalis* (Lep.: Pyralidae). *Chinese Journal of Biological Control* 7, 141.

Strand, M.R. & Vinson, S.B. (1985) *In vitro* culture of *Trichogramma pretiosum* on an artificial medium. *Entomologia Experimentalis et Applicata* 39, 203–209.

Strand, M.R., Vinson, S.B., Nettles, W.C. & Xie, Z.N. (1988) *In vitro* culture of the egg parasitoid *Telenomus heliothidis*: the role of teratocytes and medium consumption in development. *Entomologia Experimentalis et Applicata* 46, 71–78.

Takasu, K. & Yagi, S. (1992) *In vitro* rearing of the egg parasitoid, *Ooencyrtus nezarae* Ishii (Hymenoptera, Encyrtidae). *Applied Entomology and Zoology* 27, 171–173.

Thompson, S.N. (1986) Nutrition and *in vitro* culture of insect parasitoids. *Annual Review of Entomology* 31, 197–219.

Volkoff, N. & Colazza, S. (1992) Growth patterns of teratocytes in the immature stages of *Trissolcus basalis* (Woll) (Hymenoptera, Scelionidae), an egg parasitoid of *Nezara viridula* (L) (Heteroptera, Pentatomidae). *International Journal of Insect Morphology and Embryology* 21, 323–336.

Volkoff, N., Vinson, S.B., Wu, Z.X. & Nettles, W.C. (1992) *In vitro* rearing of *Trissolcus basalis* (Hym, Scelionidae) an egg parasitoid of *Nezara viridula* (Hem., Pentatomidae). *Entomophaga* 37, 141–148.

Wu, Z.X. & Qin, J.D. (1982) Ovipositional response of *Trichogramma dendrolimi* to the chemical contents of artificial eggs. *Acta Entomologica Sinica* 25, 363–371.

Wu, Z.X., Zhang, Z.P., Li, T.X. & Liu, D.M. (1980) Artificial media devoid of insect additives for rearing larvae of the endoparasitoid wasp *Trichogramma*. *Acta Entomologica Sinica* 23, 232.

Wu, Z.X., Qin, J.D., Li, T.X., Chang, Z.P. & Liu, D.M. (1982) Culturing *Trichogramma dendrolimi in vitro* with artificial media devoid of insect materials. *Acta Entomologica Sinica* 25, 128–133.

Xie, Z.N., Nettles, W.C., Morrison, R.K., Irie, K. & Vinson, S.B. (1986a) Effect of ovipositional stimulants and diets on the growth and development of *Trichogramma pretiosum in vitro*. *Entomologia Experimentalis et Applicata* 42, 119–124.

Xie, Z.N., Nettles, W.C., Morrison, R.K., Irie, K. & Vinson, S.B. (1986b) Three methods for the *in vitro* culture of *Trichogramma pretiosum* Riley. *Journal of Entomological Science* 21, 133–138.

Xie, Z.N., Nettles, W.C., Morrison, R.K. & Vinson, S.B. (1988) *In vitro* mass culture of *Trichogramma pretiosum* Riley from eggs collected using a salt solution as ovipositional stimulant. In: Voegelé, J., Waage, J.K. & van Lenteren, J.C. (eds) Trichogramma *and Other Egg Parasites. Les Colloques de l'INRA* 43, 403–406.

Xie, Z.N., Nettles, W.C., Vinson, S.B. & Li, Y.H. (1991a) Identification of ovipositional stimulants for *Anastatus japonicus*. In: Wajnberg, E. & Vinson, S.B. (eds) Trichogramma *and Other Egg Parasitoids. 3rd International Symposium. Les Colloques de l'INRA* 56, 101–104.

Xie, Z.N., Xie, Y.Q., Li, L.Y. & Li, Y.H. (1991b) A study on the oviposition stimulants of *Trichogramma neustadt. Acta Entomologica Sinica* 34, 54–59.

Xin, J.C. & Li, L.Y. (1989) Observation on the ovipositional behavior of *Anastatus japonicus* (Ashmead) by *in vitro* rearing method and the results of its continuous rearing *in vitro*. *Natural Enemies of Insects* 11, 61–64.

Xu, X., Li, K.H., Li, Y.F., Moon, Q.Z. & Li, L.Y. (1986) Culture of resistant strain of *Trichogramma japonicum* (Hym.: Trichogrammatidae) to pesticides. *Natural Enemies of Insects* 8, 155–159.

Zhang, L.W. & Cao, A.H. (1993) A study on the rearing of *Amblyseius nicholsi* (Acari: Phytoseiidae) on artificially produced *Trichogramma* pupae. *Chinese Journal of Biological Control* 9, 9–11.

Zhang, M.L., Xu, X., Han, S.C. & Chen, Q.X. (1989) Effect of trichorson on *Anastatus japonicus* Ashmead (Hymen.: Eupelmidae) *in vitro*. *Natural Enemies of Insects* 11, 10–11.

Zhong, L.S. (1987) A preliminary study on culturing *Trichogramma ostriniae in vitro. Chinese Journal of Biological Control* 11, 112–113.

Zhong, L.S. & Zhang, J.L. (1989) Influence of diet pH to the development and the efficacy of reproduction of *Trichogramma confusum* (Hym.: Trichogrammatidae) in artificial rearing. *Chinese Journal of Biological Control* 5, 101–103.

Zhong, L.S. & Zhang, J.L. (1992) Effect of food dregs in the pupal tissue fluid of *Antheraea pernyi* on the development of *Trichogramma* and a new simplified method of obtaining the fluid as a component of the artificial host eggs for *Trichogramma*. *Chinese Journal of Biological Control* 8, 145–147.

Quality Control in *Trichogramma* Production

5

Franz Bigler

Swiss Federal Research Station for Agronomy, Reckenholtzst.
191, 8046 Zürich, Switzerland

Abstract

Quality control in mass-production of *Trichogramma* is discussed in the context of the total quality aspects of the rearing programme. The overall performance of *Trichogramma* species or strains is subdivided into major components and measurable attributes. Three examples of different *Trichogramma* mass-rearing programmes are presented and major components and attributes are selected according to the specific objectives and needs of the rearing and release system. Methods to assess single attributes must be quick, simple and reliable in order to be applied in a production unit. A combination of quality traits with a high predictive value to the field performance should be established for each specific rearing system and implemented in the production and release programme. Examples of a number of quality attributes for *T. brassicae* and *T. minutum* are listed and the relationship to the field performance is discussed. It is shown that quality control encompasses more than the assessment of the parasitoid performance which is the result of the rearing system's quality. Total quality control means training of staff, adequate technical equipment and permanent production, process and product control with feedback mechanisms for corrections. A case study for *T. brassicae* demonstrating the main steps in a quality control programme is presented.

Introduction

The success of biological control is related, among other factors, to the quality of the released organisms (van Lenteren, 1991). Suboptimal success can be

indicative of low insect quality, but it tells nothing about the causes of the lack of performance. Preservation of the natural attributes of the insect colony is therefore one of the particular objectives in insect rearing for release in biological control programmes. This is often in contrast to maximum efficiency of production due to economic constraints. Artificial conditions improve the rearing success and the turnover time but they may impair the colony's quality.

Mass-rearing systems of *Trichogramma* are, in general, very artificial compared to the crop habitat where the parasitoids are released. Some common characteristics of *Trichogramma* production, relevant to quality, are: unnatural host eggs at high densities, no plants available, high intraspecific competition, very limited hosts and space, high number of generations per time, no or limited possibilities for storage and no or incomplete rejuvenation of strains. These general features of mass-production systems are critical for any *Trichogramma* population because they are subjected to genetic and/or adaptive alterations. Preservation of quality attributes in mass-reared populations is therefore crucial for success.

What Is Quality in *Trichogramma* Production?

Trichogramma are used in general for inundative or seasonal inoculative release with an expected immediate effect on the pest population. Inoculative releases are exceptional, though not impossible. Thus, high parasitoid numbers of excellent performance are required at the right time and place. Production of high quality *Trichogramma* depends on the appropriate use of suitable facilities, equipment and the accurate performance of rearing operations. This includes availability of high quality host eggs, stock colony maintenance, synchronization of emergence of the parental generations, density regulation in rearing units, optimal physical conditions and sanitation. Production requires a permanent monitoring of the relevant factors influencing the rearing success. Periodical checks of performance during production will help to detect possible quality deterioration in an early stage. Shipment and handling operations at the distributor and user level may impair product quality again. In summary, high quality production of *Trichogramma* includes optimal strain selection, suitable facility operations and appropriate shipment and release systems that ensure high performance in the field.

Relative Importance of Quality Components

According to Huettel (1976), the overall quality in the field can only be measured in terms of how well the insect population functions in its intended

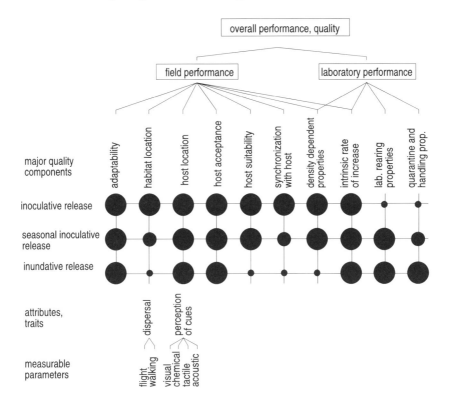

Fig. 5.1. Relative importance of quality components in relation to the objective of *Trichogramma* production and use. The size of each black circle is proportional to the importance of each trait.

role when it is released. It is certainly true that successful biological control is a clear indication of adequate insect quality, but there is need to develop and to apply test procedures that assess performance before the parasitoids are released. Assessment of overall quality seems adequate as long as no major problems are in sight. The limitations of this approach become evident as soon as problems start to appear. Performance problems may arise very rapidly because apparent success may mask concealed and gradual deterioration of quality. Then, when unfavourable conditions cause a sudden breakdown of the field efficacy, explanations must be available. It will become evident that descriptive overall quality measurement lacks the analytical power to uncover the causes.

A schematic presentation of a possible division of the overall quality of *Trichogramma* into major quality components is shown in Fig. 5.1. Each attribute can be split into single traits and measurable parameters. This division (or similar attempts to divide quality into major components) is necessary to sort out those components that might be of major importance for

field performance or are known to be especially sensitive to alteration. The selection of a few major quality components is a matter of experience and personal judgement, but it is anticipated that an acceptable consensus may emerge which will be related to the objectives of the rearing and release programme.

The major performance components consist of complex physiological and ecological properties that are not all of equal importance for successful control and depend on the use and the release strategy. The relative importance of each attribute has to be evaluated for each particular rearing programme. Thorough knowledge of ecological, physiological and biological properties of each specific pest enemy relationship is needed before evaluating major components.

Analysis of Quality Components in Three Production Systems

Three *Trichogramma* production and crop/pest systems are analysed in Table 5.1 with regard to the relative importance of quality components and measurable traits. The table shows that the crop/pest characteristics are very distinct between maize in central Europe, Asia and the Canadian forest. A rough comparison of the three production systems shows the main differences in the physical conditions (temperature, humidity), the rearing host, the possibilities of strain rejuvenation and storage. The distribution system is determined by the release strategy, the number of parasitoids released, the distribution method and the shipment to the user.

The table lists the major quality components and shows the relative importance of measurable traits. Not all of them are of the same value for the different systems and an accurate evaluation is possible only after a detailed analysis.

Adaptability to physical conditions, measured at low temperatures as parasitization rate, flight propensity and walking activity are important parameters in Switzerland and Canada because of the climatic variations in these latitudes. On the other hand, it is not an important feature in the Philippines because of the tropical climate and the production under more or less natural conditions.

Habitat location is generally not critical in *Trichogramma* used for inundative release because they are distributed in the target habitat. In the Philippines, however, where the pest is present almost all year and maize is more or less constantly planted and harvested, the egg parasitoid, once inoculated in an area, propagates naturally and migrates from one maize field to the next. Here, the property of locating the habitat is crucial. The perception of volatile chemical cues from the host plant or the host might be one of the keys for successful habitat location. It may be less important or even unnecessary if the parasitoids

Table 5.1. Analysis of three production systems: application strategies and relative importance of quality components in *Trichogramma brassicae*, *T. evanescens* and *T. minutum*. Relative importance: −, low; +, high; ++, very high. After Bigler *et al.* (1989); Bigler (1990) and Smith *et al.* (1990), respectively.

	T. brassicae	*T. evanescens*	*T. minutum*
General information			
Crop	Maize	Maize	Forest
Pest	*Ostrinia nubilalis* (European corn borer)	*Ostrinia furnacalis* (Asian corn borer)	*Choristoneura fumiferana* (Spruce budworm)
Continent	Europe	Asia	North America
Country	Switzerland	Philippines	Canada (Ontario)
Crop/pest characteristics			
Cropping season	May–October	All year	Perennial
Number of harvests per year	1	1–3	–
Pest generations per year	1	3–10	1
Location of host egg masses	All leaves	All leaves	Mainly upper part of tree
Average no. of eggs per mass	17	25	60
Density of egg masses	Low (<1 per plant)	High (1–10 per plant)	
Mass-production system			
Physical conditions	Fluctuating	± Natural	Fluctuating
Rearing host	*Ephestia kuehniella*	*Sitotroga cerealella* *Corcyra cephalonica*	*Ephestia kuehniella*
Size of rearing cages	Very small	Very small	Very small
Host and parasitoid density	Extremely high	Extremely high	Extremely high
Strain rejuvenation	Yes	Partly or completely	
Long-term storage	Diapause	No	No

Table 5.1 continued.

	T. brassicae	T. evanescens	T. minutum
Distribution system			
Release strategy	Inundative, season-long control	Inoculative (partly inundative)	Inundative
Release system	50–100 sites ha^{-1}	50 sites ha^{-1}	Bulk, broadcast
Number of releases per year	2	Varying	2
Release intervals	10 days	4–10 days	4–5 days
No. of parasitoids released	100,000 ha^{-1}	Varying	6 million ha^{-1}
Distribution method	Hands, cards	Hands, cards	Aircraft
Shipment, distribution	Express mail	Special mail	Special mail
Quality components			
Adaptability to physical conditions	++	–	++
Habitat location	–	++	–
Host location	+	+	++
Host acceptance	++	++	++
Host suitability	+	++	+
Rate of increase in mass-production	++	+	++
Rate of increase in the field	+	++	–
Examples of measurable traits			
At low temperature			
parasitization rate	++	–	++
flight and walking	++	–	++
oviposition	++	–	++

Recognition of volatile chemical cues			
from host plant	−	++	+
from host	+	++	+
Flight propensity	++	++	++
Locomotion activity	++	++	++
Recognition of eggs of the target host	++	++	++
Recognition of eggs of other hosts	−	+	−(+)
Acceptance and suitability of the target host	++	++	++
Acceptance and suitability of other hosts	−	+	−(+)
Synchronization with the host	−	+	−
Fecundity	+	+	++
Longevity	++	+	++
Sex ratio	++	+	++
Rate of emergence from factitious hosts	++	+	++
Rearing and handling properties	++	+	++

are released into the habitat close to the host eggs.

The spruce budworm lays its egg masses preferentially in the upper part of the tree. According to the release system, more or less *T. minutum* emerge in proximity to the host. For those adults emerging in the lower part of the trees or on the ground it is essential to fly upwards. In the maize crop where *T. brassicae* and *T. evanescens* are released at a few points per hectare, dispersal by flight is important as well.

If a female has located an egg mass, it must accept it and allocate the eggs in an optimal way. This is especially important if a seasonal control by natural propagation over a number of generations is expected. Natural multiplication in the field may influence the control success even in a one-generation pest situation because of the short generation time of *Trichogramma* in general. The example of corn borer control by *T. brassicae* in Switzerland may illustrate this. The oviposition period of the European corn borer lasts 4–7 weeks in central Europe (Hawlitzky, 1986; Bigler *et al.*, 1989). By releasing *T. brassicae* at the very beginning of the oviposition period and assuming a generation time of approximately 150 degree days (>10°C) (van den Heuvel, 1986), we expect the first adults of the parasitoid, developed in corn borer eggs in the field, after 16–24 days depending on the temperature. Thus, the last third of the corn borer eggs are parasitized by the naturally developed parasitoids.

A high rate of population increase is important in situations where a very high number of parasitoids must be available within a short time. This is especially critical if no procedures for long time storage are known and if high numbers of parasitoids are necessary to achieve control. A high rate of increase in the field is crucial, e.g. on the Philippines where often only a few fields of a region are inoculated with a low number of *T. evanescens* and the parasitoids build up their populations in maize fields within a short period.

Quantification of Quality

Once a thorough analysis of the production system and release strategy is made, we have to select those traits which are assumed to have a high impact on the overall quality of the parasitoids. A variety of techniques is to be developed for measuring and monitoring quality and a wealth of data will be produced.

The relatively complex structure of quality brings us to the statement that quality control procedures should be relatively simple but reliable in order to be applied widely and routinely. This breakdown of quality into innumerable parameters might indeed lead to the conclusion that quality control is so complex and sophisticated that it becomes the privilege of a few specialists. This is not the case. Thanks to the accumulating experience, improved and simplified methods and devices, and not the least to the services provided by specialized facilities, production managers should be able to analyse their problems and select or develop the proper key elements for biotests that monitor the quality of the insects they have to produce.

The question arises whether laboratory tests for monitoring insect quality can adequately assess traits that have to function under field conditions. Criticism focuses in particular on the validity of laboratory tests that measure behaviour. There is, however, a general consensus that both field and laboratory tests should be developed to complement each other because it has become obvious that both approaches have their definite advantages and shortcomings.

Advantages of the laboratory tests in general are: they are relatively cheap, quick, reproducible under standard conditions, independent of the season, comparable to a known standard and single parameters can be investigated. Field tests or tests in larger field cages have the advantage that certain components, such as dispersal, orientation mechanisms, survival rates and production of viable offspring, can be studied under the direct influence of the complex environment of the target area. Unless carried out in a highly effective manner, field experiments often produce data that are influenced by a multitude of variables that cannot be identified precisely; hence, cause and effect are obscured and analysis is difficult. Therefore, these cost- and labour-intensive field tests are the logical follow-up to preliminary laboratory experiments that provide the first indications of what direction research has to take in the field.

A number of laboratory experiments have been conducted aiming at quantifying quality attributes of *Trichogramma*. However, only a few cases are documented where laboratory findings were complemented with field tests. Female body size is often used as index of fitness or quality. It is in general correlated to traits like fecundity, longevity, rate of search and flight ability (Biever, 1972; Marston & Ertle, 1973; Bigler *et al.*, 1982; Bai *et al.*, 1992; Bourchier *et al.*, 1993; Pavlik, 1993). The significance of female body size to quality for biological control was evidenced by Kazmer & Luck (1990). These authors showed in field experiments that the largest females of *T. pretiosum* were 4.5 times more likely to arrive on a suitable host than the smallest females. They conclude that the relative fitness and quality associated with female size is substantial but not sufficiently reliable to predict the potential quality for biological control. Pavlik (1993) assessed a variety of performance attributes in the laboratory and correlated them to body size of females of different *Trichogramma* species. He concludes that body size is not a suitable parameter to predict females' performance when the parasites are reared on factitious hosts like *Ephestia kuehniella* or *Sitotroga cerealella*. Bigler *et al.* (1987) compared the quality of *T. maidis* Pintureau & Voegelé (= *T. brassicae* Bezdenko) from the two rearing hosts *Ephestia kuehniella* and *Sitotroga cerealella*. The populations reared on *E. kuehniella* from where the adults are larger showed a slightly higher field efficacy. However, the differences in body size alone did not explain sufficiently the difference in overall performance.

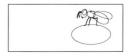

This suggests that other attributes, not correlated to body size, are modified by the rearing process.

Behavioural traits, responsible for successful host searching and acceptance, such as flight, walking and recognition of physical structures and chemical cues may be altered or lost. Salmanova (1991) demonstrated a substantial loss of females' ability for spontaneous flights in the course of mass-production on *S. cerealella*. Temperature adaptation was observed after a few generations reared under constant temperatures and the normal daily rhythm was lost after 15 generations (Shchepetilnikova & Kasinskaya, 1981). Bigler *et al.* (1988) showed that velocity and time spent walking decreased if strains were reared continuously under artificial conditions on *E. kuehniella*. Parasitism in the field was compared between these strains and a strong correlation with the locomotory activity was demonstrated. The ability to react to host cues can be changed or lost under laboratory rearing conditions (Salmanova, 1991). Neuffer (1987) and Bergeijk *et al.* (1989) showed that *T. maidis* (= *T. brassicae*), reared on factitious hosts, changed its acceptance of the natural host, *O. nubilalis*, with an increasing number of generations reared on the factitious hosts. In the same study it was demonstrated that, after 300 generations on *E. kuehniella*, the strain could not be reared again on its original host because the parasitoid larvae were not able to develop anymore.

Theoretically, a high number of single quality traits could be measured in the laboratory but, for practical purposes, only a few can be assessed. Thus, production managers should answer the following questions: (i) Which attributes are important with regard to field performance? (ii) What is the value of a single attribute for the overall quality? (iii) How are attributes correlated to each other? and (iv) What methods are quick and reliable? Greenberg (1991) evaluated quality attributes of *T. evanescens* in the laboratory, such as searching ability, emergence rate, sex ratio and fecundity, and related them to the effectiveness in the field. Taking into account these attributes he formed an integral index that is used to classify the parasitoids according to their expected performance before release. Cerutti & Bigler (1994) assessed the values of emergence rate, sex ratio, life span, walking activity, fecundity and acceptance of the natural host on 52 populations of *T. brassicae* subjected to different rearing and storage procedures. They combined these parameters to one integral quality index and compared it to field efficacy. The correlation between the quality index and the effectiveness in the field evidences the practical value of the quality assessment in the laboratory. The methods used in this investigation are described by Cerutti & Bigler (1991).

Implementation of Quality Control in Mass-rearing Programmes

Boller & Chambers (1977) defined the role of quality control to '. . . provide and coordinate a production system that ensures that the operation will produce adequate numbers of an optimum quality at minimum product costs'. Chambers & Ashley (1984) and Leppla & Fisher (1989) subdivided quality control into production, process and product control and defined these terms for insect rearing programmes in general.

Production control is the regulation of consistency, reliability and timeliness of production output. Production of high quality parasitoids depends on the appropriate use of suitable facilities, equipment and the accurate performance of rearing operations. This includes availability of high quality host eggs, stock colony maintenance, synchronization of emergence of the parental generations, density regulation in rearing units, optimal physical conditions and sanitation.

Process control is the assessment of biological parameters of the unfinished product that tell us how well the manufacturing processes are performing. Unacceptable deviations from the standard or the product specification is an alarm-signal and will ask for corrections of the rearing processes. Periodically measured parameters like parasitization rate, super-parasitism, sex ratio, fecundity, flight and walking activity, host acceptance etc., may tell us the performance of the operations and will provide feedback for corrections.

Product control is the assessment of biological parameters of the finished product that tell us how well the product is conforming to standards and specifications of quality.

The division of quality control in three primary subdivisions was adopted by Laing & Bigler (1991) and used to implement quality control in *Trichogramma* mass-rearing programmes (Fig. 5.2).

Selection of founder population and colonization

The initial use of quality and performance control for *Trichogramma* should occur during the selection of a founder population of the desired species and the first steps of colonization. The 'screening process' at this stage would be quite rigorous and time-consuming. Many tests of quality could be applied at this stage of the process; not all tests listed here would be applied to a founder population although most of them would be needed.

In an approximate order of difficulty, the tests to be used are: determination of percentage of emergence, sex ratio, fecundity, longevity, duration of emergence, developmental rate, locomotion (walking, flight propensity), host acceptance (natural and factitious hosts), host suitability (natural and

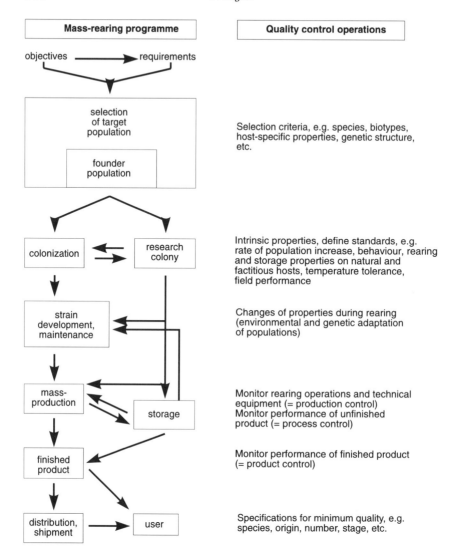

Fig. 5.2. Main elements of a *Trichogramma* mass-rearing programme and implementation of quality control.

factitious hosts), temperature tolerance (thresholds), functional response, quiescence and diapause capacity, and semi-field or field performance.

Mass-production and storage

After the selection process has been completed, one or more strains or species of *Trichogramma* will be utilized in a mass-rearing programme. A series of tests will be conducted that is less comprehensive and less time-consuming than those performed during strain selection and colonization. This is of necessity, because the rearing system itself will require a considerable proportion of the total manpower available for the programme. Thus, several of the more time-consuming tests in strain selection are omitted here. The frequency with which the following tests should be run will probably be high (once a month or more) during the start-up phase of mass-rearing, but if results do not indicate deterioration of the culture, this frequency may decline to as little as once a year. The tests to be used during mass-production are the determination of: percentage of emergence, sex ratio, fecundity, longevity, locomotion (walking, flight propensity), host acceptance (natural host), and species strain identification.

Distribution and shipment

At this time, mass-produced *Trichogramma* are being prepared for shipment to distributors and/or users. The producer must be confident that the performance characteristics of his mass-reared *Trichogramma* will be satisfied with each batch shipped. Yet, in systems without diapause storage, it is the time when the producer is trying to maximize the numbers of parasitoids produced. Thus, the time allocated to testing is necessarily restricted.

The basic information to be provided by the producer is the number of healthy female *Trichogramma* per unit shipped. In order to obtain this figure, the following information is needed: number of host eggs parasitized (number of black eggs), percentage of emergence and sex ratio. Rapid fecundity, longevity and host acceptance tests may complement the other basic tests. Each of the tests requires a minimum amount of time to complete so that the results will be known at the time of shipment. All of the tests should be run on each batch to be shipped.

In order to determine if the wasps have been affected negatively by handling and shipment, the distributor and/or user should remove a sample from the batch (a portion of cards, capsules or several hundred parasitized eggs from a bulk shipment) and hold it as near as possible at the conditions used for this test by the producer until all adults have emerged. The percentage emergence of each batch should be determined. If the *Trichogramma* do not perform well after release, this figure can be compared with that of the producer to determine if shipping has affected their quality.

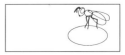

Trichogramma brassicae – A Case Study

In Switzerland, *Trichogramma brassicae* has been mass-produced since 1975 and applied commercially against the European corn borer, *Ostrinia nubilalis*, in maize since 1978. A significant loss of field efficacy was observed in 1980 (Fig. 5.3). By changing the mass-production system and the colony maintenance, it was possible to improve the performance of the strain and achieve the efficiency limit of at least 75% parasitism in the field.

A thorough analysis of the production system and the performance requirements of *T. brassicae* under the maize growing conditions in Switzerland led to the discovery of important traits which are crucial for a high efficacy. Since attributes like locomotory activity, host acceptance, host suitability and temperature tolerance were negatively affected by the former rearing system, we developed a new production unit. At the same time, risk evaluations of other deteriorations in the strain were performed and methods for measuring single traits and the field performance were developed.

In recent years, the production system of *T. brassicae* was changed from a short period production with a high daily output to a long period production with a low daily output. Improvements of the long-term storage of the

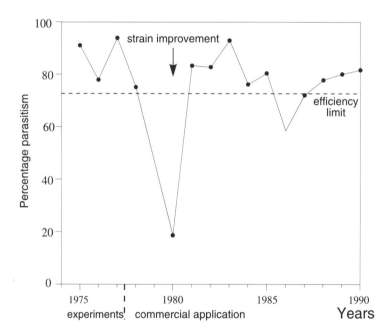

Fig. 5.3. Parasitism (%) of *Ostrinia nubilalis* eggs by *Trichogramma brassicae* in maize from 1975 to 1990 in Switzerland.

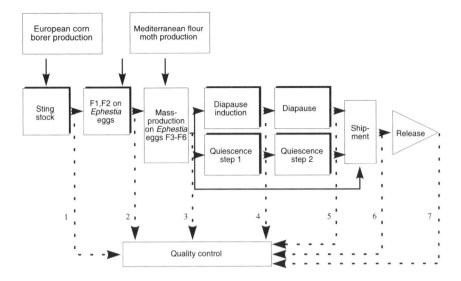

Fig. 5.4. Implementation of a quality control system in the mass-rearing programme of *Trichogramma brassicae* in Switzerland.

parasitoids (diapause) has prolonged the mass-production period from 2 to 9 months per year. The main production steps are schematically presented in Fig. 5.4. Table 5.2 lists the performance and product control procedures in the mass-rearing of *T. brassicae*.

The total production system consists of three subunits, namely the European corn borer rearing unit as natural host of the wasps, the *Ephestia kuehniella* unit as factitious (mass-rearing) host and the *T. brassicae* unit. Each of these subunits has its own quality control system. The purpose of the two host rearing units is to provide the *T. brassicae* unit at the right time with a sufficient amount of high quality eggs. Quality control procedures in the host units concentrate on: the rearing diet quality (origin, storage, grind, mixture), egg hatch, larval and pupal weight per unit rearing diet, egg production, egg sterilization efficacy, temperature and humidity regulation, ventilation control to prevent health hazards by insect scales and sanitation procedures to avoid insect diseases.

Eggs of the European corn borer are used to produce the sting stock of *T. brassicae*. This population is permanently reared under semi-field conditions, i.e. in a field insectary in summer and in a greenhouse with fluctuating temperatures in winter; corn borer egg masses are attached to corn plants and the adults of the parental generation emerge 2 to 4 m away from the plants so that egg masses must be reached by flight. A portion of the sting stock is regularly used to develop the strain on eggs of *E. kuehniella*. From our experience we know that the performance decreases with an increasing

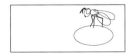

F. Bigler

Table 5.2. Performance control in the rearing of *Trichogramma brassicae* in Switzerland. Numbers correspond to those on the quality control arrows in Fig. 5.4.

Flow-chart item of Fig. 5.4	Quality control procedure	Frequency
1. Sting stock	Host acceptance Locomotion (walking) Fecundity Longevity Percentage parasitism in field	Once a year
2. F1, F2	Host acceptance Host suitability Walking speed Percentage females Fecundity Longevity	Once a year
3. F6 (no storage)	Percentage parasitism Percentage emergence Percentage females	Each batch
4. Diapause induction and quiescence (step 1)	Percentage parasitism	Each batch
5. Diapause and quiescence (step 2)	Percentage emergence Percentage females Fecundity Longevity Walking speed	Each batch
6. Shipment (transportation to user)	Percentage black eggs per release unit Percentage emergence Percentage females No. of females per release unit	Each shipment
7. Release	Percentage parasitism in field	Once a season

number of generations reared on the factitious host. Therefore, we recommend production of no more than six to seven generations before releasing the parasitoids for biological control purposes. With the possibility of diapause storage of *T. brassicae* for about 9 months, we accumulate a large number of parasitized eggs of *E. kuehniella* to have a stock ready for further rearing if needed. The long-term storage gives the production system the needed flexibility in a control programme where a high number of parasitoids must be released within 2 to 3 weeks in summer.

Since the sting stock is reared under near-natural conditions we do not expect deterioration. However, once a year we assess in the laboratory the parameters listed in Table 5.2 and, in addition, we measure the field efficacy (percentage parasitism). Since we know from our previous experiments that a change of quality attributes does not occur (or is not measurable) within the first generations on *E. kuehniella* eggs, we quantify the parameters only once a year. The sixth generation (F6) is normally sold to the farmers. The performance control programme depends on the further utilization of the parasitoids. A few rapid tests (parasitism, emergence, sex ratio) are made on each batch when the parasitized eggs are shipped immediately (without storage). Before diapause induction and quiescence (step 1), we assess only the percentage parasitism. Shortly before diapause is broken and the parasitoids are removed from quiescence storage, we conduct a number of performance controls (item 5 in Table 5.2). This is necessary because storage is a delicate procedure for *T. brassicae* and may influence its performance. The correct number of females shipped by the producer is determined by taking the corresponding figures of the previous tests (items 3–5 in Table 5.2). The final user (the farmer) is in general not able to do any performance tests. Therefore government institutes, with financial support of the *Trichogramma* producers, accomplish the tests listed under items 6 and 7.

Conclusions

Quality control in *Trichogramma* mass-rearings is one of the measures used to avoid failures in biological control with these parasitoids. The extremely artificial rearing conditions, compared to the habitat where they are released, call for the establishment of sophisticated quality control concepts. First, the objectives of mass-production must be defined clearly and the rearing system established. The shortcomings of the system must then be analysed and, based on this information, an adequate quality control system will be built up, whereby the overall performance of a species or strain may be divided into major components and measurable traits. The importance of single performance attributes has to be established and related to the field performance. The methods must be quick, simple and reliable. A single trait will never predict the overall performance accurately and therefore, the best combination of a set of laboratory methods must be developed. Whereas performance of the parasitoids in the field is the best indication for a good rearing system, low field efficacy does not tell us the causes. Regular performance control, carried out in the laboratory, will either indicate deterioration of performance and initiate corrections or make us confident to produce wasps that are within the quality specifications.

Total quality control encompasses more than the assessment of those aspects of the parasites' performance which are the result of the rearing system's quality. Perfect management, well-trained and motivated technical staff, adequate technical equipment and material, close contacts to a research

group and permanent production, process and product control with corresponding feedback for corrections and improvements are prerequisites for high quality *Trichogramma*.

References

Bai, B., Luck, R.F., Forster, L., Stephens, B. & Janssen, J.A.M. (1992) The effect of host size and quality attributes of the egg parasitoid, *Trichogramma pretiosum*. *Entomologia Experimentalis et Applicata* 64, 37–48.

Bergeijk, K.E. van, Bigler, F., Kaashoek, N.K. & Pak, G.A. (1989) Changes in host acceptance and host suitability as an effect of rearing *Trichogramma maidis* on a factitious host. *Entomologia Experimentalis et Applicata* 52, 229–238.

Biever, K.D. (1972) Effect of temperature on the rate of search by *Trichogramma* and its potential application in field releases. *Environmental Entomology* 2, 194–197.

Bigler, F. (1990) Biological control of the Asian corn borer, *Ostrinia furnacalis* in the Philippines. *Report of an Expert Consultation of the GTZ-project*, 23 pp.

Bigler, F., Baldinger, J. & Luisoni, L. (1982) L'impact de la méthode d'élevage et de l'hôte sur la qualité intrinsèque de *Trichogramma evanescens* Westw. In: Voegelé, J. (ed.) *Les Trichogrammes. Les Colloques de l'INRA* 9, 167–180.

Bigler, F., Meyer, M. & Bosshart, S. (1987) Quality assessment in *Trichogramma maidis* Pint. et Voeg. reared from eggs of the factitious hosts *Ephestia kuehniella* and *Sitotroga cerealella*. *Journal of Applied Entomology* 104, 340–353.

Bigler, F., Bieri, M., Fritschy, A. & Seidel, K. (1988) Variation in locomotion between laboratory strains of *Trichogramma maidis* and its impact on parasitism of eggs of *Ostrinia nubilalis* in the field. *Entomologia Experimentalis et Applicata* 49, 283–290.

Bigler, F., Bosshart, S. & Waldburger, M. (1989) Bisherige und neue Entwicklungen bei der biologischen Bekämpfung des Maiszünslers, *Ostrinia nubilalis* Hbn., mit *Trichogramma maidis* Pint. et Voeg. in der Schweiz. *Landwirtschaft Schweiz* 2, 37–43.

Boller, E.F. & Chambers, D.L. (1977) Concepts and approaches. In: Boller, E.F. & Chambers, D.L. (eds) *Quality Control – An Idea Book for Fruit Fly Workers. IOBC/ WPRS Bulletin* 5, 4–13.

Bourchier, R.S., Smith, S.M. & Song, S.J. (1993) Host acceptance and parasitoid size as predictors of parasitoid quality for mass reared *Trichogramma minutum*. *Biological Control* 3, 135–139.

Cerutti, F. & Bigler, F. (1991) Methods for quality evaluation of *Trichogramma evanescens* used against the European corn borer. In: Bigler, F. (ed.) *Fifth Workshop of the IOBC Global Working Group. Quality Control of Mass Reared Arthropods*. Wageningen, The Netherlands, pp. 119–126.

Cerutti, F. & Bigler, F. (1994) Quality assessment of *Trichogramma brassicae* Bezd. in the laboratory. *Entomologia Experimentalis et Applicata* (in press).

Chambers, D.L. & Ashley, T.R. (1984) Putting the control in quality control in insect rearing. In: King, E.G. & Leppla, N.C. (eds) *Advances and Challenges in Insect Rearing*. USDA, Agricultural Research Service, pp. 256–260.

Greenberg, S.M. (1991) Evaluation techniques for *Trichogramma* quality. In: Bigler, F. (ed.) *Fifth Workshop of the IOBC Global Working Group. Quality Control of Mass*

Reared Arthropods. Wageningen, The Netherlands, pp. 138–145.

Hawlitzky, N. (1986) Etude de la biologie de la pyrale du maïs, *Ostrinia nubilalis* Hbn. en région parisienne durant quatre années et recherche d'éléments prévisionnels du début de ponte. *Acta Oecologica Applicata* 1, 47–68.

Heuvel, H. van den (1986) Die biologische Bekämpfung des Maiszünslers, *Ostrinia nubilalis* Hbn. mit *Trichogramma maidis* Pint. et Voeg. *Internal Report, FAP*, 30 pp.

Huettel, M.D. (1976) Monitoring the quality of laboratory reared insects: a biological and behavioral perspective. *Environmental Entomology* 5, 807–814.

Kazmer, D.J. & Luck, R.F. (1990) Female body size, fitness and biological control quality: field experiments with *Trichogramma pretiosum*. In: Wajnberg, E. & Vinson, S.B. (eds) Trichogramma *and Other Egg Parasitoids. 3rd International Symposium. Les Colloques de l'INRA* 56, 37–40.

Laing, J.E. & Bigler, F. (1991) Quality control of mass produced *Trichogramma* species. In: Bigler, F. (ed.) *Fifth Workshop of the IOBC Global Working Group. Quality Control of Mass Reared Arthropods.* Wageningen, The Netherlands, pp. 111–118.

Lenteren, J.C. van. (1991) Quality control of natural enemies: hope or illusion? In: Bigler, F. (ed.) *Fifth Workshop of the IOBC Global Working Group. Quality Control of Mass Reared Arthropods.* Wageningen, The Netherlands, pp. 1–14.

Leppla, N.C. & Fisher, W.R. (1989) Total quality control in insect mass production for insect pest management. *Journal of Applied Entomology* 108, 452–461.

Martson, N. & Ertle, L.R. (1973) Host influence on the bionomics of *Trichogramma minutum*. *Annals of the Entomological Society of America* 5, 1155–1162.

Neuffer, U. (1987) Vergleich von Parasitierungsleistung und Verhalten zweier Ökotypen von *Trichogramma evanescens* Westw. Dissertation, Universität Hohenheim, 120 pp.

Pavlik, J. (1993) The size of the female and quality assessment of mass reared *Trichogramma* spp. *Entomologia Experimentalis et Applicata* 66, 171–177.

Salmanova, L.M. (1991) Changes of *Trichogramma* cultures in permanent rearing on the laboratory host *Sitotroga cerealella* Oliv. Dissertation, Biological Faculty, Lomonosow University, Moscow.

Shchepetilnikova, V.A. & Kasinskaya, L.V. (1981) Changes in the environmental preferences of *Trichogramma* effected through conditions of rearing. In: Pristavko, V.P. (ed.) *Insect Behaviour as a Basis for Developing Control Measures Against Pests of Field Crops and Forests.* Oxonion Press, New Delhi, pp. 225–231.

Smith, S.M., Carrow, J.R. & Laing, J.E. (eds) (1990) Inundative release of the egg parasitoid, *Trichogramma minutum* (Hymenoptera: Trichogrammatidae), against forest insect pests such as the spruce budworm, *Choristoneura fumiferana* (Lepidoptera: Tortricidae). *The Ontario Project 1982–1986. Memoirs of the Entomological Society of Canada* No. 153, 87 pp.

Methods and Timing of Releases of *Trichogramma* to Control Lepidopterous Pests

6

SANDY M. SMITH

Faculty of Forestry, University of Toronto, 33 Willcocks Street,
Toronto, Ontario, Canada M5S 3B3

Abstract

Trichogramma species are released worldwide as pupae in parasitized host eggs against a number of lepidopterous pests. The different techniques used can be separated broadly according to the distribution of the parasitized material; either from point sources or in a broadcast application. Point sources include parasitoids released from aircraft and from the ground in cards or different types of containers. Broadcast applications have been conducted with backpack sprayers or blowers from the ground as well as aerial applications of parasitoid material in water, attached to carriers such as bran or simply as unattached parasitized host eggs. The release techniques selected have depended on a number of considerations including the purpose of the study, characteristics of the crop or site, the technology available, the desired distribution of material, parasitoid behaviour and biological interactions with predators, the type of compounds to be added and the cost. Point releases have been used primarily for experimental purposes while aerial techniques have been developed for operational use. The timing of *Trichogramma* releases has varied according to factors such as host egg availability and acceptability, synchronization with the host, weather, use of other control measures, and host population dynamics. All have been important in determining when and how many releases will be made in either agricultural or forestry situations. The technology of the release mechanisms, storage parameters, and the development of artificial eggs will all have significant impact on the future approach to releasing *Trichogramma*.

Introduction

Trichogramma species are the most widely used parasitoids in biological control programmes throughout the world and a key aspect in this use is their release: how and when to get them most effectively to the field. As a biologist, Kot (1968) correctly identified the significance of factors such as the species and strain of *Trichogramma* released, host density, numbers released and behaviour of the parasitoid in modifying the results of inundative releases. However, of equal importance, is the methodology and timing of release. Methods of release can influence the behaviour of *Trichogramma* and can affect the level of pest control achieved (Keller *et al.*, 1985). Understanding those factors which influence the method and timing of release will be crucial to advance the use of *Trichogramma* and other beneficial insects.

The approach to releasing *Trichogramma* species has evolved over the past 60 years from small-scale experimental studies to large treated areas. Today, egg parasitoids are released commercially in over 12 countries with most extensive practical use in the former USSR (Anon., 1990), the People's Republic of China (Cock, 1985) and parts of Europe (Ridgway & Morrison, 1985; see Chapter 2). On an international scale, those countries where labour costs are relatively low have emphasized the development of manual release techniques where high labour inputs are not economically detrimental, while more developed countries have placed emphasis on reducing labour costs by increasing the mechanization of release. At the same time, the extent of area to be treated, from small research plots to commercial farms, has caused a shift from ground applications to aerial releases in an effort to increase accessibility and reduce the time involved.

The following review examines different methods for releasing *Trichogramma* from a historical perspective and addresses some of the factors that will influence the selection of the most appropriate method. Those factors which should be taken into account when timing releases will also be discussed, followed by a final section which examines the issues important in the future development of methodology and timing in *Trichogramma* releases.

Methods of Release

Different ways of releasing *Trichogramma* have been developed, however, all require the production of parasitized material. Morrison *et al.* (1978), Li (1982), Gusev & Lebedev (1986) and Laing & Eden (1990) report methods for mass-producing *Trichogramma* either on small egg hosts such as the Angoumois grain moth, *Sitotroga cerealella* (Olivier), the Mediterranean flour moth, *Ephestia kuehniella* Zeller, and the rice grain moth, *Corcyra cephalonica* (Stainton), or on large host eggs such as the eri-silkworm, *Philosamia cynthia* Ricini Donovan. In all cases, host eggs are exposed to female parasitoids for oviposition prior to release and the parasitoid pupae are subjected to various temperature regimes to programme adult emergence (Stinner *et al.*, 1974). The intention in programming is to obtain a high percentage of emergence from the parasitized eggs within a given period of time after they are exposed to normal ambient temperatures. It is this parasitized material which is available in various forms for release.

Ground applications

The production of large volumes of parasitized material has expanded the potential for studies on the field application of *Trichogramma*. Release of parasitoids from the ground has been the most common approach in the literature (Allen & Gonzalez, 1974; Beglyarov & Smetnik, 1977; Burbutis & Koepke, 1981; McLaren & Rye, 1983; Ridgway & Morrison, 1985; Bigler, 1986; Pak *et al.*, 1989; Neil & Specht, 1990; Prokrym *et al.*, 1992). With the exception of the People's Republic of China (Sun & Yu, 1986) and the former USSR (Anon., 1990), these studies have been research-oriented and aimed at examining parasitoid efficacy in terms of rates of release and timing as well as parasitoid behavior and dispersal rather than commercial application.

Releases from point sources

Some of the earliest records in North America report the manual release of adult *Trichogramma* (Steenburgh, 1930). In these situations, parasitized eggs were taken to the field and allowed to emerge in Petri dishes. After emergence, they were transported to the release location, the dishes opened, and the parasitoids allowed to move away freely; those parasitoids not flying were forced out by tapping the containers. Similarly, in the former USSR, early work released pharate adult parasites in rumpled paper or wilted leaves directly onto the crop (Anon., 1974, 1990), while in the United States, parasitized host eggs were allowed to emerge in the laboratory and then transported to the field for release (Gross *et al.*, 1981b). For experimental purposes, this approach is extremely practical because it ensures a precise number of adult parasitoids in a given area. However, it requires considerable time and makes large-scale work impossible. Thus, the release of parasitized host eggs rather than pharate adults has been used in later experimental work.

UNPROTECTED SOURCES

Historically, some parasitized material has been released directly onto the crop in point sources (Stinner *et al.*, 1974), however, most has been attached to a substrate of cardboard or various types of paper (wax, filter, etc.). A similar system is used today in the People's Republic of China with egg cards collected

from windows of 'rearing rooms' (Li, 1982). These substrates then have been distributed to strategic point sources and attached to the host plant by either string, staples or clips in: cotton (Allen & Gonzalez, 1974; Scholz, 1990); cole crops (Parker *et al.*, 1971; Pak *et al.*, 1989); corn (Hassan *et al.*, 1986; Ravensberg & Berger, 1986; Neil & Specht, 1990); citrus (Newton, 1988; Newton & Odendaal, 1990); apple (Dolphin *et al.*, 1972); peach and pear (Feng, 1986); and forests (Hulme & Miller, 1986; Sun & Yu, 1986). Some obvious advantages of this approach is that it requires simple technology, is easy to implement, and provides a good distribution of parasitoids according to the spacing between release points. Bigler (1986) considered that this approach in corn took about half an hour per hectare per person, however, the intensive labour required has meant that the associated costs are relatively high in developed countries (Parker *et al.*, 1971).

CONTAINERIZED MATERIAL

One of the major drawbacks of releasing *Trichogramma* from parasitized host eggs on a substrate is that the material remains exposed in the field for a longer period of time than if adult parasitoids are released directly. The likelihood of predators and bad weather reducing the number of surviving parasitoids thus increases unless measures are taken to shelter or protect the material.

Over the years, a number of different techniques have been developed to protect or contain the parasitized material and this has probably been the most standard means of releasing parasitoids to date (Gross *et al.*, 1981a). The containers have ranged from handmade prototypes (Burrell & McCormick, 1962; Ables *et al.*, 1979; Shi *et al.*, 1986) to various food packaging equipment (Stinner, 1977). Cone-shaped cups with cut tips or waxed paper have been used in orchard and forest situations (Steenburgh, 1930; Steenburgh & Boyce, 1938; Smith *et al.*, 1987); ice cream cartons in orchards (Dolphin *et al.*, 1972); cut bags of various types for orchards including Ziploc® plastic bags (Yu *et al.*, 1984b), brown paper bags (Dolphin *et al.*, 1972; Rincon-Vitova in Jones *et al.*, 1977), and nylon mesh/plastic screen bags (Hassan *et al.*, 1988); glass jars (Prokrym *et al.*, 1992) and cans (Burbutis & Koepke, 1981) in corn; plastic diet cups in cotton (Stinner *et al.*, 1974; Gross *et al.*, 1981b; Kanour & Burbutis, 1984); bamboo containers and capsules in corn (Maini *et al.*, 1986); and plastic milk bottles (Kanour & Burbutis, 1984).

The containers have been attached to the plants by ribbons, staples, glue or hung directly from plant parts. In some cases, the containers have been placed on the ground (Gross *et al.*, 1981b) or attached to wooden stakes rather than to the plants directly (Kanour & Burbutis, 1984; Smith *et al.*, 1987, 1990a), while in others a food source has been provided in the container to increase parasitoid longevity (Gross *et al.*, 1984). In most studies, the material has been distributed to specific points by hand although in some, the containers have been dropped mechanically from a moving vehicle (Ables *et al.*, 1979).

Releases in broadcast form

In order to be effective, parasitized material released from point sources necessitates that the points be evenly distributed throughout the release area and that the spacing between the points be of a minimal distance to allow parasitoid dispersal to all areas uniformly (Gusev & Lebedev, 1986). While this can be achieved effectively on small, discrete areas, even with high labour costs, it is relatively inefficient when larger hectares need to be treated. Thus, several attempts have been made to distribute parasitized material in broadcast form rather than from point sources. This has meant increased mechanization from material sprayed in various granular mixes to liquid suspension (Ables *et al.*, 1979; Anon., 1990).

The broadcast technique has been as simple as sprinkling parasitized material evenly over the target crop. Brower (1988) spread eggs parasitized by *T. pretiosum* onto the surface of peanuts in simulated storage each week to control the almond moth and the Indianmeal moth. In containerized systems such as stored nuts, this simple technology is sufficient, however, in field crops and forest applications, automated technology is required.

The technology for broadcast release of *Trichogramma* arose in North America from the availability of free-form parasitized material, and was derived, in part, from earlier programmes on applications of general insect predators (Nordlund *et al.*, 1974; Ables *et al.*, 1979). An early spray apparatus described by Schutte & Franz (1961) and Schutte (1962) applied eggs of *E. kuehniella*, parasitized by *T. embryophagum* (Hartig), in an apple orchard while an aerosol spray device was reported by Thewke & Puttler (1970) to apply lepidopterous eggs onto crops. Shands *et al.* (1972) used a compressed-air sprayer to apply eggs of *Coccinella septempunctata* L. and *C. transversoguttata* Faldermann and *Chrysopa* spp. in a 0.125% agar solution to control aphids on potatoes (Nordlund *et al.*, 1974) and this was unsuccessfully tried in Texas for the application of *Trichogramma* (Ables *et al.*, 1979). For various reasons, including problems with drip, volume and adhesive, these technologies were not suitable for large-scale applications.

During the late 1970s and early 1980s, research on broadcast techniques continued intensively in cotton against the cotton bollworm, *Heliothis zea* (Boddie). In 1974, Nordlund *et al.* (1974) developed a hand-pulled pneumatic sprayer which applied eggs of *H. zea* parasitized by *T. pretiosum*. A 15.2 m hose allowed plots as large as eight rows wide and 15.2 m long to be treated from one spray site. Subsequent work by Reeves (1975) suggested that material could be successfully attached to a carrier such as sawdust and applied. Using this technique, Jones *et al.* (1977) attached eggs of *S. cerealella* to bran flakes and applied them in a modified Cyclone® seeder on a power-driven backpack

for the first large-scale broadcast release of *Trichogramma* at rates of 176,000 to 247,000 parasitoids ha^{-1}. In 1981, Gross *et al.* (1981a) designed a system which used a peristaltic pump, conveyor belt and centrifugal fan to deliver *Trichogramma*-parasitized eggs of *H. zea* 15–25 cm above the plant canopy at 244,000 parasitoids ha^{-1}. This system was modified slightly to apply parasitized eggs at 167,000 to 835,000 parasitoids ha^{-1}, 10–15 cm above the plant canopy using a compressed-air sprayer and nozzle system (Gross *et al.*, 1984). While the majority of these studies were directed at specific research questions, the authors generally concluded that the systems developed could be used commercially for ground applications, achieving uniform distributions of undamaged parasitoids.

Research in Canada during the mid-1980s made use of modified leaf blowers to direct unattached *S. cerealella* eggs parasitized by *T. minutum* into forest stands (Smith & Wallace, 1990). The aim of these trials was to simulate aerial releases for experimental purposes rather than to develop commercial techniques for ground application although they could be used economically in localized pest problems because of their low cost and technology. Recent work in forest situations suggests that the Solo® backpack sprayer may be used to apply high quality parasitized material in a water solution on young forest stands or in speciality-use areas such as Christmas tree plantations and seed and cone orchards (S.M. Smith, unpubl.). Similar studies in the former USSR have used tractors to apply large numbers of *Trichogramma* from the ground either with sawdust or water as a carrier (Gusev & Lebedev, 1986; Anon., 1990).

Aerial applications

Ground-level releases are useful for research because the distribution of material can be tightly controlled, however, they are subject to limitations like inaccessibility to release areas during inclement weather, damage to crops by machinery and a low dissemination rate per unit time (Ables *et al.*, 1979). Also, from a commercial standpoint, target areas can be relatively large and timeliness is critical especially if natural enemies can only be released at certain times of the day (e.g. early morning or late afternoon). Thus, aerial techniques have been developed to circumvent some of these problems and are often the only practical method for commercial-sized operations (Anon., 1990). In countries such as the People's Republic of China, while some material has been distributed from the air, most is applied from the ground (Huffaker, 1977; Cock, 1985; Li, 1990). In North America and the former USSR, the aerial application of *Trichogramma* is probably the only feasible approach to widespread use.

Releases from point sources

The first aerial release methods in North America consisted of timing the distance intervals between distribution points and dropping packaged insects out of an aircraft door (Schuster *et al.*, 1971). Later, equipment used in sterile insect programmes (Boving *et al.*, 1969; Higgins, 1970) was used to mechanically eject packages from aircraft (Schuster *et al.*, 1971).

Reeves (1975) made the first grid-pattern *Trichogramma* releases in 1972 on 240 ha of cotton in Texas. Parasitoids were released in packages (8000 to 12,000) using a portable machine installed in the cabin of a Cessna 180® aircraft. The machine had a refrigerated storage compartment and automatically opened, cut the corners of the packages, and released them at a rate of 158.4 packages min^{-1} (43.2 ha^{-1}). Although parasitoid efficiency was considered equal to that of ground releases and the hours of labour were reduced 94.3% with this approach over that in ground releases, this approach was considered uneconomical. Extensive time was required to prepare and transport the material before distribution as well as the large amount of space utilized by the material; this restricted the amount of area that could be treated effectively (Jones *et al.*, 1977). As a consequence, aerial techniques which distributed parasitized material in a broadcast fashion were developed.

Releases in broadcast form

Higgins (1970) reported a few examples of automated free aerial release of insects (e.g. lacewings in sawdust, parasitic flies for sugar-cane borer) in which a machine mounted on the side of an aircraft, remotely controlled by the pilot, conveyed material for release. General reference was made in the literature to the requirement for improved technology of release for *Trichogramma* (Stinner, 1977; Ridgway *et al.*, 1977; Ables *et al.*, 1979) and simple technologies such as the use of model airplanes with automated metering devices have been tried (E. Kettela, Forestry Canada-Maritimes Region, Fredericton, New Brunswick, Canada). However, intensive studies on the aerial broadcast of *Trichogramma* in North America did not begin until the late 1970s and early 1980s when USDA researchers focused on improved parasitoid release for control of the cotton bollworm (King *et al.*, 1985).

The first prototype aerial application of *Trichogramma* was derived from the ground system designed by Jones *et al.* (1977) and made use of parasitized material attached to bran flakes. Bouse *et al.* (1980) modified a Piper PA-25® aircraft to facilitate the metering of the flakes. The plane contained a standard hopper with a Slimline venturi-type dry materials spreader, an agitator for 'fluffing' the bran, and a saw-tooth baffle metering device which was precooled. The flakes were applied in 6 m swaths flying at 3 m above the crop at 168 km h^{-1} to provide an application rate of approximately 157,000 parasitoids ha^{-1}. Jones *et al.* (1979) continued the work with releases made

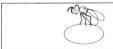

in a Model C. Piper Pawnee® aircraft with similar equipment except that bridging chains were provided to prevent clumping of the material. This system released parasitoids in a bran-flake mixture at a rate of 247,000 ha^{-1}.

While the bran-flake mixture provided a means of delivering parasitized material evenly over the crop, the quantities required to treat large areas was high and the logistics of cooling such material was difficult. In 1981, Bouse *et al.* (1981) developed a new aerial release system which applied the parasitized eggs unattached at a rate of approximately 200,000 parasitoids ha^{-1} and 3.2 ha min^{-1}. This required pre-cooling of the material between 12 and 16°C and allowed the treatment of one 20-ha block given a 9 to 12 m swath and a flight speed of 160 km h^{-1}. Although over 70 ha of cotton were treated with this technique in a 3-year period, problems with the system were encountered in terms of condensation and clumping of the eggs when temperatures exceeded 16°C, an inability of keeping the material cooler than 16°C for more than 45 min, and the lack of a consistent monitoring system to allow the pilot to determine whether the metering and dispersal equipment was working properly.

The final aerial release systems developed for use in cotton were those described by Bouse & Morrison (1985). This work arose from an USDA-ARS pilot test in 1981–1983 to examine a delivery system which would precisely meter and dispense bulk quantities of *Sitotroga* eggs containing *Trichogramma* pupae in a successful aerial broadcast (King *et al.*, 1985). A modified Cessna 172® aircraft was flown at a height of 12 m to give a swath width of 24 m. The parasitized material was treated prior to loading with an industrial starch powder (Dri Flo®; National Starch and Chemical Corp., Bridgwater, NJ 08807, USA) used to prevent clumping of the eggs and kept chilled at 15.6°C. The delivery system was refrigerated to 12°C via a solid-state thermoelectric element and the flow rate could be adjusted internally by the pilot. In 1981–1982, the hopper systems were attached externally to the aircraft wings while in 1983, a central, large-capacity (3850 ml) hopper (a modified Koolatron® P34A container) was held inside the aircraft with tubing extending to the landing gear. In all years, the hoppers were kept refrigerated at 12°C. In 1983, material was agitated and a lid added to the cold compartment to increase the flow rate. Over the 3 years of the study, 5400 ha of cotton were treated in Arkansas and North Carolina with application rates of 125,000–370,000 parasitoids ha^{-1}.

Research in forest environments has necessitated a different approach to the aerial application of *Trichogramma*. The fixed-wing (small airplane) systems developed in cotton required relatively small amounts of parasitized material (<300,000 parasitoids ha^{-1}) while application rates of 12 million female parasitoids ha^{-1} are more common in forestry (Smith *et al.*, 1987). This has meant the development of rotory-winged aircraft (helicopters) for the release of large amounts of unattached parasitized *Sitotroga* eggs on small, discrete areas. The system described by Hope *et al.* (1990) consisted of simple electrical components, mechanical components from a small grain planter and a centrifugal slinger used in aerial seeding jack pine. The equipment was mounted on a Bell® 47 helicopter and flown at 25 m above the ground. Undamaged parasitoids were distributed over a swath width of approximately 15 m using this technique (Smith *et al.*, 1990b). Because of the simplicity of the hopper construction, problems were encountered with manually feeding the material into the hopper (ca. 1.5 l capacity) at a consistent rate and with clumping of cooled eggs as they entered the ambient temperature of the hopper during the release. Despite these limitations, this technique was used to treat experimentally almost 20 ha of conifer forest between 1982 and 1985 (Smith *et al.*, 1990a) and 60 ha in 1993 for spruce budworm control (R.S. Bourchier & S.M. Smith, unpubl.).

Research in the former USSR has been directed at incorporating parasitized material in water/air formulations to apply over extensive areas (Anon., 1990). Several release mechanisms have been described with rates of 5 l ha^{-1} using Soviet AN-2 aircraft flying 5 m above the crop at a speed of 160 km h^{-1} for a swath width of approximately 60–80 m. These systems have been considered successful (capacity of 200–280 ha h^{-1}) in treating over 700,000 ha, of both agricultural and forest crops (Gusev & Lebedev, 1986; Anon., 1990).

Factors Influencing the Release Methodology

As is evidenced in the literature, parasitized host material can be released in a number of ways, either from the ground or air, in point distributions or broadcast form. A number of factors will affect the method of *Trichogramma* release to be selected in any given situation and these will be discussed in the following section.

Purpose of release

The primary goal of the release will be a major factor in determining the approach for distribution of material. Research studies focused on developing particular biocontrol strategies necessitate precise control over the amount, distribution and timing of release, and usually occur on small, localized areas (Newton & Odendaal, 1990). In this case, ground application of a known amount of material at specific points is probably the most effective approach, followed by ground application in broadcast form.

As the size of the area to be examined increases, the ability to effectively treat at a consistent rate and in a timely fashion decreases and this will require a shift to broadcast applications, first from the ground, then by aircraft. In the

majority of situations, aerial release of *Trichogramma* is the most effective approach because of the large volumes of material needed. This method ensures large, uniform applications in short periods of time at relatively low cost (Bouse & Morrison, 1985; Anon., 1990; Smith *et al.*, 1990b).

The final consideration here is the level of efficacy which is desired. In lower value crops or those where damage does not occur directly to the product (e.g. forest or orchard defoliators), extremely high levels of control may not be required. In contrast, pests causing direct damage to the crop (e.g. defoliators of cole crops) have lower economic thresholds and must be supressed at higher rates of application. If large volumes of parasitized material must be released, automated, broadcast methods are more likely to be cost-effective than point releases.

Site and crop characteristics

The characteristics of the crop where the releases are to be applied will influence the approach. First, the accessibility to the crop, either geo-graphically or seasonally, will determine whether ground or aerial releases must be considered. Specific fields tend to be difficult to reach from the ground. For example, cranberry production areas are usually located in low-lying wetland areas with high acidity (bogs) while forest stands can cover extensive areas in the tens of thousands of hectares. Similarly, fields are often inaccessible by ground machinery in early spring until they have dried or late fall until they have frozen. In these cases, aerial applications should be given priority as they are most likely to be commercially viable once the technology has been developed.

Second, the height, spacing and diversity of the crop will also affect the methodology selected. In forest situations where the crop may easily exceed 3 m (e.g. 6–12 m) and the pest eggs are located in the upper canopy, ground applications are going to be less effective than aerial releases unless the material can be directed upwards for a considerable distance. Some ground systems do achieve this, such as leaf blowers and backpack sprayers; leaf blowers have been shown to distribute material a distance of about 7 m (Smith & Wallace, 1990) and Solo® backpack sprayers a distance of about 17 m (S.M. Smith, unpubl.).

The blower systems and point releases from the ground can also be helpful in diverse crop situations where material must be applied to uneven surfaces of varying heights and shapes. When plants are tightly spaced in a stand, manoeuvrability of machines on the ground (either tractors or sprayers) will be difficult and this may necessitate aerial releases or ground releases from point sources (Hassan *et al.*, 1986; Anon., 1990; Neil & Specht, 1990).

Finally, the stage of plant development at which *Trichogramma* should be applied will be important in the selection of methodology. Knipling & McGuire (1968) hypothesized that changes in plant foliage area would affect the ability of *Trichogramma* to locate host eggs and this has been substantiated in corn and cotton systems by several authors (Need & Burbutis, 1979; Ables *et al.*, 1980; Burbutis & Koepke, 1981; Kanour & Burbutis, 1984; Keller *et al.*, 1985). As plants increase in size and architectural complexity, larger numbers of parasitoids will be required to achieve the same level of parasitism, and probably at higher levels in the canopy. Again, this leads from broadcast releases at ground level to aerial releases to be most cost-effective.

Parasitoid behaviour and biological interactions

Behavioural characteristics of *Trichogramma* species and their interactions with other biotic components will also influence how parasitoid material should be distributed in the field. These include parasitoid dispersal, probability of mating and expected levels of predation.

The dispersal rate and pattern of *Trichogramma* is species-specific and dependent on the type of crop in which they are released (Stein, 1961; Hendricks, 1967; Parker *et al.*, 1971; Stinner *et al.*, 1974). Some species such as *T. minutum* prefer the upper canopy of the stand (Kot, 1964; Keller *et al.*, 1985; Smith, 1988) or the upper regions of the crop such as *T. pretiosum* (Gonzalez *et al.*, 1970) while others prefer the lower canopy (Burbutis *et al.*, 1977) and still others show no preference (Schread, 1932; Newton, 1988). In general, *Trichogramma* disperse short distances from the point of release, usually less than 20 m (Yu *et al.*, 1984b; Keller *et al.*, 1985; Smith, 1988; Anon., 1990). Several studies have shown that the location of the release will influence the resulting distribution of parasitism in the crop in that parasitoids tend to stay in the area they are released (Hendricks, 1967; Burbutis *et al.*, 1977; Ables *et al.*, 1980; Keller *et al.*, 1985). While attempts to overcome the apparent innate tendency of some species to disperse upon release have been made including chilling of the insects before release, releasing them at night, or providing in-field food sources (Gross *et al.*, 1981b), the methodology of release provides one of the greatest means of controlling this distribution of material.

Point releases allow parasitized material to be placed precisely where it is needed to maximize the impact on host eggs (Allen & Gonzalez, 1974; Oatman & Platner, 1985; Smith *et al.*, 1987). In contrast, distribution with broadcast releases is less certain because the material must not only be placed in the desired location but also attached by some means to ensure that it remains there until parasitoid emergence (see later section). Aerial releases obviously are capable of placing material in the upper canopy while ground releases are more appropriate if the pest is located in the lower canopy (depending on crop height). One problem with aerial releases may be that substantial parasitoid movement occurs which could lead to unnecessarily high application rates (Jones *et al.*, 1979). The impact of wind on the dispersal of released material may account for some of this movement and this is obviously of more concern in aerial releases than ground releases (Yu *et al.*, 1984b; Keller *et al.*, 1985; Smith, 1988).

The requirement for mating after release and length of time available for this will also influence the type of release used. Male *Trichogramma* emerge slightly earlier than females and mate multiply with females immediately upon emergence (Waage & Ming, 1984; Lee *et al.*, 1986; Forsse *et al.*, 1992). Mated females can parasitize more host eggs than unmated females and this means that, for bisexual species/strains, successful mating is important. When these parasitoids are released in clumps or from point sources, males can readily locate females. However, in broadcast releases, the opportunity for mating is reduced because males and females may be separated by considerable distances in the crop and males must first locate receptive females. While point releases appear to be more conducive to mating and high levels of parasitism, most female parasitoids from broadcast releases appear to mate in the field (Smith *et al.*, 1990a) as in nature (Keller *et al.*, 1985; Bai & Smith, 1994); this is probably achieved through sex pheromones.

Andow & Prokrym (1991) estimated that the disappearance rate of *Trichogramma* released in corn was high and that 40% of the females disappeared daily. The cause of this 'disappearance' was not known although dispersal, hot dry weather and predation were all considered important factors. Several studies have indicated that predation has a significant impact on parasitoid survival and thus, efficacy of release. Losses due to predation have been observed when point releases were made in cabbage (Parker *et al.*, 1971), corn (Burbutis & Koepke, 1981), cotton (Nordlund *et al.*, 1974; Jones *et al.*, 1977), peach (Steenburgh, 1930), young spruce plantations (Smith, 1991) and in storage (Tran & Hassan, 1986). The primary predators in these studies included ants, chrysopid larvae, and coccinellid adults and larvae. While material released from point sources is undoubtedly easier to find and more attractive to predators than material scattered in a broadcast form, point sources are easier to protect by using containers than broadcast material. Thus, although a containerized point source is probably the technique to use in areas where predation is known or anticipated to be high, further research is needed to determine the true level of predation on broadcast material.

Additives to the parasitized material

In determining the methodology for release, consideration must also be given to other substances which may need to be or are desirable to add to the released parasitoids. In some situations, a food source (sugar/honey solution), kairomones or compounds which reduce predation may be considered while unparasitized host eggs, stickers and extenders may also be desired. The characteristics of the substance added and its volume (bulk) will influence the type of release method that is best suited.

Trichogramma are known to have poor longevity when no sugar source is available (Anunciada & Voegelé, 1982; Voronin, 1982; Newton & Odendaal, 1990) and unfed females parasitize fewer hosts than those provided with a sugar source (Ashley & Gonzalez, 1974; Yu *et al.*, 1984a). This suggests that greater levels of parasitism can be achieved if a food source is provided to released parasitoids, and situations where few releases can be made but in which parasitoids must remain active for a long period of time should consider this approach. To date, no studies have examined the consequences of adding a sugar source to parasitoids released in the field. It is worth noting that while the food source provides energy for the parasitoid in foraging, it also acts as an attractant to other animals, both vertebrate and invertebrate (e.g. ants), and these are likely to remove parasitized host eggs as well as the sugar source. Obviously then, such an approach is better suited to point releases at ground level, where the material can be protected from predators, than to broadcast releases.

With material that is broadcast for release, inert stickers or extenders are often added to the parasitized eggs. These have various functions, primarily to ensure that the material remains attached to the crop and to provide more accurate metering and a uniform distribution of material (Jones *et al.*, 1977; Ables *et al.*, 1979). Both dry (granular) materials have been used as well as liquids. Jones *et al.* (1979) attached parasitized eggs to a bran-flake carrier (mucilage : water, 1 : 1) for distribution in cotton. In the former USSR, sawdust has been used as a carrier for broadcast releases from the ground (Anon., 1990) while Smith *et al.* (1990a) used timothy seed as an extender for the aerial application of *T. minutum*. In all cases, the extenders allowed a relatively small amount of material to be spread more evenly over a larger area than would otherwise be possible if it had been applied unattached (neat). Nordlund *et al.* (1974) used a solution of Plantgard® (Polymetrics Intl, New York, USA), a retainer for insecticides and water (1 : 2) to carry the parasitized eggs and stick them to cotton plants in the field. Water alone has been used in ground and aerial broadcast systems developed in the former USSR (Anon., 1990) and in Canada (S.M. Smith, unpubl.). Because the factitious host eggs (*Ephestia* or *Sitotroga* spp.) are covered in a substance which becomes naturally sticky when wet, these systems help attach eggs to their location without additives.

The dispersal capabilities of *Trichogramma* suggests that there is a need in some situations to retain parasitoids in the release area and stimulate continued foraging when pest populations are low (Tumlinson *et al.*, 1993). Kairomones have been shown to elicit searching in *Trichogramma* (Lewis *et al.*, 1979; Gross *et al.*, 1981b; Keller *et al.*, 1985; see Chapter 8) and several studies have attempted to incorporate kairomones into released material.

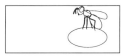

Nordlund *et al.* (1974) and Jones *et al.* (1973) first added kairomones from *Heliothis zea* moth scales to parasitized eggs for hand applications. Beevers *et al.* (1981) absorbed the kairomones on to particles of diatomaceous earth to aid in their distribution, while Nordlund *et al.* (1974) and Gross *et al.* (1984) later applied the kairomone tricosane in a hexane spray at 75 μg ml^{-1} with a pneumatic sprayer. These studies were considered successful and suggest that broadcast applications are better suited to the addition of kairomones than point releases where some dispersal away from the point is desirable.

To date, no work has been done to identify compounds which could be added to parasitized material to reduce predation in the field. The high levels of predation observed in the field, however, suggests that this is a fertile area of investigation particularly in situations where the emergence of parasitoids needs to be extended over a long period of time. The addition of such compounds, if available, would apply to both point source and broadcast releases. A recent study by Bjorkman & Larsson (1991) on plant compounds which reduce ant predation on leaf-feeding sawflies may provide future possibilities in this direction.

In some situations, it is desirable to add unparasitized host eggs to the parasitoid releases. This was first suggested by Knipling & McGuire (1968) and Thewke & Puttler (1970) and enables *Trichogramma* to cycle and continue to build up populations in the field when natural levels of target eggs are not available. Nordlund *et al.* (1974) suggested that these unparasitized eggs retained female parasitoids in the area to increase egg parasitism (Keller *et al.*, 1985); hand application experiments were undertaken (Gross *et al.*, 1981b). In this case, the increase in the amount of material to be released may require a shift from point sources to broadcast methods to ensure that material is uniformly distributed. Parker *et al.* (1971) and Parker & Pinnell (1972) released fertile adult moths concomitantly with *Trichogramma* to obtain similar results. In this case, adult moths are unlikely to be incorporated into the parasitized material for release and both point source and broadcast methods would be equally acceptable.

Cost of release

Of obvious consideration in the method to be used for release is the relative cost. Stinner (1977) concluded in his comprehensive review of literature from the 1960s that the major obstacle in the use of beneficial arthropods in biological control programmes was economic rather than ecological. Despite continued work, this remains true: the major block to achieving cost-effectiveness is our inability to produce and deliver large amounts of material in the field. Those factors affecting cost include the technology available, labour involved, and the value of the crop or stand to be protected.

Point releases made by hand from the ground require relatively low technology input (unless elaborate containers are constructed) and provided labour costs are kept low, offer an extremely attractive option. This approach has worked very successfully, even over large areas, in countries like the People's Republic of China (Li, 1990). Here, the cost was estimated to be approximately 3 yuan ha^{-1} (US\$0.60 ha^{-1}). Similarly, in the former USSR, ground releases were considered to take one man 1 day to treat 10 ha (excluding prerelease preparation time) (Anon., 1974). Where labour costs are higher, this approach to distribution will be more expensive (Newton & Odendaal, 1990) and other methods must be considered.

In developed countries, labour costs are high and represent a major factor limiting the use of *Trichogramma*. Substantial savings would be achieved by shifting to increased mechanization in broadcast releases (Jones *et al.*, 1977). Broadcast releases, either ground or aerial, reduce significantly the time and number of people required. For example, in the former USSR, one man (driving a tractor) is able to distribute parasitoids over 10 ha h^{-1} (Anon., 1990). With this approach, the more material to release, the less the overall operation will cost because of the diminishing marginal costs.

Finally, the value of the crop or stand to be protected will have some influence on the type of release used relative to the cost (see earlier section on purpose of release). In most cases, growers will select the cheapest option, however, if the crop has relatively high value, they may opt for a slightly more expensive option (in terms of technology or labour) in order to achieve a better level of control. Thus, the cost of the release method is also linked to the level of efficacy desired and that will vary according to factors such as purpose of release, site/crop characteristics, parasitoid behaviour and additives.

Timing of Releases

The timing of parasitoid release is a crucial aspect in their successful use. Incorrect timing can result in failure of the release, either totally or in part, independent of the method of release. Some important considerations in achieving the correct timing include the availability and acceptability of host eggs, the ability or necessity to programme parasitoid emergence, the weather conditions at the time of and immediately following the release, the requirement to integrate the releases with other control measures, and the population dynamics of the target host, both in the short and long term.

Host egg availability and acceptability

The most important aspect in the timing of parasitoid releases is the availability (oviposition period) and acceptability (ability to be parasitized by *Trichogramma*) of host eggs. These factors will determine not only the timing

and frequency of parasitoid releases, but as well, the length of time parasitoids must remain actively foraging.

Inundative releases of *Trichogramma* are aimed at synchronizing the parasitoid with the egg-laying period of the target host. If the host is multivoltine, then a different number of releases or rates may be appropriate (Bigler, 1986; Newton & Odendaal, 1990). For example, in Switzerland, the corn borer, *Ostrinia nubilalis* Hbn., completes one generation in the north and two generations in the south each year. This has resulted in two different strategies for parasitoid release with three consecutive releases in the same generation made against the univoltine form and one release at the beginning of each generation for the multivoltine form (Bigler, 1986). Similarly, in order to achieve synchronization with the host, Hassan *et al.* (1988) proposed four to six releases in apple for control of codling moth; Gusev & Lebedev (1986) reported using two or three releases against most pests in the Ukraine; and Brower (1988) successfully used triweekly releases of *T. pretiosum* against the almond moth in peanuts. Lawrence *et al.* (1985) proposed only two releases of *T. minutum* against the spruce budworm in Maine because progeny from the first release would be present in the field at the end of budworm oviposition and simulate the effects of a third release.

The length of time *Trichogramma* remains active in the field will influence the number of releases required in order to achieve continuous parasitism throughout the oviposition period of the host. Unfortunately, only inferred information is available on parasitoid activity in the field (Dolphin *et al.*, 1972; Gusev & Lebedev, 1986; Prokrym *et al.*, 1992). Kanour & Burbutis (1984) believed that *T. nubilale* adults lived about 2 days, based on laboratory data, however, they found levels of field activity ranging from 4–15 days. Similarly, *T. minutum* lived 1.8 days when held in the laboratory but maintained parasitism levels on sentinel egg masses over an 18-day period following release (see Fig. 7d in Smith *et al.*, 1990a). In order to avoid unreliable estimates of parasitoid activity in apple orchards, Hassan *et al.* (1988) conducted successive releases a few days before the parasitoids from the previous application had died based on subsamples kept at field temperatures.

One of the critical aspects of synchronization is the timing of the first release. Different approaches have been used to ensure that parasitoids are active in the field at the time of first oviposition by the host. Most successful has been traps to collect adult hosts, either regular lights (Hassan *et al.*, 1986; Maini *et al.*, 1986), black lights (Neil & Specht, 1990; Prokrym *et al.*, 1992) or pheromone traps (Maini *et al.*, 1986; Smith *et al.*, 1987; Hassan *et al.*, 1988; Smith *et al.*, 1990a). Observations have also been made on the timing of the first host egg mass in the field (Tran & Hassan, 1986), although experience suggests that parasitoids need to be released several days before the first egg mass is detected to be truly effective (Hassan 1984; Kanour & Burbutis 1984; Smith *et al.* 1990a). A detailed method of predicting oviposition for the corn borer was developed by both Hawlitzky & Voegelé

(1991) and Prokrym *et al.* (1992) using degree-day accumulations and this is the basis for the predictive regression line derived for spruce budworm development and oviposition used in Smith *et al.* (1990a).

Finally, the length of time the host egg is susceptible to parasitism will influence the number and timing of *Trichogramma* releases. The correct assessment of host acceptability is important in determining the impact of the parasitoids (Lopez & Morrison, 1985), and this is generally done prior to any type of field trial (Houseweart *et al.*, 1982; Pak *et al.*, 1986; Smith & Strom, 1993). Host eggs usually become less susceptible to attack by *Trichogramma* as they age, and are rarely acceptable beyond 6–10 days (depending on temperature) after oviposition unless overwintering or diapause occurs in the egg stage of the host such as the spruce budmoth, *Zeiraphera canadensis* (Lep.: Olethreutidae) (R.S. Bourchier & S.M. Smith, unpubl.) and the blackheaded fireworm, *Rhopobota naevana* (Lep.: Tortricidae) (D. Henderson, E.S. Cropconsult, Vancouver, British Columbia, Canada). The longer host eggs are available for parasitism, the more releases which can be made or the more selective can be the programming of the released material.

Programming parasitoid emergence

Parasitoid emergence is varied according to the temperature at which the parasitoids are reared (Morrison *et al.*, 1978; Voegelé *et al.*, 1986; Laing & Eden, 1990). Several field studies have deliberately used mixtures of different developmental stages of *Trichogramma* in order to achieve longer periods of activity in the field (Steenburgh & Boyce, 1938; Kanour & Burbutis, 1984; Bigler, 1986; Hassan *et al.*, 1988; Hawlitzky & Voegelé, 1991; Prokrym *et al.*, 1992). In other cases, extended emergence of parasitoids, while not planned, occurred with basically the same effect (Smith *et al.*, 1990a). This is important if the life span of field-released parasitoids is expected to be short or if they have a high rate of disappearance in the field (Andow & Prokrym, 1991; Prokrym *et al.*, 1992). The mixture of programmed material may even be specified so that the largest proportion of parasitoids emerge during peak oviposition by the host (Smith & You, 1990; Prokrym *et al.*, 1992).

By programming parasitoids to emerge over 2- to 20-day periods, the number of releases required could be reduced as well as the application rate because fewer parasitoids are actively seeking hosts over a continuous period rather than larger numbers in a relatively short time. Lund (1938) was the first to suggest a colonization schedule based on the fact that *Trichogramma* have an expected adult life span of 4–6 days and they deposit the majority of their eggs (75%) within the first 48 h of emergence. This staggered emergence may also allow for parasitoid replication in the target host and *Trichogramma* emerging from these natural hosts are often superior in quality than those from the factitious host (Bigler, 1986; Smith *et al.*, 1987).

Weather conditions

Weather is comprised of a number of environmental conditions including temperature, rainfall, humidity (and leaf wetness), solar radiation (or hours of sunshine) and wind speed and direction, which can influence the outcome of parasitoid releases. Often these components act in conjunction with one another and are difficult to separate (e.g. cool, wet, windy, overcast days). They are extremely important, however, in the timing of parasitoid releases because they cannot be controlled, only avoided.

As a poikilotherm, the activity of *Trichogramma* is highly dependent on temperature and other environmental conditions. Under laboratory conditions, lower temperatures result in lower levels of parasitism or activity (Biever, 1972; Boldt, 1974; Forsse *et al.*, 1992). This has been verified in field studies (Parker *et al.*, 1971; Ravensberg & Berger, 1986). Parker *et al.* (1971) observed a linear relationship between minimum daily temperatures and egg parasitism and R.S. Bourchier & S.M. Smith (unpubl.) have recently found low levels of parasitism by *T. minutum* in Canadian boreal forests when daily temperatures dropped below 17°C. Higher temperatures have also been shown to be detrimental to *Trichogramma* activity and have restricted their use in some crop situations (e.g. cotton, Fye & Larsen, 1969).

Several authors have noted that rainy weather reduced the effectiveness of parasitoid releases (Yu *et al.*, 1984b; Smith *et al.*, 1986, 1990a; Tran & Hassan, 1986). Humidity, however, is unlikely to play an important role in parasitism because levels in the field rarely drop below 30–40% or 60% on the foliage surface (Keller *et al.*, 1985). Steenburg (1930) observed reduced flight initiation by parasitoids on overcast days or when clouds passed in front of the sun while Ashley *et al.* (1973) reported no activity by *Trichogramma* during the night. Finally, wind speeds greater than 30 cm s^{-1} appear to inhibit parasitoid flight (Keller, 1985) although dispersal by *Trichogramma* has been often linked to wind direction (Hendricks, 1967; Yu *et al.*, 1984b).

It has been suggested that improved parasitoid efficacy could be achieved in field releases during inclement weather (particularly temperature) through the collection and selection of tolerant strains or species (Lopez & Morrison, 1980; Pak & Van Heiningen, 1985; Neil & Specht, 1990). A more short-term approach would be to avoid releasing parasitoids during unfavourable temperature conditions. In crops where temperatures reach relatively high levels during the day, releases should be conducted in early morning or late evening (Allen & Gonzalez, 1974; Anon., 1990; Smith *et al.*, 1990a), whereas in situations with low temperatures, releases could be scheduled just at midday so that parasitoids experience maximum temperatures. Similarly, releases should be timed to avoid conditions that are overcast, rainy or windy. Unfortunately, programmed material held during periods of inclement weather (usually at low temperatures in the dark) may suffer reductions in quality (viability, longevity, fecundity or host acceptance) and these losses must be traded-off against the expected effects of releasing parasitoids in unfavourable weather (Gross *et al.*, 1981a; Voegelé *et al.*, 1986).

Other control measures

Timing releases of *Trichogramma* for successful programmes requires careful consideration of other activities in the crop, especially other control measures. This is particularly true in situations where there is a complex of pests with multiple generations on the same crop, all requiring different approaches to management.

Previous work in the United States during the 1970s showed that parasitoid releases were incompatible with the regular insecticide applications being conducted on cotton (Bull & House, 1983; Lopez & Morrison, 1985). In China, the use of *Trichogramma* in cotton and rice is limited by the use of insecticides (Li, 1990) while in Germany, Hassan *et al.* (1988) reported that fungicide and insecticide use in apple orchards interfered with the success of the release trials. Obviously, releasing parasitoids during periods of broad-spectrum pesticide application is to be avoided as such releases have an extremely low potential for success, depending on the timing relative to the timing of the pesticide application.

Navarajan (1986) demonstrated that carbaryl, methyl parathion, mala-thion and triazophos are extremely toxic to *Trichogramma* while the synthetic pyrethroids fenvalerate and flucythrinate and the traditional insecticides monocrotophos, endrin, phosalone and endosulfan were less so. Recent studies have identified several pesticides which are more compatible with biological control agents such as *Trichogramma* including the insecticides Spanon, Rogor, Metaphos + BHC (57), Dipel, Dimilin 25WP, Pirimor-Granulat, the acaricide Shell Torrque, the fungicides Nimrod, Ronilan, Orthocid 83, Ortho Difolatan, Bayleton flussig and the herbicides Gesatop 50 and Kerb 50W (Hassan, 1984; Hassan *et al.*, 1987).

Cultural controls may also influence the timing of parasitoid releases. In China, vegetables are often planted adjacent to cotton fields or interplanted with cotton to attract moths in before the pests of cotton start to lay eggs (Li, 1990). The timing of parasitoid releases in these situations would be earlier than if cotton was planted alone because host eggs in the vegetables would allow for the build-up of *Trichogramma* before pest populations arrived in the cotton and the releases would be aimed at inoculating the vegetable crop for later suppression of the cotton pests. In the former USSR, agroecosystems have been enriched with nectariferous plants for augmenting *Trichogramma* releases (Voronin, 1982).

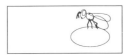

Host population dynamics

The final factor which will influence the timing of parasitoid releases is the population dynamics of the target pest. *Trichogramma* releases are aimed at the egg stage of the host but their impact or success is measured by how well they reduce damage in subsequent larval stages. Most of the previous discussion on timing has focused on annual crops with pest populations that must be controlled yearly. The timing of releases in these situations is tightly linked to the period of host oviposition in each year, the goal being to provide parasitoids in the field from just before the first egg is laid, throughout oviposition. If conducted properly (i.e. with sufficient numbers, appropriate location, good weather, good parasitoid quality, etc.), this will achieve maximum levels of egg parasitism, depending on the degree to which egg parasitism and larval mortality are density-dependent (May *et al.*, 1981; Hamburg & Hassell, 1984; Prokrym *et al.*, 1992).

In contrast, pest populations in perennial crops such as orchard, vineyard or berry systems and forests, may fluctuate naturally over several years ranging in any given area from low, endemic levels to outbreak, epidemic numbers. The major factor here is time and this will vary depending on the system under examination. For example, spruce budworm populations fluctuate over periods of 20–30 years (Royama, 1984) while the management of strawberries provides only about a 3-year period for pest populations to fluctuate. The timing of *Trichogramma* releases in such situations has long-term implications for suppression.

One of the questions which arises in the use of parasitoid releases against pests in perennial crops is whether they should be applied annually or only at specific phases in the pest's population cycle. Traditionally, releases in these situations have been applied annually with little concern for the long-term population dynamics of the system (Sun & Yu, 1986; Newton & Odendaal, 1990; Smith *et al.*, 1990a). As more information becomes available on the components in pest management, however, the latter option requires serious consideration (Ridgway & Morrison, 1985; Sun & Yu, 1986; You & Smith, 1990).

The best time for the release of an egg parasitoid such as *Trichogramma* in the population cycle of perennial hosts will depend on the impact of all natural enemies as well as the interaction of the pest with its host plant; that is the effect of the components driving the population levels. A simple life systems analysis model developed by You & Smith (1990) to examine the timing of *T. minutum* releases against the spruce budworm in terms of its population cycling suggested that the most successful releases would be made when the pest was at the end of its plateau phase or at the beginning of its declining phase. This provided a period of about 5–6 years for release and predicted that populations would remain low for several years beyond the regular endemic phase of the budworm.

A recent approach to spruce budworm management may allow us to test this prediction in the field. R.S. Bourchier & S.M. Smith (unpubl.) are currently investigating the potential for integrating the use of *Bacillus thuringiensis kurstakii* Berliner against the larval stage with releases of *T. minutum* against the egg stage. The intention here is to suppress the budworm population later in its epidemic phase so that the other natural enemies in the system (which are thought to cause the collapse of the host) will have an immediate and significant impact. If this works, fewer releases of *Trichogramma* would be required and these could be timed for the most susceptible stage in the population cycle of the budworm. Similar work has been conducted in the former USSR against the diamondback moth on cabbage (Voronin, 1982) and in sugarbeet (Gusev & Lebedev, 1986).

In some cases, *Trichogramma* released in a field during one year may carry over into subsequent years. In perennial crop systems, this may reduce the necessity of annual applications, depending on the carryover levels. In other cases, the natural background levels of *Trichogramma* may be higher in some years than others and careful monitoring of these levels prior to releases will determine whether inundative releases need to be conducted in any given year, and if so, where. This will ensure that releases are synchronized with host cycles in specific locations.

Future Issues in Releasing *Trichogramma*

Improved methods of releasing *Trichogramma* and a better understanding of the factors to be considered in the timing of releases will be needed to make this biological approach a feasible alternative to present-day controls. Fortunately, considerable research has already been conducted in these areas as most programmes to develop release strategies have addressed timing and methodology. Three aspects remain, however, which will become increasingly important in the commercial development of *Trichogramma* releases and affect the timing and methodology used: the technology available in the future, the ability to store parasitized material, and the development of artificial eggs.

Ables *et al.* (1979) suggested that automated techniques which would allow rapid and efficient releases of insect parasitoids and predators would be needed to ensure the economic feasibility of this type of biological control. They reviewed 10 years of research into release techniques from the 1960s to the 1970s and identified three phases of augmentative releases which would have to be mechanized in order to improve the economic feasibility of this approach: rearing, transport/holding and distribution. In the past 20 years, we have seen significant advances in the technology available for distributing *Trichogramma*, however, this automation is not available for all crop systems and where it has been developed, there remain problems which

need to be overcome before the systems can be truly commercial.

Both biological and engineering aspects need to be examined in the successful development of improved technology for parasitoid distribution. From the biological perspective, issues such as how to ensure minimal injury to the released material, the optimal spacing within the target area to maintain uniformity, the timeliness of the release(s), the potential for reducing predation and other losses, and the ability of the material to adhere to the foliage (or wherever it is targeted) must be considered. From the engineering view, questions as to the most appropriate mechanism to use (e.g. ground or aircraft), the range and manoeuvrability of the equipment, the hopper space required, the speed and area which can be covered effectively, the ability to maintain the material cool and unclumped, and whether liquid or dry applications are most appropriate, need to be answered. Obviously, whether from the ground or from aircraft, economic feasibility and overall effectiveness will be assured if the technology shifts to lightweight, compact units with minimal loading and fuel consumption while at the same time maintaining a high release capacity and no effect on the biological parameters of the system (Ables *et al.*, 1979).

Ables *et al.* (1979) also identified transport/holding as an important area for future research. High temperatures during shipment and holding are known to reduce parasitoid efficacy (Tran & Hassan, 1986) and previous work has shown that cold temperatures are effective in storing parasitoids, even with a minimum of technology input (Bouse & Morrison, 1985; Voegelé *et al.*, 1986; Hawlitzky & Voegelé, 1991). Unfortunately, in those instances where specific biological parameters were examined, cold storage also resulted in reduced quality (Ridgway *et al.*, 1973) and this must be factored in when making recommendations for the storage of such material, either at the production facility or in the field.

Holding/storage and transport of parasitized host eggs has been clearly linked with production efficiency, however, they have rarely been extrapolated to include the shipment and containment of material after it leaves the facility (Ridgway & Morrison, 1985). Successful holding/transport will be increasingly important as *Trichogramma* releases become commercially viable. To ensure that the material is held and released in the most optimal way will require improved understanding and communication of the biological requirements of each system. Realistic 'shelf-lives' will have to be developed and instructions on how to hold the material as well as the consequences of holding it provided for applicators in the field. The technology of release is integrally linked to the condition and requirements of the material to be released and any effective development in the former must take into consideration the limitations imposed by the latter. For example, material which arrives at the field at one level of moisture is of little value in calibrating equipment when the next shipment will arrive at a different level and there is no way of standardizing the condition of the material. Previous successes in pest control can be directly attributed to this smooth flow of technology from the producer to the manager in the field (Hawlitzky & Voegelé, 1991).

The final aspect affecting the methods and timing of *Trichogramma* releases in the future is the potential for developing artificial eggs. Various approaches are now being investigated (Strand & Vinson, 1985; Xie *et al.*, 1986; Li, 1990; see Chapter 4) and if successful, will provide a different type of product for release. Current methodology is based on applying factitious host eggs either alone or in combination with other substances. Often the problems associated with storing and releasing this material are related to the fact that it is a pure biological product (e.g. high humidity causing clumping of material before/during release), and in a more artificial system, some of the restrictions for distribution may be reduced or eliminated by the design of the egg capsule. Egg size, in particular, will be subject to manipulation and this will impact on the design and calibration of release equipment as well as on the timing because of changes in parasitoid size, longevity and fecundity. The ability to encapsulate eggs in various substances will not only affect the flowability of the material but also affect the level of predation in the field and thus, the programming of parasitoid emergence and timing of releases.

The future of *Trichogramma* releases appears to be bright. As research identifies more pest situations in which this approach can have an impact on reducing crop damage, more emphasis will be placed on producing large numbers of parasitoids for commercial use. This, in turn, will provide sufficient material to develop improved systems for distribution and the abundance of parasitized eggs will allow field programmes to improve their strategies for timing releases. The 'catch-22 situation' which we found ourselves in in the past, where large numbers of parasitoids were needed to demonstrate field efficacy and yet large production was not justified until this success was demonstrated, limited the development of release technology. It will be a very different scenario over the next 40 years.

References

Ables, J.R., Reeves, B.G., Morrison, R.K., Kinzer, R.E., Jones, S.L., Ridgway, R.L. & Bull, D.L. (1979) Methods for the field release of insect parasites and predators. *Transactions of the American Society of Agricultural Engineers* 18, 59–62.

Ables, J.R., McCommas, D.W., Jones, S.L. & Morrison, R.K. (1980) Effects of cotton plant size, host egg location, and location of parasite release on parasitism by *Trichogramma pretiosum. Southwestern Entomologist* 5, 261–264.

Allen, J.C. & Gonzalez, D. (1974) Spatial attack patterns of *Trichogramma pretiosum* around release sites compared with a random diffusion model. *Environmental Entomology* 3, 647–652.

Andow, D.A. & Prokrym, D.R. (1991) Release density, efficiency and disappearance of *Trichogramma nubilale* for control of European corn borer. *Entomophaga* 36, 105–113.

Anon. (1974) Methodological directions on mass rearing and utilization of *Tricho-gramma* for the control of pests of farm crops. In: *Proceedings of the 1972 All-Union Seminar of Specialists of Productive Biolaboratories.* Kolos Publishing House, Moscow.

Anon. (1990) Application of *Trichogramma* against complexes of enemies in agricul-tural crops (recommendations). *National Research Institute of Biological Protection and National Research Communications in Agriculture for the USSR.* Agropromizdat, Moscow, 45 pp.

Anunciada, L. & Voegelé, J. (1982) L'importance de la nourriture dans le potentiel biotique de *Trichogramma maidis* Pint. et Voeg. et *T. nagarkattii* Voeg. et Pint. (Hym.: Trichogrammatidae) et l'oosorption dans les femelles en contention ovarienne. In: Voegelé, J. (ed.) *Les Trichogrammes. Les Colloques de l'INRA* 9, 79–84.

Ashley, T.R. & Gonzalez, D. (1974) Effect of various food substances on longevity and fecundity of *Trichogramma. Environmental Entomology* 3, 169–171.

Ashley, T.R., Gonzalez, D. & Leigh, T.F. (1973) Reduction in effectiveness of laboratory reared *Trichogramma. Environmental Entomology* 2, 1069–1073.

Bai, B.B. & Smith, S.M. (1994) Patterns of host exploitation by the parasitoid wasp, *Trichogramma minutum* (Hymenoptera: Trichogrammatidae), when attacking eggs of the spruce budworm, *Choristoneura fumiferana* (Lepidoptera: Tortricidae), in Canadian forests. *Annals of the Entomological Society of America* (in press).

Beevers, M., Lewis, W.J., Gross, H.R. & Nordlund, D.A. (1981) Kairomones and their use for management of entomophagous insects. X. Laboratory studies on manipulation of host-finding behavior of *Trichogramma pretiosum* Riley with kairomone extracted from *Heliothis zea* (Boddie) moth scales. *Journal of Chemical Ecology* 7, 635–648.

Beglyarov, G.A. & Smetnik, A.I. (1977) Seasonal colonization of entomophages in the U.S.S.R. In: Ridgway, R.L. & Vinson, S.B. (eds) *Biological Control by Augmentation of Natural Enemies.* Plenum Press, New York, pp. 283–328.

Biever, K.D. (1972) Effect of temperatures on rate of search by *Trichogramma* and its potential application in field releases. *Environmental Entomology* 1, 194–197.

Bigler, F. (1986) Mass production of *Trichogramma maidis* Pint. et Voeg. and its field application against *Ostrinia nubilalis* Hbn. in Switzerland. *Journal of Applied Entomology* 101, 23–29.

Bjorkman, C. & Larsson, S. (1991) Pine sawfly defence and variation in host plant resin acids: a trade-off with growth. *Ecological Entomology* 16, 283–289.

Boldt, P.E. (1974) Temperature, humidity, and host: effect on rate of search of *Trichogramma evanescens* and *T. minutum* Auctt. (not Riley, 1871). *Annals of the Entomological Society of America* 67, 706–708.

Bouse, L.F. & Morrison, R.K. (1985) Transport, storage, and release of *Trichogramma pretiosum. Southwestern Entomologist* 8, 36–48.

Bouse, L.F., Carlton, J.B., Jones, S.L., Morrison, R.K. & Ables, J.R. (1980) Broadcast aerial release of an egg parasite for lepidopterous insect control. *Transactions of the American Society of Agricultural Engineers* 23, 1359–1368.

Bouse, L.F., Carlton, J.B. & Morrison, R.K. (1981) Aerial application of insect egg parasites. *Transactions of the American Society of Agricultural Engineers* 24, 1093–1098.

Boving, P.A., Winterfield, R.G., Butt, B.A. & Donier, C.E. (1969) A box ejector for insect distribution from a helicopter. *USDA-ARS Report* No. 42–156, 8 pp.

Brower, J.H. (1988) Population suppression of the almond moth and the Indianmeal moth (Lepidoptera: Pyralidae) by release of *Trichogramma pretiosum* (Hymenoptera: Trichogrammatidae) into simulated peanut storage. *Journal of Economic Entomology* 81, 944–948.

Bull, D. & House, V. (1983) Effects of different pesticides on parasitism of host eggs by *Trichogramma pretiosum* Riley. *Southwestern Entomologist* 8, 46–53.

Burbutis, P.P. & Koepke, C.H. (1981) European corn borer control in peppers by *Trichogramma nubilale. Journal of Economic Entomology* 74, 246–247.

Burbutis, P.P., Curl, G.D. & Davis, C.P. (1977) Host searching behavior by *Trichogramma nubilale* on corn. *Environmental Entomology* 6, 400–402.

Burrell, R.W. & McCormick, W.J. (1962) Effect of *Trichogramma* releases on parasitism of sugarcane borer eggs. *Journal of Economic Entomology* 55, 880–882.

Cock, M.J.W. (1985) The use of parasitoids for augmentative biological control of pests in the People's Republic of China. *Commonwealth Institute of Biological Control, Biocontrol News and Information* 6, 213–223.

Dolphin, R.E., Cleveland, M.L., Mouzing, T.E. & Morrison, R.K. (1972) Releases of *Trichogramma minutum* and *T. cacoeciae* in apple orchard and the effects on populations of codling moth. *Environmental Entomology* 1, 481–484.

Feng, J.-G. (1986) Studies on the biological control of insect pests in fruit tree and oak tree with *Trichogramma dendrolimi* Matsumura. In: Voegelé, J., Waage, J.K. & van Lenteren, J.C. (eds) Trichogramma *and Other Egg Parasites – Les Trichogrammes et autres parasitoïdes oophages. Les Colloques de l'INRA* 43, 461–467.

Forsse, E., Smith, S.M. & Bourchier, R. (1992) Flight initiation in the egg parasitoid *Trichogramma minutum*: effect of temperature, mates, food, and host eggs. *Entomologia Experimentalis et Applicata* 62, 147–154.

Fye, R.E. & Larsen, D.J. (1969) Preliminary evaluation of *Trichogramma minutum* as a released regulator of lepidopterous pests of cotton. *Journal of Economic Entomology* 62, 1291–1296.

Gonzalez, D., Orphanides, G., van den Bosch, R. & Leigh, T.F. (1970) Field cage assessment of *Trichogramma* as parasites of *Heliothis zea*: development of methods. *Journal of Economic Entomology* 63, 1292–1296.

Gross, H.R., Harrell, E.A., Lewis, W.J. & Nordlund, D.A. (1981a) *Trichogramma* spp.: concurrent ground application of parasitized eggs, supplemental *Heliothis zea* host eggs, and host-seeking stimuli. *Journal of Economic Entomology* 74, 227–229.

Gross, H.R., Lewis, W.J. & Nordlund, D.A. (1981b) *Trichogramma pretiosum*: effect of prerelease parasitization experience on retention in release areas and efficiency. *Environmental Entomology* 10, 554–556.

Gross, H.R., Lewis, W.J., Beevers, M. & Nordlund, D.A. (1984) *Trichogramma pretiosum* (Hymenoptera: Trichogrammatidae): effects of augmented densities and distributions of *Heliothis zea* (Lepidoptera: Noctuidae) host eggs and kairomones on field performance. *Environmental Entomology* 13, 981–985.

Gusev, G.V. & Lebedev, G.I. (1986) Present state of *Trichogramma* application and research. In: Voegelé, J., Waage, J.K. & van Lenteren, J.C. (eds) Trichogramma *and Other Egg Parasites – Les Trichogrammes et autres parasitoïdes oophages. Les Colloques de l'INRA* 43, 477–481.

Hamburg, H. Van & Hassell, M.P. (1984) Density dependence and the augmentative

release of egg parasitoids against graminaceous stalk borers. *Ecological Entomology* 9, 101–108.

Hassan, S.A. (1984) Massenproduktion und anwendung von *Trichogramma*. 4. Feststellung der gunstigsten Freilassungstermine fur die Bekampfung des Maiszunslers *Ostrinia nubilalis* Hubner. *Gesunde Pflanzen* 36, 40–45.

Hassan, S.A., Stein, E., Dannemann, K. & Reichel, W. (1986) Mass-production and utilization of *Trichogramma*: 8. Optimizing the use to control the European corn borer *Ostrinia nubilalis* Hbn. *Journal of Applied Entomology* 101, 508–515.

Hassan, S.A., Bigler, F., Bogenschutz, H., Boller, E., Brun, J., Chiverton, P., Edwards, P., Mansour, F., Naton, E., Oomen, P., Overmeer, W., Polgar, L., Rieckmann, W., Samsoe-Petersen, L., Staubli, A., Sterk, G., Tavares, K., Tuse, J., Viggiani, G. & Vivas, A. (1987) Results of the third joint pesticide testing programme by the IOBC/WPRS-working group 'Pesticides and beneficial organisms'. *Journal of Applied Entomology* 103, 92–107.

Hassan, S.A., Kohler, E. & Rost, W.M. (1988) Mass production and utilization of *Trichogramma*: 10. Control of the codling moth *Cydia pomonella* and the summer fruit tortrix moth *Adoxophyes orana* (Lep.:Tortricidae). *Entomophaga* 33, 413–420.

Hawlitzky, N. & Voegelé, J. (1991) Démarche utilisée pour élaborer une stratégie de lutte biologique par lâchers d'entomophages contre un ravageur du maïs. Problèmes apparus lors de la pratique et solutions apportées. *Bulletin de la Société Zoologique de France* 116, 319–329.

Hendricks, D.E. (1967) Effect of wind on dispersal of *Trichogramma semifumatum*. *Journal of Economic Entomology* 60, 1367–1373.

Higgins, A.H. (1970) A machine for free aerial release of sterile pink bollworm moths. *USDA-ARS Report* No. 81–40.

Hope, C.A., Nicholson, S.A. & Churcher, J.J. (1990) Aerial release system for *Trichogramma minutum* Riley in plantation forests. In: Smith, S.M., Carrow, J.R. & Laing, J.E. (eds) *Inundative Release of the Egg Parasitoid, Trichogramma minutum (Hymenoptera: Trichogrammatidae), against Forest Insect Pests such as the Spruce Budworm, Choristoneura fumiferana (Lepidoptera: Tortricidae): The Ontario Project 1982–1986. Memoirs of the Entomological Society of Canada* No. 153, 38–44.

Houseweart, M.W., Southard, S.G. & Jennings, D.T. (1982) Availability and acceptability of spruce budworm eggs to parasitism by the egg parasitoid, *Trichogramma minutum* (Hymenoptera: Trichogrammatidae). *Canadian Entomologist* 114, 657–666.

Huffaker, C.B. (1977) Augmentation of natural enemies in the People's Republic of China. In: Ridgway, R.L. & Vinson, S.B. (eds) *Biological Control by Augmentation of Natural Enemies*. Plenum Press, New York, pp. 329–339.

Hulme, M.A. & Miller, G.E. (1986) Potential for control of *Barbara colfaxiana* (Kearfott) (Lepidoptera: Olethreutidae) using *Trichogramma* sp. In: Voegelé, J., Waage, J.K. & van Lenteren, J.C. (eds) Trichogramma *and Other Egg Parasites – Les Trichogrammes et autres parasitoïdes oophages. Les Colloques de l'INRA* 43, 483–488.

Jones, S.L., Lewis, W.J., Beroza, M., Bierl, B.A. & Sparks, A.N. (1973) Host-seeking stimulants (kairomones) for the egg parasite *Trichogramma evanescens*. *Environmental Entomology* 2, 593–596.

Jones, S.L., Morrison, R.K., Ables, J.R. & Bull, D.L. (1977) A new and improved technique for the field release of *Trichogramma pretiosum*. *Southwestern. Entomologist* 2, 210–215.

Jones, S.L., Morrison, R.K., Ables, J.R., Bouse, L.F., Carlton, J.B. & Bull, D.L. (1979) New techniques for the aerial release of *Trichogramma pretiosum*. *Southwestern Entomologist* 4, 14–19.

Kanour, W.W. & Burbutis, P.P. (1984) *Trichogramma nubilale* (Hymenoptera: Trichogrammatidae) field releases in corn and a hypothetical model for control of European corn borer (Lepidoptera: Pyralidae). *Journal of Economic Entomology* 77, 102–107.

Keller, M.A. (1985) The role of movement in the population dynamics of *Trichogramma* species. PhD Dissertation, North Carolina State University, 161 pp.

Keller, M.A., Lewis, W.J. & Stinner, R.E. (1985) Biological and practical significance of movement by *Trichogramma* species: a review. *Southwestern Entomologist* 8, 138–155.

King, E.G., Bull, D.L., Bouse, L.F. & Phillips, J.R. (1985) Introduction: biological control of *Heliothis* spp. in cotton by augmentative releases of *Trichogramma*. *Southwestern Entomologist* 8, 1–10.

Knipling, E.F. & McGuire, J.U. (1968) Population models to appraise the limitations and potentialities of *Trichogramma* in managing host insect populations. *USDA Technical Bulletin* No. 1387.

Kot, J. (1964) Experiments in the biology and ecology of species of the genus *Trichogramma* Westw. and their use in plant protection. *Ekologia Polska* 12, 243–303.

Kot, J. (1968) Factors affecting the efficiency of *Trichogramma* introduction. *Proceedings of the XIII International Congress of Entomology* 2, 155–158.

Laing, J.E. & Eden, G.M. (1990) Mass-production of *Trichogramma minutum* Riley on factitious host eggs. In: Smith, S.M., Carrow, J.R. & Laing, J.E. (eds) *Inundative Release of the Egg Parasitoid,* Trichogramma minutum *(Hymenoptera: Trichogrammatidae), against Forest Insect Pests such as the Spruce Budworm,* Choristoneura fumiferana *(Lepidoptera: Tortricidae): The Ontario Project 1982–1986. Memoirs of the Entomological Society of Canada* No. 153, 10–24.

Lawrence, R.K., Houseweart, M.W., Jennings, D.T., Southard, S.G. & Halteman, W.A. (1985) Development rates of *Trichogramma minutum* (Hym.: Trichogrammatidae) and implications for timing augmentative releases for suppression of egg populations of *Choristoneura fumiferana* (Lep.: Tortricidae). *Canadian Entomologist* 117, 556–563.

Lee, K.-Q., Jiang, F.-L. & Guo, J.-J. (1986) Preliminary study on the reproductive behaviour of the parasitic wasps (*Trichogramma*). In: Voegelé, J., Waage, J.K. & van Lenteren, J.C. (eds) Trichogramma *and Other Egg Parasites – Les Trichogrammes et autres parasitoïdes oophages. Les Colloques de l'INRA* 43, 215–219.

Lewis, W.J., Beevers, M., Nordlund, D.A., Gross, H.R. & Hagen, K.S. (1979) Kairomones and their use for management of entomophagous insects. IX. Investigation of various kairomone treatment patterns for *Trichogramma* spp. *Journal of Chemical Ecology* 5, 673–680.

Li, Li-Ying (1982) *Trichogramma* sp. and their utilization in Peoples' Republic of China. In: Voegelé, J. (ed.) *Les Trichogrammes. Les Colloques de l'INRA* 9, 23–29.

Li, Li-Ying (1990) Research and utilization of *Trichogramma* in China. Unpublished report from Guangdong Entomological Institute distributed at the 3rd International Symposium on *Trichogramma* and Other Egg Parasitoids in San Antonio, Texas, USA, 23–27 September, 21 pp.

Lopez, J.D. & Morrison, R.K. (1980) Effects of high temperatures on *Trichogramma pretiosum* programmed for field release. *Journal of Economic Entomology* 73, 667–670.

Lopez, J.D. & Morrison, R.K. (1985) Parasitization of *Heliothis* spp. eggs after augmentative releases of *Trichogramma pretiosum* Riley. *Southwestern Entomologist* 8, 110–138.

Lund, H.O. (1938) Studies on longevity and productivity in *Trichogramma evanescens*. *Journal of Agricultural Research* 56, 421–440.

Maini, S., Burchi, C., Gattavecchia, C., Celli, G. & Voegelé, J. (1986) *Trichogramma maidis* Pint. Voeg. in northern Italy: augmentative releases against *Ostrinia nubilalis* (Hbn.). In: Voegelé, J., Waage, J.K. & van Lenteren, J.C. (eds) *Trichogramma and Other Egg Parasites – Les Trichogrammes et autres parasitoïdes oophages. Les Colloques de l'INRA* 43, 515–517.

May, R.M., Hassell, M.P., Anderson, R.M. & Tonkyn D.W. (1981) Density dependence in host–parasitoid models. *Journal of Animal Ecology* 50, 855–865.

McLaren, I.W. & Rye, W.J. (1983) The rearing, storage and release of *Trichogramma ivelae* Pang and Chen (Hymenoptera: Trichogrammatidae) for control of *Heliothis punctiger* Wallengren (Lepidoptera: Noctuidae) on tomatoes. *Journal of the Australian Entomology Society* 22, 119–124.

Morrison, R.K., Jones, S.L. & Lopez, T.D. (1978) A unified system for the production and preparation of *Trichogramma pretiosum* for field release. *Southwestern Entomologist* 3, 62–68.

Navarajan, P. (1986) Toxicity of different pesticides to parasitoids of the genus *Trichogramma*. In: Voegelé, J., Waage, J.K. & van Lenteren, J.C. (eds) *Trichogramma and Other Egg Parasites – Les Trichogrammes et autres parasitoïdes oophages. Les Colloques de l'INRA* 43, 423–432.

Need, J.T. & Burbutis, P.P. (1979) Searching efficiency of *Trichogramma nubilale*. *Environmental Entomology* 8, 62–68.

Neil, K.A. & Specht, H.B. (1990) Field releases of *Trichogramma pretiosum* Riley (Hymenoptera: Trichogrammatidae) for suppression of corn earworm, *Heliothis zea* (Boddie) (Lepidoptera: Noctuidae), egg populations on sweet corn in Nova Scotia. *Canadian Entomologist* 122, 1259–1266.

Newton, P.J. (1988) Movement and impact of *Trichogrammatoidea cryptophlebiae* Nagaraja (Hymenoptera: Trichogrammatidae) in citrus orchards after inundative releases against the false codling moth, *Cryptophlebia leucotreta* (Meyrick) (Lepidoptera: Tortricidae). *Bulletin of Entomology Research* 78, 85–99.

Newton, P.J. & Odendaal, W.J. (1990) Commercial inundative releases of *Trichogrammatoidea cryptophlebiae* (Hym.: Trichogrammatidae) against *Cryptophlebia leucotreta* (Lep.: Tortricidae) in citrus. *Entomophaga* 35, 545–556.

Nordlund, D.A., Lewis, W.J., Gross, H.R. & Harrell, E.A. (1974) Description and evaluation of a method for field application of *Heliothis zea* eggs and kairomones for *Trichogramma*. *Environmental Entomology* 3, 981–984.

Oatman, E.R. & Platner, G.R. (1985) Biological control of two avocado pests. *California Agriculture* 39, 21–23.

Pak, G.A. & Van Heiningen, T.G. (1985) Behavioural variations among strains of *Trichogramma* spp.: adaptability to field-temperature conditions. *Entomologia Experimentalis et Applicata* 38, 3–13.

Pak, G.A., Buis, H., Heck, I. & Hermans, M. (1986) Behavioural variations among strains of *Trichogramma* spp.: host-age selection. *Entomologia Experimentalis et Applicata* 40, 247–258.

Pak, G.A., Van Heiningen, T.G., Van Alebeek, F., Hassan, S.A. & van Lenteren, J.C. (1989) Experimental inundative releases of different strains of the egg parasite *Trichogramma* in brussel sprouts. *Netherlands Journal of Plant Pathology* 95, 129–142.

Parker, F.D. & Pinnell, R.E. (1972) Further studies on the biological control of *Pieris rapae* using supplemental host and parasite releases. *Environmental Entomology* 1, 150–157.

Parker, F.D., Lawson, F.R. & Pinnell, R.E. (1971) Suppression of *Pieris rapae* using a new control system: mass releases of both the pest and its parasites. *Journal of Economic Entomology* 64, 721–735.

Prokrym, D.R., Andow, D.A., Ciborowski, J.A. & Sreenivasam, D.D. (1992) Suppression of *Ostrinia nubilalis* by *Trichogramma nubilale* in sweet corn. *Entomologia Experimentalis et Applicata* 64, 73–85.

Ravensberg, W.J. & Berger, H.K. (1986) Biological control of the European corn borer (*Ostrinia nubilalis* Hbn., Pyralidae) with *Trichogramma maidis* Pint. Voeg. in Austria in 1980–1985. In: Voegelé, J., Waage, J.K. & van Lenteren, J.C. (eds) *Trichogramma and Other Egg Parasites – Les Trichogrammes et autres parasitoïdes oophages. Les Colloques de l'INRA* 43, 557–564.

Reeves, B.G. (1975) Design and evaluation of facilities and equipment for mass production and field release of an insect parasite and an insect predator. PhD Dissertation, Texas A&M University, 180 pp.

Ridgway, R.L. & Morrison, R.K. (1985) Worldwide perspective on practical utilization of *Trichogramma* with special reference to control of *Heliothis* on cotton. *Southwestern Entomologist* 8, 190–195.

Ridgway, R.L., Morrison, R.K., Kinzer, R.E., Stinner, R.E. & Reeves, B.G. (1973) Programmed releases of parasites and predators for control of *Heliothis* spp. on cotton. In: *Proceedings of the Beltwide Cotton Research Conference*, pp. 92–94.

Ridgway, R.L., King, E.G. & Carrillo, J.R. (1977) Augmentation of natural enemies for control of plant pests in the western hemisphere. In: Ridgway, R.L. & Vinson, S.B. (eds) *Biological Control by Augmentation of Natural Enemies*. Plenum Press, New York, pp. 379–428.

Royama, T. (1984) Population dynamics of the spruce budworm *Choristoneura fumiferana*. *Ecological Monographs* 54, 429–462.

Scholz, B.C.G. (1990) Evaluation and selection of native egg parasitoids for bollworm management in Australian cotton. In: Wajnberg, E. & Vinson, S.B. (eds) *Trichogramma and Other Egg Parasitoids. 3rd International Symposium. Les Colloques de l'INRA* 56, 235–238.

Schread, J.C. (1932) Behavior of *Trichogramma* in field liberations. *Economic Entomologist* 25, 370–374.

Schuster, M.F., Boling, J.C. & Marony, J.J. (1971) Biological control of rhodesgrass scale by airplane releases of an introduced parasite of limited dispersal ability. In: Huffaker, C.B. (ed.) *Biological Control*. Plenum Press, New York, pp. 227–250.

Schutte, F. (1962) Aktuelle Fragen Des *Trichogramma*-Einsatzes. *Journal of Applied Entomology* 50, 131–137.

Schutte, F. & Franz, J.M. (1961) Utersuchungen Zur Apfelwickler-Bekampfung

(*Carposapsa pomonella* L.) Mit Hilfe von *Trichogramma* Hartig. *Entomophaga* 6, 237–247.

Shands, W.G., Gordon, C.C. & Simpson, G.W. (1972) Insect predators for controlling aphids on potatoes. 6. Development of a spray technique for applying eggs in the field. *Journal of Economic Entomology* 65, 1099–1103.

Shi, G.-Z., Zhou, Y.-N., Zhao, J.-S., Li, T., Lian, M.-L., Chang, R.-Q., Li, T.-X., Chen, S.-J., Yang, X.-L. & Niou, J.-Q. (1986) The techniques of protection of the *Trichogramma* population in the fields. In: Voegelé, J., Waage, J.K. & van Lenteren, J.C. (eds) *Trichogramma and Other Egg Parasites – Les Trichogrammes et autres parasitoïdes oophages. Les Colloques de l'INRA* 43, 581–583.

Smith, S.M. (1988) Pattern of attack on spruce budworm egg masses by *Trichogramma minutum* (Hymenoptera: Trichogrammatidae) released in forest stands. *Environmental Entomology* 17, 1009–1015.

Smith, S.M. (1991) Predation on released parasitoid material: implications for spruce budworm suppression. In *Proceedings of the Eastern Spruce Budworm Research Work Conference* 27–30 January, Sault Ste. Marie, Ontario, Canada.

Smith, S.M. & Strom, K. (1993) Oviposition by the forest tent caterpillar (Lep.: Lasiocampidae) and acceptability of its eggs to parasitism by *Trichogramma minutum* (Hym.: Trichogrammatidae). *Environmental Entomology* 22, 1375–1382.

Smith, S.M. & Wallace, D.R. (1990) Ground systems for releasing *Trichogramma minutum* Riley in plantation forests. In: Smith, S.M., Carrow, J.R. & Laing, J.E. (eds) *Inundative Release of the Egg Parasitoid,* Trichogramma minutum *(Hymenoptera: Trichogrammatidae), against Forest Insect Pests such as the Spruce Budworm,* Choristoneura fumiferana *(Lepidoptera: Tortricidae): The Ontario Project 1982–1986. Memoirs of the Entomological Society of Canada* No. 153, 31–37.

Smith, S.M. & You, M. (1990) A life system simulation model for improving inundative releases of the egg parasitoid, *Trichogramma minutum* (Hymenoptera: Trichogrammatidae) against the spruce budworm (Lepidoptera: Tortricidae). *Ecological Modelling* 51, 123–142.

Smith S.M., Hubbes, M. & Carrow, J.R. (1986) Factors affecting inundative releases of *Trichogramma minutum* against the spruce budworm. *Journal of Applied Entomology* 101, 29–39.

Smith, S.M., Hubbes, M. & Carrow, J.R. (1987) Ground releases of *Trichogramma minutum* (Hym.: Trichogrammatidae) against epidemic populations of eastern spruce budworm. *Canadian Entomologist* 119, 251–263.

Smith, S.M., Wallace, D.R., Howse, G. & Meating, J. (1990a) Suppression of spruce budworm populations by *Trichogramma minutum*. In: Smith, S.M., Carrow, J.R. & Laing, J.E. (eds) *Inundative Release of the Egg Parasitoid,* Trichogramma minutum *(Hymenoptera: Trichogrammatidae), against Forest Insect Pests such as the Spruce Budworm,* Choristoneura fumiferana *(Lepidoptera: Tortricidae): The Ontario Project 1982–1986. Memoirs of the Entomological Society of Canada* No. 153, 46–81.

Smith, S.M., Wallace, D.R., Laing, J.E., Eden, G.M. & Nicholson, S.A. (1990b) Deposit and distribution of *Trichogramma minutum* Riley following aerial release. In: Smith, S.M., Carrow, J.R. & Laing, J.E. (eds) *Inundative Release of the Egg Parasitoid,* Trichogramma minutum *(Hymenoptera: Trichogrammatidae), against Forest Insect Pests such as the Spruce Budworm,* Choristoneura fumiferana *(Lepidoptera: Tortricidae): The Ontario Project 1982–1986. Memoirs of the Entomological Society of Canada* No. 153, 45–55.

Stein, W. Von (1961) Die Verteilung des Eiparasiten *Trichogramma embryophagum cacoeciae* (Htg.) in den Baumkronen nach seiner Massenfreilassung zur Bekampfung des Apfelwicklers. *Zeitschrift für Pflanzenkrankheiten und Pflanzenschute* 68, 502–508.

Steenburgh, W.E. Van (1930) Notes on the natural and introduced parasites of the oriental peach moth (*Laspeyresia molesta* Busck) in Ontario. *Annual Report of the Canadian Entomology Society of Ontario* 1930, 124–130.

Steenburgh, W.E. Van & Boyce, H.R. (1938) Biological control of the oriental fruit moth *Laspeyresia molesta* Busck in Ontario: a review of ten years' work. *Annual Report of the Canadian Entomological Society of Ontario* 1938, 65–74.

Stinner, R.E. (1977) Efficacy of inundative releases. *Annual Review of Entomology* 22, 414–531.

Stinner, R.E., Ridgway, R.L., Coppedge, J.R., Morrison, R.K. & Dickerson, W.A. (1974) Parasitism of *Heliothis* eggs after field releases of *Trichogramma pretiosum* in cotton. *Environmental Entomology* 3, 497–500.

Strand, M.R. & Vinson, S.B. (1985) *In vitro* culture of *Trichogramma pretiosum* Riley on an artificial medium. *Entomologia Experimentals et Applicata* 39, 203–209.

Sun, X. & Yu, E. (1986) Use of *Trichogramma dendrolimi* in forest pest control in China. In: Voegelé, J., Waage, J.K. & van Lenteren, J.C. (eds) Trichogramma *and Other Egg Parasites – Les Trichogrammes et autres parasitoïdes oophages. Les Colloques de l'INRA* 43, 591–596.

Thewke, S.E. & Puttler, B. (1970) Aerosol application of lepidopterous eggs and their susceptibility to parasitism by *Trichogramma. Journal of Economic Entomology* 65, 1033–1034.

Tran, L.C. & Hassan, S.A. (1986) Preliminary results on the utilization of *Trichogramma evanescens* Westw. to control the Asian corn borer *Ostrinia furnacalis* Guenee in the Philippines. *Journal of Applied Entomology* 101, 18–23.

Tumlinson, J.H., Turlings, T.C.J. & Lewis, W.J. (1993) Semiochemically mediated foraging behavior in beneficial parasitic insects. *Archives of Insect Biochemistry and Physiology* 22, 385–391.

Voegelé, J., Pizzol, J., Raynaud, B. & Hawlitzky, N. (1986) La diapause chez les trichogrammes et ses avantages pour la production de masse et la lutte biologique. *Mededelingen van de Faculteit Landbouwwetenschappen Rijksuniversiteit Gent* 51, 1033–1039.

Voronin, K.E. (1982) Biocenotic aspects of *Trichogramma* utilization in integrated plant protection control. In: Voegelé, J. (ed.) *Les Trichogrammes. Les Colloques de l'INRA* 9, 269–274.

Waage, J.K. & Ming, N.S. (1984) The reproductive strategy of a parasitic wasp. I. Optimal progeny and sex allocation in *Trichogramma evanescens. Journal of Animal Ecology* 53, 401–415.

You, M. & Smith, S.M. (1990) Simulated management of an historical spruce budworm population using inundative parasite release. *Canadian Entomologist* 122, 1167–1176.

Yu, D.S.K., Hagley, E.A.C. & Laing, J.E. (1984a) Biology of *Trichogramma minutum* Riley collected from apples in southern Ontario. *Environmental Entomology* 13, 1324–1329.

Yu, D.S.K., Laing, J.E. & Hagley, E.A.C. (1984b) Dispersal of *Trichogramma* spp. (Hymenoptera: Trichogrammatidae) in an apple orchard after inundative

releases. *Environmental Entomology* 13, 371–374.

Xie, Z.-N., Nettles, W.C., Morrison, R.K. & Vinson, S.B. (1986) *In vitro* mass culture of *Trichogramma pretiosum* Riley from eggs collected using a salt solution as ovipositional stimulant. In: Voegelé, J., Waage, J.K. & van Lenteren, J.C. (eds) Trichogramma *and Other Egg Parasites – Les Trichogrammes et autres parasitoïdes oophages. Les Colloques de l'INRA* 43, 403–406.

Biological Control with Egg Parasitoids Other Than *Trichogramma*

7

Ferdinando Bin

Agricultural Entomology Institute, University of Perugia, Borgo XX
Giugno, 06121 Perugia, Italy

Abstract

Representatives of six families, Encyrtidae, Eulophidae, Eupelmidae, Mymaridae, Platygastridae and Scelionidae, have been employed for biological control of crop and forest pests. A few attempts have also been made in greenhouses, against blood-sucking insects (triatomid bugs – *Triatoma* spp. – and other reduvids) and a poisonous spider (black widow – *Latrodectus mactans* (F.)). Some attempts have been made with conservation and augmentation strategies but most projects concern importation which has been used against over 40 pests, a quarter of which have been economically controlled. Failures, potential and investigative needs for these egg parasitoids are briefly discussed; the combined or alternative uses are also considered.

Introduction

Egg exploitation by parasitoids occurs in more than a dozen Hymenoptera families, among them *Parasitica* and *Aculeata*, which exert a natural control on two orders of spiders and 15 of insects (F. Bin *et al.*, unpubl.). Among the *Parasitica* and besides the Trichogrammatidae, seven families have been used for biological control, two of which are entirely composed of egg parasitoids (Mymaridae and Scelionidae), while five include scattered genera and species (Encyrtidae, Eulophidae, Eupelmidae, Platygastridae and Tetracampidae), a few of which as immatures or as adults (host feeding) are egg predators. An eighth family, Pteromalidae, includes both parasitoids and predators but because only the latter have been used (against scales; Luck, 1981) it is not considered here.

Most of these parasitoids are effective in situations where the *Tricho-gramma* impact is nil or economically negligible on account of host morpho-logical traits opposing physical resistance to the parasitoid ovipositor or rendering most of the eggs unaccessible. These can be hardness or thickness of chorion, protective coatings and coverings composed of scales, setae, silk, faeces, chewed plant material, mud or spumaline which sometimes form oothecae, egg mass architecture or location on plant or a combination of some of these (Gross, 1993; F. Bin *et al.*, unpubl.). Internal physiological traits may also play a protective role. However, they seem to affect negatively both *Trichogramma* and *Telenomus* species (Bin *et al.*, 1991).

A previous attempt to compare different roles in egg parasitoids (efficacy, competition, interchangeable or complementary use and potential) was restricted to Telenominae and *Trichogramma* (Bin & Johnson, 1982). Now this has been extended to all other groups. Most of the literature on the subject is summarized in the reference section (Anderson, 1976; Greathead, 1976, 1986; Clausen, 1978; Luck, 1981; Orr, 1988). Biological control is tradition-ally subdivided into conservation, augmentation and importation.

Conservation

Wild plants surrounding agroecosystems help in various ways to maintain a natural population of parasitoids which can be managed for a greater impact on pests. The wheat bug *Eurygaster integriceps* Put. was much more attacked by Telenominae when fields were near nectar-bearing plants. This carbohy-drate source together with honeydew permits the Scelionidae to live longer and increases fecundity (Orr, 1988). Corn may work as a trap crop for a sugar-cane pest, *Eldana saccharina* Wlk., and at the same time as a mass-breeding support for its parasitoid *Telenomus* sp. nr. *dignus* Gahan (Cochereau, in litt.). Grape leafhoppers can be effectively controlled by Mymaridae when plants nearby the commercial vineyards support the overwinter of alternative hosts. This has been reported for *Anagrus epos* Gir. which breeds throughout the year in the embedded eggs of the *Rubus* leafhopper and switches to the grape leafhopper eggs in spring (Doutt & Nakata, 1973). Instead, *Anagrus atomus* (L.) uses the same host, *Empoasca vitis* Goethe, on different plants (Cerruti *et al.*, 1989).

In forest ecosystems, alternative unrelated hosts are exploited by general-ists belonging to Encyrtidae and Eupelmidae between pest outbreaks. Pres-ence of heteropteran eggs appears to be crucial for the development of *Anastatus bifasciatus* Fonsc. females because only males are produced in certain lepidopterans (Tiberi *et al.*, 1991). *Ooencyrtus pityocampae* (Mercet) develops successfully in the eggs of several heteropterans and switches to defoliator lepidopterans as they become abundant (Battisti *et al.*, 1988).

Augmentation

Augmentative releases have been carried out several times in an attempt to increase the natural parasitism rate using natural hosts for specialists, factitious hosts for generalists, different storage techniques (Orr, 1988) and in one case an artificial diet (Liu *et al.*, 1988; see Chapter 4).

Several crops are damaged by heteropterans. Against the exposed eggs of coreids and pentatomids some Scelionidae and one Eupelmidae have been released with a significant positive response in the following systems: hazelnut – *Gonocerus acuteangulatus* (Goeze) – *Gryon muscaeformis* (Nees); soyabean – *Nezara viridula* (L.) – *Trissolcus basalis* Woll.; rice – *Scotinophora lurida* Burm. – *Telenomus gifuensis* Ashm.; wheat – *Eurygaster integriceps* Put. – *Trissolcus* spp. (Orr, 1988); litchi – *Tessarotoma papillosa* Drury – *Anastatus* sp. nr. *japonicus* (Nanta, 1988). The embedded eggs of the mirid *Lygus hesperus* Knight, laid in an artificial substrate, served to produce the mymarid *Anaphes iole* Gir. (Jones & Jackson, 1990) to be released in small cotton field plots (King & Powell, 1992).

Other crops are attacked by lepidopterans which lay egg masses in different ways and which are exploited by Scelionidae. Corn borers like *Sesamia* spp. and *Mithymna unipuncta* insert their egg masses between stem and leaf sheath where a good percentage of control is reached respectively by *Telenomus busseolae* Gahan (Alexandri & Tsitsipis, 1990; Hassan, 1992) and *T. cirphivorus* Liu (Liu *et al.*, 1960), the former crawling in the narrow space and the latter piercing through the leaf blade. Corn and sorghum, attacked by *Spodoptera frugiperda* (J.E.S.), have been treated with *Telenomus remus* Nixon in Honduras on a 2000 ha area since 1990 (Hassan, 1992). Cauliflower and cabbage have been partially protected from *Spodoptera litura* (F.) with *Telenomus remus* Nixon which can attack egg masses covered with hairs and *Telenomus* sp. nr. *dignus* Gahan has been used against *Eldana saccharina* Wlk. on sugar-cane (Orr, 1988).

A greenhouse leafhopper, *Hauptidia maroccana*, harmful to tomato and ornamentals, has been controlled with the mymarid *Anagrus atomus* (L.) in Great Britain on a 10 ha area since 1989 (Hassan, 1992). In protected aubergine cultivations, attacked by Colorado potato beetle, encouraging results have been obtained with the eulophid *Edovum puttleri* Griss. (Maini *et al.*, 1989–1990).

Forest defoliator lepidopterans have been the focus of some attempts. Small-scale releases have shown that two pine processionary moths that lay egg masses protected with scales, *Thaumetopoea pityocampa* (Den. & Schiff) and *T. wilkinsoni* Tams, can be suppressed with *Tetrastichus servadeii* Dom. (Tiberi, 1980; Halperin, 1990) and with *Ooencyrtus pityocampae* (Merc.) (Halperin, 1990). Large-scale releases of *Telenomus laeviusculus* Ratz. and *T. terebrans* Ratz. were employed respectively against two defoliators with exposed egg masses, *Malacosoma neustrium* L. and *Dendrolimus sibiricus* Tschtv. (Orr, 1988).

Livestock pests, i.e. tabanid flies, which lay multi-layered egg masses, were controlled by releasing large quantities of eggs naturally parasitized by *Telenomus olsenni* Johns. with questionable results (Orr, 1988). Blood-sucking Reduviidae, such as the triatomid bug vectors of Chagas disease, might be controlled with several Scelionidae (Masner, 1975).

Classical Biological Control

Importation has been employed by reassociating a parasitoid with its coevolved host (old associations used in classical biological control) or associating a parasitoid with a non-coevolved one (new associations used in an innovative approach). Some attempts fall into the category of intra-areal transfers (Greathead, 1976).

Most of these attempts have been world-reviewed (Clausen, 1978), updated and economically assessed (Luck, 1981) by conventional success rates (P, partial; S, substantial; C, complete). The second of these reviews gives data for quantifying results up to 1981.

There were 92 attempts made, 59% of which resulted in either no or uncertain parasitoid establishment, whereas 41% acclimatized. Of this last percentage 42% failed, 24% gave partial economic control and 34% substantial to complete control.

There were 54 parasitoid species employed in these attempts, some of which were used more than once. Only about a quarter of species proved to be economically relevant and were distributed thus in the various families: Mymaridae, 5; Scelionidae, 3; Eulophidae and Encyrtidae, 2; Eupelmidae and Platygastridae, 1. In contrast, the most frequently used family was Scelionidae with 20 species, followed by Mymaridae, 15; Eulophidae and Encyrtidae, 8; Eupelmidae, 2; Platygastridae, 1; Tetracampidae, 1. Economic control was partial for eight species (15%): 3 Mymaridae, 2 Encyrtidae, 2 Eulophidae, 1 Eupelmidae; and substantial to complete for six (11%): 3 Scelionidae, 2 Mymaridae, 1 Platygastridae.

There were 40 pest species scattered in seven orders. Economic control was only obtained for a quarter: 1 grasshopper, 1 pentatomid bug, 3 homopterans, 3 lepidopterans and 2 coleopterans as listed below. Crop pests have been controlled with the following results (success rate abbreviations as above): (S) sugar-cane chinese grasshopper *Oxya chinensis* (Thun.) – *Scelio pembertoni* Tim.; (S) sugar-cane top borer *Tryporyza nivella* (F.) – *Telenomus beneficiens* var. *elongatus* Ish.; (P) sugar-cane leafhopper *Perkinsiella saccharicida* Kirk. – *Ootetrastichus beatus* (Perk.), *O. formosanus* Tim., *Anagrus optabilis* and *A. frequens*; (P) canna – arrowroot leaf roller *Calpodes ethlius* – *Ooencyrtus* sp.; (S) apple leafhopper *Edwardsiana froggatti* (Baker) – *Anagrus armatus nigriventris* Gir.; (P) cereal leaf beetle *Oulema melanopus* (L.) – *Anaphes flavipes* Foer.; (S–C) various crops – *Nezara viridula* (L.) – *Trissolcus basalis*; (C) citrus, coffee, mango, other plants – torpedo bug *Siphanta acuta* (Wlk.) – *Aphanomerus pusillus* Perk. (Luck, 1981). Apple, grape and many other plants are no longer frequently damaged by the planthopper *Stictocephala bisonia* after the importation of *Polynema striaticorne* Gir. (Alma *et al.*, 1987).

Forest and ornamental trees have been protected from several defoliators: (P) gypsy moth – *Anastatus disparis* Rus. and *Ooencyrtus kuwanae* How.; (S–C) eucalyptus snout beetle *Gonipterus scutellatus* (Gyl.) – *Patasson nitens* Grit. (Luck, 1981). Elm leaf beetle, *Pyrrhalta luteola* (Muell.), has been regulated by *Tetrastichus gallerucae* (Fonsc.) and has successfully recovered 50 years after its introduction (Hall & Johnson, 1983). *Oxydia trichiata* suppression has been an outstanding success obtained by a new association of this South American pest with a North American polyphagous species, *Telenomus alsophilae* (Drooz *et al.*, 1977). A unique attempt was that effected against a poisonous spider, the black widow, *Latrodectus mactans* (F.), with the scelionid *Baeus latrodecti* Doz. which was successfully established (Clausen, 1978).

Intra-areal transfers (Greathead, 1976) have been attempted in several instances but, in spite of the fact that the parasitoid was only moved to a new location within the same pest area, results are different. The scelionid *Eumicrosoma benefica* Gahan, previously reported only on the chinch bug *Blissus leucopterus* Say west of the Mississippi, showed a parasitization over 50% of the hairy chinch bug *B. hirtus* Montd. east of the river where it was incidentally transferred (Dicke, 1937). However, planned transfers of *Telenomus* spp. against the same pyralids within the tropics, but at much greater distances, have failed (Polaszek & Kimani, 1990).

Success has been obtained by introducing from the same area as the host or a different area, purposely or accidentally, one or several species against exposed, protected or embedded eggs or by introducing a species which is able to exploit host eggs at any stage of development.

Unexpected results are reported for the beet leafhopper, *Circulifer tenellus* (Bak.). The native parasitoids were so competitive that establishment of exotic species was probably adversely affected (Clausen, 1978). Fortuitous biological control might therefore have occurred without any introduction. Although the sugar-cane leafhopper was the intended target of *Anagrus frequens* Perk. in later years this was found most frequently on the corn leafhopper (Clausen, 1978).

Uncertain results or failures have been reported for all other pests which belong to the same groups for which economic control was obtained and for four species of forest defoliator sawflies (Luck, 1981). Some instances of failure to establish or be effective are due to: (i) lack of resistance to cold winters, e.g. *Edovum puttleri* Griss. a subtropical eulophid which specializes on *Leptinotarsa* species (Colazza & Bin, 1992); (ii) a too delicate chorion of the target pest, *Epilachna varivestis* Muls., which does not withstand parasitization by *Tetrastichus ovulorum* Ferr. introduced from another country where it develops in a different coccinellid (Angalet *et al.*, 1968); (iii) lack of adaptability to

different hosts within the same family, e.g. *Ooencyrtus kuwanae* (How.); (iv) host migratory habit, e.g. the beet leafhopper, involving flights of several hundred miles that cannot be accomplished by parasitoids; or (v) inability of egg parasitoids to compete with a mirid egg predator subsequently introduced, e.g. taro leafhopper (Clausen, 1978).

Conclusions

Attempts to use egg parasitoids other than *Trichogramma* have included a wide range of pests: phytophages in agroecosystems, forests and greenhouses, blood-sucking bugs which carry diseases, livestock tabanid flies and a poisonous spider. However, their potential and limits are still largely unexplored and several basic and applied aspects must be investigated.

Conservation can be attained by environmental manipulation which considers obligate and facultative hosts together with food sources for adults in simplified agroecosystems and complex forests.

Augmentative release strategies lag far behind when compared to the results obtained with *Trichogramma*, especially as regards *in vitro* rearing. Promising attempts are in progress with general parasitoids (Masutti *et al.*, 1992), but for specialist parasitoids, contact semiochemicals are probably indispensable for inducing recognition and acceptance of an artificial diet. Behavioural manipulation could also be effected in the environment since the host pheromones used for mating disruption do not affect parasitoid efficiency (Brown & Cameron, 1979; Orr, 1988). Field releases have proved the validity of the augmentative approach but several failures, probably due to parasitoid dispersal outside the target area, still make it somewhat unpredictable.

Classical biological control, i.e. application of coevolved associations, is by far the most commonly used method. Its success rate is comparable with that obtained with non-coevolved associations (Conti *et al.*, 1991) but in both cases some of the basic mechanisms underlying host specialization or host switching are only now being understood. Furthermore, even what has been considered a landmark example of classical approach, may need adequate evaluation of some important aspects such as parasitoid identification and post-release assessment (Clarke, 1990).

Economic assessment with standardized rates, although difficult, has to be pursued. Many attempts to introduce parasitoids, unless they are an outstanding success or a total failure, have in fact been reported as an increase in parasitism rate without necessarily assessing the economic control. Recently proposed parameters based on parasitoid behaviour (Bin & Vinson, 1991) could help to better define efficacy.

Since there are associations of more than one parasitoid species with the same host and complexes of pests laying eggs by different strategies on the same plants, using the most effective parasitoid(s) in alternative or combined strategies should be considered to achieve better economic control.

Acknowledgements

Research supported by MURST 40%.

References

Alexandri, M.P. & Tsitsipis, J.A. (1990) Influence of the egg parasitoid *Platytelenomus busseolae* (Hym.: Scelionidae) on the population of *Sesamia nonagrioides* (Lep.: Noctuidae) in central Greece. *Entomophaga* 35, 61–70.

Alma, A., Arno, C. & Vidano, C. (1987) Particularities on *Polynema striaticorne* as egg parasite of *Stictocephala bisonia* (Rhynchota Auchenorrhyncha). In: *Proceedings of the 6th Auchenorrhyncha Meeting, Turin, Italy, 7–11 Sept. 1987.* National Research Council, Rome, pp. 597–603.

Anderson, J.F. (1976) Egg parasitoids of forest defoliating Lepidoptera. In: Anderson, J.F. & Kaya, H.K. (eds) *Perspectives in Forest Entomology.* Academic Press, New York, pp. 233-249.

Angalet, G.V., Coles, L.W. & Stewart, J.A. (1968) Two potential parasites of the Mexican bean beetle from India. *Journal of Economic Entomology* 61, 1073–1075.

Battisti, A., Colazza, S., Roversi, P.F. & Tiberi, R. (1988) Alternative hosts of *Ooencyrtus pityocampae* (Mercet) (Hymenoptera Encyrtidae) in Italy. *Redia* 71, 321–328.

Bin, F. & Johnson, N.F. (1982) Potential of Telenominae in biocontrol with egg parasitoids (Hym., Scelionidae). In: Voegelé, J. (ed.) *Les Trichogrammes. Les Colloques de l'INRA* 9, 275–287.

Bin, F. & Vinson, S.B. (1991) Efficacy assessment in egg parasitoids: a proposal for a unified terminology. In: Wajnberg, E. & Vinson, S.B. (eds) *Trichogramma and Other Egg Parasitoids. 3rd International Symposium. Les Colloques de l'INRA* 56, 175–179.

Bin, F., Colazza, S. & Ricci, C. (1991) Efficacia e limiti dei parassitoidi oofagi e pupali dell' Ifantria americana, *Hyphantria cunea* Drury (Lep. Arctiidae). In: *Atti Convegno Lotta biologica. Ministero dell' Agricoltura e delle Foreste, Acireale, 28 novembre 1991.* Ministry of Agriculture and Forestry, Rome.

Brown, M.W. & Cameron, E.A. (1979) Effects of disparlure and egg mass size on parasitism by the gypsy moth egg parasite, *Ooencyrtus kuwanai. Environmental Entomology* 8, 77–80.

Cerruti, F., Delucchi, V., Baumgaertner, J. & Rubli, D. (1989) Ricerche sull'ecosistema 'vigneto' nel Ticino: II. La colonizzazione dei vigneti da parte della cicalina *Empoasca vitis* Goethe (Hom., Cicadellidae, Typhlocybinae) e del suo parassitoide *Anagrus atomus* Haliday (Hym., Mymaridae) e importanza della flora circostante. *Mitteilungen der Schweizerischen Entomologischen Gesellschaft* 62, 253–267.

Clarke, A.R. (1990) The control of *Nezara viridula* L. with introduced egg parasitoids in Australia. A review of a landmark example of classical biological control. *Australian Journal of Agricultural Research* 41, 1127–1146.

Clausen, C.P. (ed.) (1978) Introduced parasites and predators of Arthropod pests and

weeds: a world review. *Agricultural Research Service, USDA, Agriculture Handbook* No. 480, 545 pp.

Colazza, S. & Bin, F. (1992) Introduction of the oophage *Edovum puttleri* Griss. (Hymenoptera: Eulophidae) in Italy for the biological control of Colorado potato beetle. *Redia* 75, 203–225.

Conti, E., Colazza, S. & Bin, F. (1991) New associations in classical biological control: the case of oophage parasitic Hymenoptera. In: Wajnberg, E. & Vinson, S.B. (eds) *Trichogramma and Other Egg Parasitoids. 3rd International Symposium. Les Colloques de l'INRA* 56, 1183–186.

Dicke, F.F. (1937) *Eumicrosoma benefica* Gahan as an egg parasite of the hairy chinch bug. *Journal of Economic Entomology* 30, 376.

Doutt, R.L. & Nakata, J. (1973) The *Rubus* leafhopper and its egg parasitoid: an endemic biotic system useful in grape-pest management. *Environmental Entomology* 2, 381–386.

Drooz, A.T., Bustillo, A.E., Fedde, G.F. & Fedde, V.H. (1977) North American egg parasite successfully controls a different host genus in South America. *Science* 197, 390–391.

Greathead, D.J. (1976) A review of biological control in western and southern Europe. *Commonwealth Institute of Biological Control Technical Communication* No. 7, 182 pp.

Greathead, D.J. (1986) Parasitoids in classical biological control. In: Waage, J. & Greathead, D. (eds) *Insect Parasitoids*. Academic Press, New York, pp. 289–318.

Gross, P. (1993) Insect behavioral and morphological defenses against parasitoids. *Annual Review of Entomology* 38, 251–273.

Hall, R.W. & Johnson, N.F. (1983) Recovery of *Tetrastichus gallerucae* (Hymenoptera: Eulophidae), an introduced egg parasitoid of the elm leaf beetle (*Pyrrhalta luteola*) (Coleoptera: Chrysomelidae). *Journal of the Kansas Entomological Society* 56, 297–298.

Halperin, J. (1990) Mass breeding of egg parasitoids (Hym., Chalcidoidea) of *Thaumetopoea wilkinsoni* Tams (Lep., Thaumetopoeidae). *Journal of Applied Entomology* 109, 336–340.

Hassan, S.A. (ed.) (1992) Trichogramma *News*, vol. 6. Federal Biological Research Centre for Agriculture and Forestry, 46 pp.

Jones, W.A. & Jackson, C.G. (1990) Mass production of *Anaphes iole* for augmentation against *Lygus hesperus*: effects of food on fecundity and longevity. *Southwestern Entomologist* 15, 463–468.

King, E.G. & Powell, J.E. (1992) Propagation and release of natural enemies for control of cotton insect and mite pests in the United States, including status and prospects for using *Microplitis croceipes*. *Cotton-Integrated Pest Management: Proceedings of a Symposium. USDA-ARS Report* No. 106, 47–60.

Liu, Chung-Lo, Fu, Yi-Ling & Chen, Tai-Lu (1960) Biological studies and field liberation of an armyworm egg-parasite *Telenomus cirphivorus* Liu (Hymenoptera: Scelionidae). *Acta Entomologica Sinica* 10, 283–288.

Liu, Zhi-Cheng, Wang, Zhi-Yon, Sun, Yi-Ren, Liu, Jiang-Feng & Yang, Wu-Hong (1988) Studies on culturing *Anastatus* sp. a parasitoid of litchi stink bug, with artificial host eggs. In: Voegelé, J., Waage, J.K. & van Lenteren, J.C. (eds) *Trichogramma and Other Egg Parasites. 2nd International Symposium. Les Colloques de l'INRA* 43, 351–360.

Luck, R.F. (1981) Parasitic insects introduced as biological control agents for arthropod pests. In: Pimentel, D. (ed.) *Handbook of Pest Management in Agriculture*, vol. 2. CRC Press, Boca Raton, Florida, pp. 125–284.

Maini, S., Nicoli, G. & Manzaroli, G. (1989–1990) Evaluation of the egg parasitoid *Edovum puttleri* Grissel (Hym. Eulophidae) for biological control of *Leptinotarsa decemlineata* (Say) (Col. Chrysomelidae) on eggplant. *Bollettino dell' Istituto di Entomologia 'Guido Grandi' Universita'di Bologna* 44, 161–168.

Masner, L. (1975) Two new sibling species of *Gryon* Haliday (Hymenoptera, Scelionidae), egg parasites of blood-sucking Reduviidae (Heteroptera). *Bulletin of Entomological Research* 65, 209–213.

Masutti, L., Battisti, A., Milani, N. & Zanata, M. (1992) First success in the *in vitro* rearing of *Ooencyrtus pityocampae* (Mercet) (Hym., Encyrtidae). Preliminary note. *Redia* 75, 227–232.

Nanta, P. (1988) Biological control of longan stink bug, *Tessarotoma papillosa* Drury in Thailand. In: Voegelé, J., Waage, J.K. & van Lenteren, J.C. (eds) *Trichogramma and Other Egg Parasites. 2nd International Symposium. Les Colloques de l'INRA* 43, 525–526.

Orr, D.B. (1988) Scelionid wasps as biological control agents: a review. *Florida Entomologist* 71, 506–528.

Polaszek, A. & Kimani, S.W. (1990) *Telenomus* species (Hymenoptera: Scelionidae) attacking eggs of pyralid pests (Lepidoptera) in Africa: a review and guide to identification. *Bulletin of Entomological Research* 80, 57–71.

Tiberi, R. (1980) Modificazioni della distribuzione dei parassiti oofagi in ovature di processionaria del pino conseguenti al potenziamento artificiale di *Tetrastichus servadeii* Dom. (Hymenoptera, Chalcidoidea). *Redia* 63, 307–321.

Tiberi, R., Roversi, P.F. & Bin, F. (1991) Egg parasitoids of pine and oak processionary caterpillars in central Italy. *Redia* 74, 249–250.

Habitat Location by *Trichogramma* 8

Donald A. Nordlund

USDA-ARS, Subtropical Agricultural Research Laboratory, Biological Control of Pests Research Unit, 2413 East Highway 83, Weslaco, Texas 78596, USA

Abstract

Members of the genus *Trichogramma* are some of the most widely used biological control agents in the world. Thus, considerable effort has been expended to decipher the host selection behaviour of some of the more important species. This behaviour (host-habitat location, host location and host acceptance) is mediated by a number of stimuli. Chemical stimuli, however, appear to dominate the process. Host-habitat location is the important first step in the process and these minute insects exhibit preferences for certain habitat types. A variety of plants, including *Amaranthus*, tomato and corn, produce chemical stimuli that mediate host-habitat location. In addition, some *Trichogramma* species respond to the sex pheromone released by adult hosts. Host selection, with particular emphasis on habitat location, of *Trichogramma* spp. will be reviewed.

Introduction

The genus *Trichogramma* (Chalcidoidea, Trichogrammatidae) is comprised of almost 100 species (Hung *et al.*, 1985) and is the most widely used genus of parasitoids for biological control of insect pests in the world (Stinner, 1977; King *et al.*, 1985). The former USSR and China lead the world in area of application, with a combined total of more than 20 million ha (Li, 1984; Ridgway & Morrison, 1985; Gusev & Lebedev, 1988; see Chapter 2 for a review). Members of the genus have been studied extensively over the years, because of their importance, and considerable effort has been extended to decipher the host selection behaviour of some important species of this genus (Noldus, 1989).

The success of a parasitoid is, to a great extent, dependent on the ability of the females to parasitize hosts. Successful parasitization has been divided into several steps: host-habitat location, host location, host acceptance and host regulation (Laing, 1937; Flanders, 1953; Doutt, 1964; Vinson, 1975, 1981; Nordlund *et al.*, 1981a,b). The first three of these steps are referred to as host selection (Nordlund *et al.*, 1988). The host selection process is a continuum of behaviours mediated by a variety of stimuli. The host-habitat location, host location and host acceptance steps were devised to help us in studying and describing the process, not because the parasitoids recognize these steps. Thus, it is difficult and not particularly useful to attempt to identify distinct dividing lines between the steps (see Arthur, 1981; Vinson, 1981; Weseloh, 1981). There is a large body of literature related to the stimuli mediating the behaviours involved in the host selection process (Nordlund *et al.*, 1981a,b; Vet & Dicke, 1992; Tumlinson *et al.*, 1993), yet, we are only beginning to develop an understanding of the process.

We recognized quite early that when an adult parasitoid emerges, it may or may not find itself in a habitat populated by its host. If it is to reproduce, it must have the capacity to locate a habitat in which hosts might be present. Salt (1935) stated: 'Now it is certain that some parasites, and probable that many more, are first attracted not to a particular host but to a certain type of environment.' Laing (1937) recognized the importance of habitat-related stimuli when saying: 'The movement of a parasite that is outside the area of distribution of its host, must be either entirely random with respect to that area, or be influenced by the environmental factors characteristic of it.'

Thus, host-habitat location is the first step in host selection. This step often places a greater limitation on the number of host species actually attacked by a parasitoid than does host suitability (Picard & Rabaud, 1914; Flanders, 1962; Townes, 1962). Picard & Rabaud (1914), for example, reported that many parasitic Hymenoptera would attack larvae from different families and even different orders, if they were found on the same plant. Many parasitoids exhibit a high degree of specificity in their choice of habitats in which to search for hosts (Nordlund, 1987; Nordlund *et al.*, 1988). Despite the recognized importance of host-habitat location to successful parasitization, relatively little is known about this behaviour for parasitoids in general and for *Trichogramma* spp. in particular (Vinson, 1981)

Habitat Preferences

That *Trichogramma* species exhibit various habitat preferences was recognized in the 1930s. Flanders (1937), for example, noted that *Trichogramma evanescens* Westwood preferred a field habitat, *Trichogramma embryophagum* (Hartig) was found most frequently in arboreal habitat and *Trichogramma semblidis* (Aurivillius) preferred a marsh habitat. Steenburg (1934) demon-

Table 8.1. Percentage parasitization of *Helicoverpa zea* eggs by *Trichogramma* spp. in monoculture or polyculture plots of corn, beans and tomato (Nordlund *et al.*, 1984).

Treatment	Mean*
Tomato	42.88 a
Corn–tomato	33.78 ab
Bean–tomato	28.54 bc
Corn–bean–tomato	28.49 bc
Bean	18.62 cd
Corn–bean	13.42 d
Corn	1.49 e

* Means followed by different letters are significantly different ($P <$ 0.05), as determined by Duncan's multiple range test.

strated the importance of habitat preferences by releasing large numbers of *T. semblidis* and *T. evanescens* in peach orchards where *T. embryophagum* was the only species occurring naturally. Both of the released species quickly disappeared from the orchard environment. *Trichogramma minutum* Riley parasitizes *Manduca sexta* (L.) eggs on tomato and jimsonweed but not on tobacco (Rabb & Bradley, 1968). *Helicoverpa virescens* (Fabricius) eggs are also parasitized on tomato but not on tobacco (Martin *et al.*, 1981). Bar *et al.* (1979) found that *Trichogramma semifumatum* Riley parasitized *Heliothis armigera* (Hübner) eggs on tomato but not on cotton.

Data from polyculture studies are ambiguous. Parasitism of *Choristoneura fumiferana* (Clemens) by *T. minutum* increased as the proportion of balsam fir in the stand decreased (Kemp & Simmons, 1978). Altieri *et al.* (1981) reported that parasitization of *Helicoverpa zea* (Boddie) eggs by *Trichogramma* spp. was significantly higher in soyabean–corn intercropping and weedy soyabean plots than in weed-free soyabean monocultures. Nordlund *et al.* (1984) reported that *Trichogramma* spp. were much more active in plots of tomatoes than in plots of corn and that parasitism of *H. zea* eggs was higher in a corn–bean–tomato polyculture than in corn but lower than in tomato monocultures (Table 8.1). Andow & Risch (1987) observed higher parasitism rates in corn monocultures than in corn–bean–squash or corn–clover polycultures.

A variety of factors characteristic of the habitat may, and probably do, play a role in observed habitat preferences of *Trichogramma* spp. Temperature, humidity and light are three factors that undoubtedly play a role in limiting the distribution of these parasitoids (Pak, 1988). However, what habitat characteristics might actually attract a female parasitoid from a distance? Factors that might be involved include sound, visual cues, other forms of

electromagnetic radiation and volatile chemicals. Chemical stimuli appear to play a major role in the process.

The Role of Synomones

Trichogramma spp. use a variety of host-produced chemical stimuli (kairomones) in their host location and host acceptance behaviours (see Noldus, 1989, for a review). Thus, it should not be surprising to find that chemical stimuli from plant (synomones) also play a role in the host-habitat location behaviour of these parasitoids.

Altieri *et al.* (1981, 1982) demonstrated that application of water extracts of *Amaranthus* spp. and corn resulted in increased parasitization of *H. zea* eggs by *Trichogramma* spp. in a variety of cropping systems, including soyabean. Parasitization increased even when the application consisted of cotton wicks, impregnated with the extract, suspended from the plants. Kaiser (1988) and Kaiser *et al.* (1989) documented, using a four-arm airflow olfactometer, that *Trichogramma maidis* Pintureau & Voegelé (= *T. brassicae* Bezdenko) females are arrested by volatiles from corn leaves.

The findings by Nordlund *et al.* (1984) (see above) led to a series of studies of the role that semiochemicals from plants play in the host-habitat location behaviour of *Trichogramma*. Nordlund *et al.* (1985a) demonstrated, in a Petri dish experiment, that the presence of extracts of tomato plants resulted in increased parasitization by *Trichogramma pretiosum* Riley females while extracts of corn did not. These studies did not demonstrate that *T. pretiosum* females are attracted to tomato, though they did indicate that *T. pretiosum* should be arrested by contact with stimuli from tomato. Nordlund *et al.* (1985b), demonstrated, using a 'jar olfactometer', that volatiles from the hexane extract of tomato also stimulated search behaviour, resulting in more than a twofold increase in parasitization. A Y-tube olfactometer was then used to demonstrate attraction by volatiles in the extract. The olfactometer was a basic Y-tube made of 2 cm (diam.) glass tubing, with each arm 20 cm long and the leg 33 cm long (Nordlund *et al.*, 1983). Females (10 per replication for 24 replications) were introduced into the leg of the olfactometer and after 10 min the number of females in the treated and control arms were recorded. The mean number of females found in the treated arm (5) was significantly greater ($P<0.05$) than the mean number found in the control arm (2). In field studies the response to chemicals in tomato extract was elicited, even when cotton rolls, treated with hexane extract of tomato leaves, were suspended in field plots of corn (Table 8.2).

Not all responses to plant-produced volatiles are positive. Bar *et al.* (1979) and Cabello & Vargas (1985) reported that *T. semifumatum* and *T. sp.p. buesi* are repelled by cotton volatiles in olfactometer studies.

From the above we see that plant-produced stimuli do elicit behavioural

Table 8.2. Parasitization of *Helicoverpa zea* eggs by *Trichogramma pretiosum* in plots of 'Silver Queen' corn in which cotton rolls, treated with a hexane extract of 'Floridade' tomato leaves, or hexane were suspended (Nordlund *et al.*, 1984).

Exposure			Mean ($\pm$ SE) percentage* parasitization in:	
Reading	Time (h)	*n*	Extract plots	Hexane plots
1	4	30	49.5 ($\pm$ 4.3) a	28.8 ($\pm$ 4.1) b
2	18	28[†]	44.7 ($\pm$ 4.8) a	19.7 ($\pm$ 3.3) b
3	4	20[‡]	30.3 ($\pm$ 3.0) a	16.4 ($\pm$ 3.0) b

* Means for each reading followed by different letters are significantly different as determined by analysis of variance.
[†] In two treated plots no eggs were recovered so these replications were eliminated from the analysis.
[‡] The third reading of one test was destroyed by rain.

responses from *Trichogramma* females. However, because of their minute size, *Trichogramma* are not particularly strong flyers, and long distance movement is probably via wind currents. Orientation to a stimulus source by these minute insects over any significant distance is unlikely. Thus, it is not clear what behaviours are elicited in nature that result in host-habitat location by *Trichogramma*. One possibility is that they float on the wind until they detect an appropriate stimulus and then simply drop.

The Role of Kairomones

The above discussion relates to stimuli that attract *Trichogramma* spp. to and/ or stimulate their search in a habitat regardless of the presence or absence of hosts. This has been traditionally thought of as host-habitat location. However, it has recently been demonstrated that *T. pretiosum*, *T. maidis* (= *T. brassicae*) and *T. evanescens* also respond to components of the sex pheromone released by female moths (Lewis *et al.*, 1982; Kaiser *et al.*, 1989; Noldus, 1989). Obviously orientation to such chemicals would tend to lead *Trichogramma* to habitats with a high probability of containing host eggs and thus, to bring the parasitoids closer to potential hosts. Noldus (1989) reported that the sex pheromone of *Mamestra brassicae* L. females was adsorbed on to the leaf surface of Brussels sprouts to such an extent that it was subsequently capable of eliciting a behavioural response from conspecific males and from *T. evanescens*. These kairomones are not a direct indication of the presence of host eggs, only of adult moths. This illustrates the continuum of behavioural responses involving host-habitat cues to close range host acceptance cues. How these long range host-produced stimuli interact with plant-produced materials is not yet understood.

Conclusion

Host-habitat location is an important first step in host selection. Successful parasitization requires that the female parasitoid be capable of locating a habitat in which its host may be found. Unfortunately, relatively little is known about the stimuli that mediate host-habitat location.

From the stand point of biological control, the need for a selected parasitoid to be effective in the target habitat is obvious. Thus, the need for consideration of habitat preferences of selected parasitoids. However, with proper knowledge other options may be fruitful. For example, it may be possible to breed plants that produce the necessary synomones, or to design intercropping systems that attract and retain important parasitoids. Finally, it may be possible to apply the appropriate synomone to a crop, attracting and retaining the parasitoids necessary to protect the crop.

We need a greater understanding of the specific behavioural mechanisms that allow *Trichogramma* and other entomophagous insects to use the vast array of stimuli in their environment to locate hosts and prey. Unfortunately, we have only just begun to scratch the surface of this problem. Much remains to be done before we have a basic understanding of the host-habitat location behaviour of *Trichogramma*, not to mention the host selection process as a whole. It is important to remember that we have a behavioural continuum, which, for our convenience, we divide into a number of steps, which the insects do not necessarily recognize. Also, the host selection process is complicated by the ability of some parasitoids to learn and by the fact that, in some cases, plants that have been damaged by potential hosts can be much more attractive or stimulatory to parasitoids (Vet & Dicke, 1992; Tumlinson *et al.*, 1993).

References

Altieri, M.A., Lewis, W.J., Nordlund, D.A., Gueldner, R.C. & Todd, J.W. (1981) Chemical interactions between plants and *Trichogramma* wasps in Georgia soybean fields. *Protection Ecology* 3, 259–263.

Altieri, M.A., Annamalai, S., Katiyar, K.P. & Flath, R.A. (1982) Effects of plant extracts on rates of parasitization of *Anagasta kuehniella* (Lep.: Pyralidae) eggs by *Trichogramma pretiosum* (Hym.: Trichogrammatidae) under greenhouse conditions. *Entomophaga* 27, 431–438.

Andow, D.A. & Risch, S.J. (1987) Parasitism in diversified agroecosystems: phenology of *Trichogramma minutum* (Hymenoptera: Trichogrammatidae). *Entomophaga* 32, 255–260.

Arthur, A.P. (1981) Host acceptance by parasitoids. In: Nordlund, D.A., Jones, R.L. & Lewis, W.J. (eds) *Semiochemicals: Their Role in Pest Management*. Wiley, New York, pp. 97–120.

Bar, D., Gerling, D. & Rossler, Y. (1979) Bionomics of the principal natural enemies attacking *Heliothis armigera* in cotton fields in Israel. *Environmental Entomology* 8, 468–474.

Cabello, T. & Vargas, P. (1985) Estudio con olfactometro de la influencia de la planta y del insecto huesped en la actividad de busqueda de *Trichogramma cordubensis* Vargas Y Cabello y de *T.* sp. p. *buesi* (Hyme. Trichogrammatidae). *Boletin del Servicio de Defensa contra Plagas e Inspección Fitopatológia* 11, 237–241.

Doutt, R.L. (1964) Biological characteristics of entomophagous adults. In: DeBach, P. (ed.) *Biological Control of Insect Pests and Weeds*. Reinhold, New York, pp. 145–167.

Flanders, S.E. (1937) Habitat selection by *Trichogramma. Annals of the Entomological Society of America* 30, 208–210.

Flanders, S.E. (1953) Variation in susceptibility of citrus-infesting coccids to parasitization. *Journal of Economic Entomology* 46, 266–269.

Flanders, S.E. (1962) The parasitic Hymenoptera: specialists in population regulation. *Canadian Entomologist* 94, 1133–1147.

Gusev, G.V. & Lebedev, G.I. (1988) Present state of *Trichogramma* application and research. In: Voegelé, J., Waage, J.K. & van Lenteren, J.C. (eds) *Trichogramma and Other Egg Parasites. 2nd International Symposium. Les Colloques de l'INRA* 43, 477–481.

Hung, A.C.F., Vincent, D.L., Lopez, J.D. & King, E.G. (1985) *Trichogramma* (Hymenoptera: Trichogrammatidae) fauna in certain areas of Arkansas and North Carolina. *Southwestern Entomologist* 8, 11–20.

Kaiser, L. (1988) Plasticité comportementale et rôle des médiateurs chimiques dans la sélection de l'hôte par *Trichogramma maidis* Pint. et Voeg. (Hym.: Trichogrammatidae). PhD Thesis, Université de Paris-Sud, Paris, 190 pp.

Kaiser, L., Pham-Delegue, M.H., Bakchine, E. & Masson, C. (1989) Olfactory responses of *Trichogramma maidis* Pint. et Voeg.: effects of chemical cues and behavioral plasticity. *Journal of Insect Behavior* 2, 701–712.

Kemp, W.P. & Simmons, G.A. (1978) The influence of stand factors on parasitism of spruce budworm eggs by *Trichogramma minutum. Environmental Entomology* 7, 685–688.

King, E.G., Hopper, K.R. & Powell, J.E. (1985) Analysis of systems for biological control of crop arthropod pests in the U.S.A by augmentation of predators and parasites. In: Hoy, M.A. & Herzog, D.C. (eds) *Biological Control in Agricultural IPM Systems*. Academic Press, Orlando, pp. 201–227.

Laing, J. (1937) Host-finding by insect parasites. I. Observations on the finding of hosts by *Alysia manducator, Mormoniella vitripennis* and *Trichogramma evanescens. Journal of Animal Ecology* 6, 298–317.

Lewis, W.J., Nordlund, D.A., Gueldner, R.C., Teal, P.E.A. & Tumlinson, J.H. (1982) Kairomones and their use for management of entomophagous insects. XIII. Kairomonal activity for *Trichogramma* spp. of abdominal tips, excretion, and a synthetic sex pheromone blend of *Heliothis zea* (Boddie) moths. *Journal of Chemical Ecology* 8, 1323–1331.

Li, Li-ying (1984) Research and utilization of *Trichogramma* in China. In: *Proceedings of the Chinese Academy of Sciences and United States National Academy of Sciences Joint Symposium on Biological Control of Insects Sept. 25–28, 1982, Beijing.* Science Press, Beijing, pp. 204–223.

Martin, P.B., Lingren, P.D., Greene, G.L. & Grissel, E.E. (1981) The parasitoid complex

of three noctuids (Lep.) in a northern Florida cropping system: seasonal occurrence, parasitization, alternate hosts, and influence of host-habitat. *Entomophaga* 26, 401–419.

Noldus, L.P.J.J. (1989) Chemical espionage by parasitic wasps. PhD Thesis, Wageningen Agricultural University, Wageningen, The Netherlands, 252 pp.

Nordlund, D.A. (1987) Plant produced allelochemics and their involvement in the host selection behavior of parasitoids. In: Labeyrie, V., Fabres, G. & Lachaise, D. (eds) *Insects – Plants*. Dr. W. Junk Publishers, Dordrecht, pp. 103–107.

Nordlund, D.A., Jones, R.L. & Lewis, W.J. (eds) (1981a) *Semiochemicals: Their Role in Pest Control*. Wiley, New York.

Nordlund, D.A., Lewis, W.J. & Gross, H.R. Jr. (1981b) Elucidation and employment of semiochemicals in the manipulation of entomophagous insects. In: Mitchell, E.R. (ed.) *Management of Insect Pests with Semiochemicals*. Plenum Press, New York, pp. 463–475.

Nordlund, D.A., Lewis, W.J. & Gueldner, R.C. (1983) Kairomones and their use for management of entomophagous insects XIV. Response of *Telenomus remus* to abdominal tips of *Spodoptera frugiperda*, (Z)-9-tetradecene-1-ol acetate and (Z)-9-dodecene-1-ol acetate. *Journal of Chemical Ecology* 9, 695–791.

Nordlund, D.A., Chalfant, R.B. & Lewis, W.J. (1984) Arthropod populations, yields and damage in monocultures and polycultures of corn, beans and tomatoes. *Agriculture, Ecosystems and Environment* 11, 353–367.

Nordlund, D.A., Chalfant, R.B. & Lewis, W.J. (1985a) Response of *Trichogramma pretiosum* females to extracts of two plants attacked by *Heliothis zea*. *Agriculture, Ecosystems and Environment* 12, 127–133.

Nordlund, D.A., Chalfant, R.B. & Lewis, W.J. (1985b) Response of *Trichogramma pretiosum* females to volatile synomones from tomato plants. *Journal of Entomological Science* 20, 372–376.

Nordlund D.A., Lewis, W.J. & Altieri, M.A. (1988) Influences of plant-produced allelochemicals on the host/prey selection behavior of entomophagous insects. In: Barbosa, P. & Letourneau, D.K. (eds) *Novel Aspects of Insect–Plant Interactions*. Wiley, New York, pp. 65–90.

Pak, G.A. (1988) Selection of *Trichogramma* for inundative biological control. PhD Thesis, Wageningen Agricultural University, Wageningen, The Netherlands, 224 pp.

Picard, F. & Rabaud, E. (1914) Sur le parasitime externe des Braconides (Hym.). *Bulletin de la Societé Entomologique de France*, 266–269.

Rabb, R.L. & Bradley, J.R. (1968) The influence of host plants on parasitism of eggs of the tobacco hornworm. *Journal of Economic Entomology* 61, 1249–1252.

Ridgway, R.L. & Morrison, R.L. (1985) Worldwide perspective on practical utilization of *Trichogramma* with special reference to control of *Heliothis* on cotton. *Southwestern Entomologist* 8, 190–198.

Salt, G. (1935) Experimental studies in insect parasitism. III. Host selection. *Proceedings of the Royal Society* 107, 450–454.

Steenburgh, W.E. van (1934) *Trichogramma minutum* Riley as a parasite of the oriental fruit moth (*Laspeyresia molesta* Busck) in Ontario. *Canadian Journal of Research* 10, 287–314.

Stinner, R.E. (1977) Efficacy of inundative releases. *Annual Review of Entomology* 22, 513–531.

Townes, H. (1962) Host selection patterns in some nearctic Ichneumonids (Hymenoptera). *Verhandlungen XI. International Kongress für Entomologie (Wein)* 2, 738–741.

Tumlinson, J.H., Lewis, W.J. & Vet, L.E.M. (1993) How parasitic wasps find their hosts. *Scientific American* 268, 100–106.

Vet, L.E.M. & Dicke, M. (1992) Ecology of infochemical use by natural enemies in a tritrophic context. *Annual Review of Entomology* 37, 141–172.

Vinson, S.B. (1975) Biochemical coevolution between parasitoids and their hosts. In: Price, P. (ed.) *Evolutionary Strategies of Parasitic Insects and Mites.* Plenum, New York, pp. 14–48.

Vinson, S.B. (1981) Habitat location. In: Nordlund, D.A., Jones, R.L. & Lewis, W.J. (eds) *Semiochemicals: Their Role in Pest Control.* Wiley, New York, pp. 51–77.

Weseloh, R.M. (1981) Host location by parasitoids. In: Nordlund, D.A., Jones, R.L. & Lewis, W.J. (eds) *Semiochemicals: Their Role in Pest Control.* Wiley, New York, pp. 79–95.

Host Recognition and Acceptance by *Trichogramma* 9

JONATHAN M. SCHMIDT

*Department of Environmental Biology, University of Guelph, Guelph,
Ontario, Canada N1G 2W1*

Abstract

After contacting a potential host, female trichogrammatid wasps examine its surface and internal contents to assess its suitability as an oviposition site. The number and sex of the eggs allocated to a host during oviposition varies with the wasp's assessment of host size, age, nutritional suitability and previous parasitization. This chapter describes the processes of host examination and oviposition and examines the sensory and behavioural mechanisms that underlie host recognition. The chemical and physical factors influencing host acceptance, clutch size and sex ratio adjustment are surveyed and discussed in the context of host preference studies. The ability of these wasps to recognize previously parasitized hosts and to selectively superparasitize such hosts is also examined. The chapter focuses on the opportunistic aspects of trichogrammatid behaviour and the experimental evidence for the use of simple behavioural rules by these wasps to solve complex foraging and progeny allocation problems.

Introduction

The life cycles of the trichogrammatid wasps alternate between parasitic and free-living stages. The immature stages develop as parasites of a single host egg, and eventually destroy their hosts by feeding on the contents. In contrast, adult *Trichogramma* are free-living, and the females behave as predators, seeking and identifying potential hosts for their progeny. This chapter will focus on the behaviour of the adult female after it finds a host, and examine the processes of host recognition, progeny allocation and the determination

of sex ratio. The physiological mechanisms underlying these activities will be described and some of the factors influencing the expression of these behaviours will be discussed.

Whereas most parasitoids are limited to a few host species, female *Trichogramma* have been reported to exploit an enormous diversity of insect eggs (Hase, 1925; Quednau, 1955; Taylor & Stern, 1971). The eggs of over 300 species belonging to at least eight insect orders are parasitized by members of the genus *Trichogramma*, and the list of known hosts continues to expand (Hase, 1925). However, caution must be exercised in order to avoid confusing the apparent polyphagy of most *Trichogramma* species with an inability to discriminate between host species. At least part of the reported polyphagy of *Trichogramma* females may be experimental artefact. Many tests of host suitability have been conducted as single choice tests in which the wasps had no alternative hosts available for oviposition (Quednau, 1955; Taylor & Stern, 1971; Pak, 1988). Two or multiple choice tests have often revealed distinct preferences for a particular host species or developmental stage (De Jong & Pak, 1984; Pak, 1988). Furthermore, the list of known host species includes *Galleria melonella* L. (Pyralidae) and numerous factitious rearing hosts, often stored product pests such as *Corcyra cephalonica* Stainton (Pyralidae), *Ephestia kuehniella* Zeller (Pyralidae), and *Sitotroga cerealella* (Oliver) (Gelechidae). It is unlikely that these hosts are frequently encountered in the field, and, indeed, the restriction of *Trichogramma* species to specific habitats probably greatly limits their actual host range (Flanders, 1937).

The ability of *Trichogramma* species to recognize and utilize a variety of host species is closely coupled to their multivoltine life cycles and limited capacity to control their dispersal in the field (Sachtleben, 1929). The occurrence of several generations each year probably favours greater poly-phagous flexibility than is generally associated with univoltine parasitoids. Successive generations of *Trichogramma* are simply unlikely to locate the same host species especially in temperate climate regions.

Polyphagy in *Trichogramma* species is associated with the ability of adult females to identify a variety of insect eggs as suitable oviposition sites and the capacity of the developing larvae to utilize the nutritionally diverse contents of these hosts. Rather than relying on highly specific chemical cues to limit host selection to a few species, female *Trichogramma* species appear to rely on more general physical features, especially size and shape, to identify egg-like objects.

Historical Review

The use of general physical cues for host recognition by *Trichogramma* species was first recognized by Holloway (1912), who found that the wasps attempted to oviposit into minute beads of plant gum and other objects of similar size and shape to insect eggs. Comparable observations were later reported by Salt (1935) and Marchal (1936). In a series of pioneering behavioural studies, Salt presented experimental evidence identifying the sensory basis of host recognition by *T. evanescens* Westwood, and demonstrated the ability of *T. evanescens* to discriminate between parasitized and unparasitized hosts (Salt, 1934, 1935, 1937, 1958). The work of Salt, Flanders and Marchal also demonstrated the pronounced influence of the host on *Trichogramma* species development and survival, providing a basis for subsequent theoretical and experimental analyses of host suitability (Salt, 1934, 1935, 1937, 1958; Flanders, 1935; Marchal, 1936). German researchers, including Hase, Maier and Quednau have contributed considerable information on host selection by *Trichogramma* species and focused attention on the effects of the abiotic environment (Hase, 1925; Quednau, 1955, 1960; Maier, 1960). The behavioural investigations undertaken by Klomp & Teerink (1962, 1967), using *T. embryophagum* Hartig, greatly extended the earlier work by Salt.

Subsequent research on various *Trichogramma* species has addressed issues associated with the use of these parasitoids in various biological control programmes. They have also been studied as model systems for investigating several progeny and sex allocation hypotheses (Iwasa *et al.*, 1984; Parker & Courtney, 1984; Waage & Godfray, 1984; Charnov & Skinner, 1985; Waage, 1986).

Much recent research on *Trichogramma* species has focused on host preference and host suitability (Pak, 1988). Attempts have also been made to quantify behavioural traits associated with host acceptance and oviposition as quality control and strain selection criteria (De Jong & Pak, 1984; Pak & van Heiningen, 1985; Pak, 1988). More detailed studies of specific, heritable behavioural traits have been reported by Chassain & Boulétreau (1987), Chassain *et al.* (1988), Wajnberg (1989), Wajnberg *et al.* (1989) and Schmidt (1991) (see Chapter 12 for a recent detailed synthesis).

Studies of the behaviour and sensory physiology of *Trichogramma* species have been continued by Schmidt & Smith (1985a,b; 1987a,b,c), Noldus & van Lenteren (1985) and Pak (1988). Discrimination of previously parasitized hosts has been an area of special attention and the focus of research by Klomp *et al.* (1980), Suzuki *et al.* (1984), Waage & Lane (1984), Waage & Ng (1984) and Dijken & Waage (1987). The oviposition of *T. chilonis* Ishii was analysed in detail by Suzuki *et al.* (1984), whose experiments identified the specific behavioural events associated with fertilization and sex determination.

Based on this research effort that spans more than 70 years, it is now possible to develop a reasonably comprehensive description of the host selection and oviposition behaviour of *Trichogramma*. In addition, many of the factors influencing host acceptance and progeny allocation have been identified. The selection of appropriate quality control criteria and successful artificial rearing of *Trichogramma* species for biological control programmes

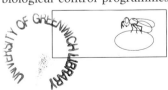

are dependent on a such a detailed understanding of these parasitoids (see Chapters 4 and 5).

Host Examination and Oviposition

The process of parasitization consists of a series of interconnected stages (Vinson, 1976). Location of the appropriate host habitat and associated community generally precedes detection of the host itself. Once a host is found, it may be inspected to evaluate its identity, condition and suitability as an oviposition site. Based on sensory cues acquired prior to and during contact, the parasitoid determines the acceptability of the host, the number of eggs to lay and, in many instances, the sex of the eggs. Before attempting to oviposit into a potential host, female *Trichogramma* examine the host surface by walking back and forth over the host while drumming it continuously with its antennae (Salt, 1935; Klomp & Teerink, 1962; Schmidt & Smith, 1989). De Jong & Pak (1984) have shown that most rejections occur before completion of the examination walk. If the host is accepted, the wasp assumes a stereotypical drilling posture and begins to probe the host chorion with its ovipositor. In some cases hosts may be rejected prior to penetration but, in general, drilling bouts are prolonged even on unsuitable objects such as glass beads (Salt, 1935; Quednau, 1955). After penetrating the chorion, the wasp usually deposits the full complement of eggs allocated to the host during a single insertion of the ovipositor (Klomp & Teerink, 1962, 1967; Suzuki *et al.*, 1984). If a host is rejected, the ovipositor is withdrawn prior to egg deposition. Marker substances and factors that regulate the metabolism of the host may also be introduced during oviposition (Salt, 1937; Klomp *et al.*, 1980; Vinson, 1985).

After contacting a host, a female *Trichogramma* briefly antennates the surface before stepping on to the chorion and making its initial transit over the host surface. During its initial transit the wasps usually follow a straight path that brings them close to the highest point of the host above the substrate (Schmidt & Smith, 1989). The initial transit continues until the wasp contacts the substrate on the opposite side of the host from its starting point. Following antennal contact with the substrate the wasp stops briefly and turns to continue its inspection of the surface. In some cases the wasps may reject the host and simply walk off the host after completing the initial transit (Schmidt & Smith, 1989).

The duration of host examination is correlated with host curvature (not volume) (Klomp & Teerink, 1962; Schmidt & Smith, 1985a, 1987a,c). On small hosts such as *E. kuehniella*, inspection of the host rarely lasts longer than 10 s at 25°C. In contrast, large hosts such as *Manduca sexta* (L.) (Sphingidae) eggs may be examined for more than 40 s (Schmidt & Smith, 1987a,c). The duration of host examination is also temperature dependent, and is prolonged at low temperatures (Schmidt & Pak, 1991). The durations of host examinations by *Trichogramma* species are comparable to those observed for other parasitoids such as *Telenomus heliothidis* Ashmead (Scelionidae) and *Caraphractus cinctus* Walker (Mymaridae) that attack sessile hosts (Jackson, 1966; Strand & Vinson, 1983).

The duration of oviposition is also variable. The hardness of the chorion largely determines the time spent drilling. Hosts with thin chorions, for example *E. kuehniella* and *Mamestra brassicae* L. (Noctuidae), can be penetrated in less than 60 s at 25°C. Large hosts generally have thick chorions, and drilling is correspondingly prolonged (Rajendram, 1978a,b). Drilling through the chorion of *Papilio xuthus* L. (Papilionidae) by *T. chilonis* requires an average of about 400 s (Suzuki *et al.*, 1984). A similar period is required by *T. minutum* Riley drilling into *M. sexta* eggs (Schmidt & Pak, 1991). The duration of egg laying is proportional to the number of eggs laid. At 27.5°C the time required to lay a single female egg averages about 20 s (Suzuki *et al.*, 1984). The duration of drilling and oviposition are both temperature dependent (Schmidt & Pak, 1991).

The host examination behaviour of *T. minutum* on *M. sexta* has been described in detail by Schmidt & Smith (1989). The paths taken by the wasps over the host surface vary greatly between individuals. Most (82%) of the paths observed were distributed evenly over the host surface without longitudinal preference or concentrated areas of examination. Approximately half the paths were evenly distributed with respect to latitude, whereas about one-third of wasps showed an obvious preference for the lower region of the host surface. A few (about 10%) wasps restricted their examination to small regions of the host surface. The paths of these wasps were characterized by repeated, frequent turning. Despite the marked differences between individuals, all observed examinations resulted in attempts at drilling and oviposition (Schmidt & Smith, 1989).

The latitudinal distribution of the paths over the surface of *M. sexta* eggs is not random. Schmidt & Smith (1989) found that the lower third of the host surface included only 16% of the examination path, while the middle third of the surface area contained 54% of the path. In general the wasps avoided the region at the top of the host. The height distribution of the paths was independent of wasp size (Schmidt & Smith, 1989).

This non-random latitudinal distribution may be controlled by the turning behaviour of the wasps as they explore the surface. After crossing the host surface, the wasp's antennae usually contact the substrate on which the host is resting. In response to this contact the wasp stops walking and turns so that it remains on the host surface. The angle of these turns determines the vertical distribution of the examination path. Few of these turns result in paths that carry the wasp over the highest point on the host surface. Instead the examination path is largely confined to a preferred area or band on the host surface (Schmidt & Smith, 1989). The distribution of the examination

path has important consequences for the application and detection of surface marking substances by the wasps. It also establishes the non-random distribution of oviposition sites (Schmidt & Smith, 1989).

Several other parasitoids of sessile hosts have been shown to use non-random, structured examination walks (Vinson, 1985). Adjustment of the examination duration to the relative size of the host increases the amount of sensory information that can be gathered by the parasitoid. In particular, adequate coverage of very large host surfaces such as *M. sexta* eggs is ensured by the wasp circumambulating the host several times (Schmidt & Smith, 1987a).

In addition to adjusting the duration and spatial distribution of the examination, *Trichogramma* species also regulate their walking speeds while on the host surface. The velocity of *T. minutum* females examining *M. sexta* hosts at 24°C was determined by Schmidt & Smith (1989). Average walking speed was 0.64 mm s^{-1} or about 2.3 m h^{-1}. Walking speed during host examination was independent of wasp size. Instantaneous walking speed varied during individual examination walks, especially when the wasps slowed or stopped briefly after crossing the host surface and contacting the substrate. The wasps also slowed slightly towards the end of the examination walk. *Trichogramma* on non-host substrates walk more rapidly than during host examination. On leaves *T. exiguum* Pinto & Platner can travel at 2.4 mm s^{-1} (Keller, 1987). On a level, clean plastic surface, *T. minutum* walk at 8–12 mm s^{-1} at 24°C (Schmidt & Smith, 1989), 12 to 20 times faster than during host examination. On non-host substrates walking speed is proportional to body size and stride length (Schmidt & Smith, 1989).

The reduction of walking speed by *T. minutum* during host examination is accompanied by an increased frequency of antennal drumming and stereotypical body postures (Schmidt & Smith, 1986). Slower walking speeds and increased frequency of antennal contact lead to a greater intensity of sensory sampling of the substrate, increasing the likelihood that a patchily distributed contact chemical such as a marker will be detected. Reduced walking speeds during host examination have been reported for many other parasitoids which attack sessile hosts (Edwards, 1954; Jackson, 1966; van Lenteren & DeBach, 1981; Strand & Vinson, 1983).

Trichogramma species display strong klinokinetic (velocity) and ortho-kinetic (turning) responses to chemical cues, especially compounds associated with the accessory glands and wing scales of various host species (Laing, 1938; Jones *et al.*, 1973; Lewis *et al.*, 1975; Noldus & van Lenteren, 1985; Nordlund *et al.*, 1987). In general, the locomotory responses to these materials act to confine searching to a limited area. Similar contact chemical cues acting as arrestment factors could be involved in the reduction of walking speed during host examination. However, other non-chemical factors must also act as klinokinetic cues, since *Trichogramma* species also reduce their walking speed while examining glass beads, droplets of gum resin and mercury globules, surfaces unlikely to have contact kairomones (Salt, 1935, 1958; De Jong & Pak 1984; Schmidt & Smith, 1985a). In these cases, and perhaps also in the case of insect eggs, host shape and size are more likely cues for the reduction of walking speed and initiation of antennal drumming (Salt, 1935, 1958; Schmidt & Smith, 1985a, 1986, 1987a).

Details of the oviposition behaviour of *T. embryophagum* were reported by Klomp *et al.* (1980). Similar observations were made by Salt on *T. evanescens* (Salt, 1935, 1937). Suzuki *et al.* (1984) analysed the behaviour of *T. chilonis* females lacking previous oviposition experience attacking the eggs of *P. xuthus*. They identified four major stages: (i) arrestment and adoption of the characteristic oviposition posture with the head elevated, the abdomen depressed, the antennae folded against the head and the ovipositor removed from its sheath and extended downwards, perpendicular to the chorion surface; (ii) drilling; (iii) penetration of the chorion by the ovipositor, followed by rhythmic jerking of the abdomen lasting about 30 s (the 'initial movement'); and (iv) oviposition.

The oviposition of female and male eggs are behaviourally distinct processes (Suzuki *et al.*, 1984). The process of laying female eggs consists of a discontinuous movement pattern. To and fro wagging of the abdomen lasting about 3 s is followed by an 8-s interval of relaxing during which there is little abdominal movement. This pause is followed by a second 3–4-s bout of abdominal wagging, during which the ovipositor is extended and inserted to its full length, while the abdomen moves back and forth. Egg laying is terminated by a period of rapid, small amplitude trembling of the abdomen lasting 5–7 s. There is no relaxing phase during the laying of male eggs. Instead, the process is reduced to only two distinct components: a period of prolonged abdominal wagging lasting about 9 s and a subsequent bout of trembling similar to that observed during the laying of female eggs (Suzuki *et al.*, 1984). Fertilization is thought to occur during the relaxing phase, resulting in the production of a diploid (female) zygote (Suzuki *et al.*, 1984). The period of trembling corresponds with the passage of the egg through the ovipositor. Each cycle of wagging, relaxing (when present) and trembling corresponds to the deposition of one egg. In general the process is repeated without intervening withdrawal of the ovipositor until the full complement of eggs is laid (Klomp & Teerink, 1962; Suzuki *et al.*, 1984). In some cases the female may remove her ovipositor and subsequently reinsert it into the same hole (Suzuki *et al.*, 1984).

Suzuki *et al.* (1984) determined that there is a fairly consistent pattern to egg laying by inexperienced wasps attacking *P. xuthus*. The first egg laid is usually (95%) female. Over 91% of the females laid one male egg among the first three eggs oviposited. At least one, and on average seven, consecutive depositions of female eggs follow each deposition of a single male egg, with the result that male eggs are laid into a single host at approximately eight egg intervals. As a result, sex ratio changes as a function of the number of eggs

laid per host (Schmidt & Smith, 1985a, 1987c). Dijken & Waage (1987) and Waage & Ng (1984) observed similar initial patterns of male and female egg laying by *T. evanescens* parasitizing patches of smaller hosts, and this has also been observed for *T. brassicae* Bezdenko (Wajnberg, 1993; see Chapter 12).

Withdrawal of the ovipositor is preceded by continuous jerking movements of the abdomen as it is gradually pulled from the host. When only the tip of the ovipositor remains in the chorion, the entire ovipositor is abruptly reinserted and then rapidly removed completely from the host (Suzuki *et al.*, 1984). The function of these movements has not been fully elucidated. However, Klomp *et al.* (1980) suggest that the deposition of an internal marker occurs only after the completion of egg laying.

Factors Affecting Host Acceptance and Progeny Allocation

Female *Trichogramma* accept single, clustered and stalked hosts. *Trichogramma* species will also attempt to parasitize the eggs of many insect species that they are unlikely to encounter in the field, including *Tinea* spp. (Tineidae), *Rhodnius prolixus* Stal (Reduviidae) and *Cimex lectularius* L. (Cimicidae) (Quednau, 1955). Even unsuitable inorganic objects such as grains of sand, droplets of plant resin, glass beads and mercury globules are subjected to persistent drilling attempts (Salt, 1935). However, the acceptance of objects as potential oviposition by *Trichogramma* species is by no means entirely indiscriminate. For example, by adjusting the size of a mercury droplet during the examination walk of *T. evanescens*, Salt (1958) was able to show that host acceptance is limited by upper and lower thresholds for diameter. Furthermore numerous choice and preference tests have demonstrated distinct preferences for selected host species based on differences in size, surface odour, colour and contents (Taylor & Stern, 1971; De Jong & Pak, 1984; Dijken *et al.*, 1986; Pak, 1988).

The oviposition behaviour of *Trichogramma* species is plastic. Based on oviposition experience thresholds for the acceptance of hosts are actively adjusted by the wasps to maximize reproductive success under varying conditions of host availability and suitability (Waage, 1986; Pak, 1988). The number of progeny allocated to a host varies with the quantity and nutritional quality of the host contents (Klomp & Teerink, 1967; Schmidt & Smith, 1987c). In addition, *Trichogramma* species have been shown to adjust their progeny allocation in order to maximize their reproductive success when exploiting aggregations of hosts (Waage & Lane, 1984; Waage & Ng, 1984; Schmidt & Smith, 1985b, 1987b; Waage, 1986).

The number of progeny allocated to a host is adjusted according to the wasp's external measure of host volume (Klomp & Teerink, 1962, 1967; Schmidt & Smith, 1985a, 1987c). Host curvature plays an important role in host recognition and acceptance (Salt, 1935; De Jong & Pak, 1984). In addition to size and shape (Klomp & Teerink, 1962), surface odour (Salt, 1937; De Jong & Pak, 1984; Nordlund *et al.*, 1987), host age (Pak, 1986), chorion thickness (Quednau, 1955), chorion hardness (Rajendram, 1978a,b) and chemical cues from the host interior (Wu & Qin, 1982; Nettles *et al.*, 1983) can also affect host acceptance and cause variations in clutch size. Differences among individual *Trichogramma* based on nutritional status and genotype have also been reported (Flanders, 1935; Walter, 1983; Wajnberg *et al.*, 1989; see Chapter 12). Waage & Ng (1984) found that *T. evanescens* reduced the number of eggs laid with successive ovipositions. Previous ovipositional experience, the interval between successive ovipositions, depletion of egg supplies and the density and spacing of hosts can all influence acceptance and clutch size (Klomp *et al.*, 1980; Waage, 1986; Schmidt & Smith, 1987b). These factors must be controlled in any analysis of other factors.

Visual factors

Like most other hymenopteran species studied, *Trichogramma* can detect differences in light intensity and wavelength. *Trichogramma* species may also be capable of limited form perception. Laing (1938) reported that *T. evanescens* females are attracted by the visual profiles of 0.3 mm high *E. kuehniella* hosts at a distance of about 1–2 mm. Using video tracking of the walking path of isolated *T. brassicae* females, Wajnberg (see Chapter 12) provides another estimation of this behavioural trait. Salt (1937) and Schmidt & Smith (1985a) have shown that *T. evanescens* and *T. minutum* can parasitize successfully in complete darkness. Although the rate of host finding may be reduced in the dark, the absence of visual cues has no effect on the assessment of host volume (Schmidt & Smith, 1985a). Furthermore, the absence of light does not interfere with the discrimination of previously parasitized hosts (Salt, 1937).

De Jong & Pak (1984) were able to demonstrate some colour (wavelength) preferences for *Trichogramma* species. Female *T. buesi* attacked white or transparent glass bead models more frequently than yellow beads. This pattern of colour preference corresponds with limited acceptance of yellow *P. brassicae* eggs by this species (De Jong & Pak, 1984). In contrast, female *T. maidis* Pintureau & Voegelé (= *T. brassicae* Bezdenko), which readily accept *P. brassicae* eggs, attack yellow and transparent glass beads with similar frequencies. It has been suggested that black hosts are rejected more frequently because they appear to have been previously parasitized or killed (Lewis & Redlinger, 1969). However, Rajendram (1978a) reported that *T. californicum* females readily attacked the empty blackened shells of parasitized *E. kuehniella* eggs.

Measurements of the ommatidia of *T. minutum* by Schmidt & Smith (1987a) suggest that the visual acuity of these wasps is very limited. On the

basis of lens size and measurements of ommatidial angles, it has been estimated that *T. minutum* can detect objects subtending a visual angle of 9° (Schmidt & Smith, 1987a). These results correspond well with the observations of Laing (1938).

Physical factors

The hardness, thickness and permeability of the host chorion are often cited as factors affecting host acceptance and suitability (Salt, 1937; Quednau, 1955; Pak, 1988). Quednau (1955) found that *Trichogramma* species females could not penetrate the 'porcelain-like' chorion of *Orgyia antiqua* (Hübner) (Lymantriidae), despite prolonged drilling. Measurements by Quednau (1955) indicate that several *Trichogramma* species have difficulty penetrating chorions thicker than 20 μm. The chorions of *Bombyx mori* L. (Bombycidae) (22 μm), *C. lectularius* and *R. prolixus* (28 μm) all present at least partial mechanical barriers to parasitization by *Trichogramma*.

Rajendram (1978a) found that artificial eggs made of parowax (melting point (m.p.) 70°C) and beeswax (m.p. 80°C) were not penetrated by the ovipositor even after prolonged drilling. Waxes with lower melting points such as ceresin (m.p. 50°C) and paraffin (m.p. 47°C) were penetrated. Using combinations of varying proportions of Vaseline and ceresin or paraffin, Rajendram (1978a) was able to demonstrate a significant correlation between melting point and the time required for ovipositor penetration. The time required to penetrate the host, or sensory inputs from stress detectors such as campaniform sensilla present on the ovipositor may be important factors controlling the persistence of drilling attempts.

Pak (1988) implicates chorion structure as a factor responsible for the reduced success of *T. maidis* (= *T. brassicae*) developing in *Pieris brassicae* L. (Pieridae) at low ambient humidities. Although the chorions of *P. brassicae* and *M. brassicae* are similar in thickness, *M. brassicae* eggs have a denser internal chorion layer than *P. brassicae* eggs. Larval survival in the eggs of *M. brassicae* was largely unaffected by ambient humidity (Pak, 1988).

Chemical factors

Chemicals present on the host surface can promote or inhibit acceptance by *Trichogramma* species. Quednau (1955) attributed the rejection of *Agelastica almi* L. (Chrysomelidae), *Coccinella* species (Coccinellidae), and *Pyrrhocoris apterus* (L.) (Pyrrhocoridae) eggs to repellent chemicals present on the chorion. His results also suggested that the surface chemistry of *Argynnis paphia* L. (Nymphalidae), *P. brassicae* and *P. rapae* (Pieridae) eggs deterred some species of *Trichogramma*.

Pak (1988) found that acceptance of *P. brassicae* eggs varied among *Trichogramma* strains, whereas *M. brassicae* eggs were readily accepted by almost all the strains tested. A significant factor governing the acceptance or rejection of these hosts was the composition of the accessory gland secretion coating the chorion (De Jong & Pak, 1984; Pak, 1988). Ovarian eggs of *P. brassicae* dissected from gravid females lack the accessory gland coating. Such eggs are more attractive to female *T. buesi* Voegelé than normally laid eggs coated with the accessory gland secretion (Pak, 1988). Glass beads coated with a methanol wash of *P. brassicae* eggs were rejected by *T. buesi* prior to drilling attempts more frequently than clean glass beads. About 40% of the treated eggs were rejected following initial contact, compared to an acceptance rate of greater than 80% for clean beads following initial contact (Pak, 1988). Although the accessory gland secretion of *P. brassicae* is a deterrent for *T. buesi*, this effect is not consistent. Pak (1988) found that acceptance of *P. brassicae* by *T. maidis* (= *T. brassicae*) is promoted by the same accessory gland secretion. Ovarian eggs of *P. brassicae* are rejected more frequently by *T. maidis* than normally laid eggs, and clean glass beads are less attractive than beads coated with the accessory gland secretion. The accessory gland secretion of *M. brassicae* was a positive recognition for both *T. buesi* and *T. maidis* (Pak, 1988). Ovarian eggs of *M. brassicae* were rejected by more than 50% of the females tested, whereas over 90% of normally laid eggs induced oviposition.

Nordlund *et al.* (1977) first suggested that the accessory gland secretions of *Heliothis zea* (Boddie) (Noctuidae) and *Trichoplusia ni* (Hübner) (Noctuidae) increased rates of parasitization. Subsequent studies of *Telenomus pretiosum* demonstrated that aqueous homogenates of the accessory glands and gland reservoirs of *H. zea* did stimulate drilling and ovipositor probing when applied to glass bead models (Nordlund *et al.*, 1987). In contrast, accessory gland extracts from *Spodoptera frugiperda* Smith (Noctuidae) were ineffective. Strand & Vinson (1983) have shown that specific accessory gland proteins are required for host recognition by the oligophagous parasitoid *T. heliothidis*. Although opportunistic, generalist parasitoids like *Trichogramma* species are unlikely to rely on specific proteins as recognition factors, the results of Pak (1988) and Nordlund *et al.* (1987) do indicate that these wasps can discriminate between the secretions of different host species, and that this sensitivity differs between *Trichogramma* strains.

Although chemicals on the host surface can influence host acceptance or rejection, they are not required for acceptance (Salt, 1935; Marchal, 1936). Glass beads and mercury droplets induce antennal drumming and drilling in the absence of organic stimuli (Salt, 1958; De Jong & Pak, 1984).

Host marking and superparasitism

Superparasitism occurs when a parasitoid oviposits on or into a host that has been previously parasitized (Dijken & Waage, 1987). Such behaviour can have deleterious consequences if competition between larval parasitoids for limited host nutritional resources leads to mortality or reduced fitness (Salt, 1937; Klomp *et al.*, 1980). However, under some conditions superparasitism may be advantageous, for example, if there are few unparasitized hosts available and there is some likelihood that the superparasitizing female's offspring will survive despite greater competition for host nutrients (Bakker *et al.*, 1985; Alphen *et al.*, 1987; Wajnberg *et al.*, 1989). Waage (1986) and Dijken & Waage (1987) have suggested that superparasitism may also be used to adjust clutch size to local host densities. In view of the potential advantages conferred by the ability to discriminate between parasitized and unparasitized hosts, it is not surprising that many parasitoids have evolved methods of marking the hosts that they have attacked (Klomp *et al.*, 1980). These markers can be employed subsequently to regulate the occurrence of superparasitism in response to the conditions of host and conspecific density encountered by the parasitoid.

Salt (1934) restricted his definition of superparasitism to 'the occurrence of more parasitoids of a single species on a host than that host can support'. As a result he considered primarily those situations in which 'some or all of the parasitoids feeding upon (the host) are insufficiently nourished, are unable to complete their development, and die' (Salt, 1934). Salt predicted that under these conditions, parasitoids should discriminate between previously parasitized and unparasitized hosts, provided that there is some likelihood of encountering unparasitized hosts (Salt, 1934). The hosts used in Salt's experiments, the eggs of *S. cerealella*, are relatively small and generally support the development of a single larva of *T. evanescens*. Superparasitism of these hosts usually does result in the death of all the larvae prior to the completion of development (Salt, 1934). In this case, Salt's definition of superparasitism is generally valid. The likelihood of larval survival increases appreciably only if hosts containing older conspecific larvae are superparasitized (Salt, 1934). The use of recently parasitized hosts in Salt's experiments precluded the investigation of host discrimination under these conditions.

Salt (1934) found that *T. evanescens* exposed to recently parasitized hosts rarely superparasitize them if unparasitized hosts were also available. In addition, the wasps avoided superparasitizing hosts that they had previously parasitized themselves (self-superparasitism). When the number of hosts available is limited, so that the wasps encounter many previously parasitized hosts while still having a large complement of their own eggs, superparasitism occurred much more frequently than when large numbers of unparasitized hosts were available (see also Burbutis *et al.*, 1983; Wajnberg *et al.*, 1989). Under these circumstances of limited host availability, the total number of eggs laid by the wasps was also reduced (see also Waage & Ng, 1984; Wajnberg *et al.*, 1989).

Salt (1934) concluded that the wasps have the ability to identify previously parasitized hosts and can refrain from ovipositing into them. Furthermore, the wasps generally selected larger hosts for superparasitism and could distinguish between large and small hosts. Host discrimination was not based on memory of previous ovipositions and depends on a marker (Salt, 1934). This marker is not specific for individuals, although Salt's data suggest that the wasps may superparasitize hosts marked by other individuals more frequently than hosts marked by themselves (Salt, 1934).

In a series of classic experiments, Salt (1937) identified the sensory basis of host discrimination by *T. evanescens*. Visual cues were not required for either oviposition or avoidance of previously parasitized hosts: both occurred in complete darkness. The avoidance of previously parasitized hosts was also not based on the detection of penetration holes, since if these were concealed by reorientation of the hosts, the wasps continued to discriminate between hosts (Salt, 1937). Furthermore, female *T. evanescens* can distinguish between clean eggs and hosts that have only been walked on by other wasps, demonstrating that the wasps respond to specific materials deposited by the feet of the parasitoids themselves prior to oviposition. The external marker is volatile and appreciable amounts are lost in moving air after 2.5 h (Salt, 1937). The material is also water soluble, and can be washed from the host surface and artificially reapplied to clean hosts, resulting in increased rejection of the treated hosts. The response of female *T. evanescens* to the marker is dose-dependent and the amount of material deposited on the host is correlated with the duration of the examination walk (Salt, 1937).

Based on these observations, some suggestions can be made concerning host examination behaviour (Schmidt & Smith, 1989). Adjustment of the duration of host examination to host size (curvature) ensures that the concentration of the marker remains high despite increased host surface area (Schmidt & Smith, 1987a). Furthermore, confinement of the examination to a limited range of heights above the substrate concentrates the marking substance in a band surrounding the host (Schmidt & Smith, 1989). Such a distribution of marker is advantageous, since it will be encountered at the beginning of the host examination, usually during the initial antennation that precedes mounting the host. A patchily distributed marker would probably be less effective and could increase the time invested by the wasp on an unsuitable host (Klomp *et al.*, 1980). Furthermore, the detection of a marking substance during initial contact with the host also prevents the ambiguity that arises if the wasp cannot differentiate between a marking substance already present on the host and the new marker it applies during its examination walk. The wasp should reject a marked host prior to application of its own marker. If a marker is not detected during the early stages of the examination, the wasp must treat any marker detected subsequently as its own (Schmidt & Smith, 1989).

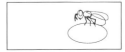

Removal of the external marker by washing the hosts revealed that *T. evanescens* detect internal chemical cues left during previous ovipositions (Salt, 1937). Klomp *et al.* (1980) determined that *T. embryophagum* also detected external and internal markers left during previous ovipositions. Using the eggs of *E. kuehniella*, they found that inexperienced wasps that have not oviposited previously accepted both parasitized and unparasitized hosts for oviposition. Experienced females that had oviposited once into unparasitized hosts continued to accept unparasitized hosts for oviposition, but rejected hosts that had already been parasitized shortly before the second encounter. Based on these results, Klomp *et al.* (1980) concluded that the wasps learnt to discriminate by ovipositing into unparasitized hosts. This conclusion has been subsequently disputed (Alphen *et al.*, 1987; Dijken & Waage, 1987). These later studies have shown that female *Trichogramma* species can identify previously parasitized hosts prior to any oviposition experience (Alphen *et al.*, 1987; Dijken & Waage, 1987). Rather than learning to discriminate between parasitized hosts, the female wasp chooses to superparasitize until it encounters unparasitized hosts (Alphen *et al.*, 1987).

The tendency to reject previously parasitized hosts depends upon the availability of unparasitized hosts (Morrison *et al.*, 1980; Burbutis *et al.*, 1983). Klomp *et al.* (1980) found that the proportion of experienced females rejecting a previously parasitized host is inversely correlated with the length of time between ovipositing into an unparasitized host and encountering a parasitized host. When the interval is less than 10 min, all the experienced females rejected the previously parasitized hosts. After 90 min, even experienced *T. embryophagum* would accept both unparasitized and parasitized hosts.

Klomp *et al.* (1980) also found that in many cases the amount of external marker left on the chorion was insufficient to prevent oviposition attempts by *T. embryophagum* females. In these cases the host was rejected only after insertion of the ovipositor. Incompletely attacked hosts were rejected by experienced females only if the incomplete attack included insertion of the ovipositor. This suggests that there may be significant differences in the efficacy of the external marker used by *T. embryophagum* and that observed by Salt (1937) for *T. evanescens*. Female *T. embryophagum* that were disturbed during drilling or initial ovipositor insertion into an unparasitized host did not subsequently discriminate between unparasitized and parasitized hosts, demonstrating that the tendency to accept only unparasitized hosts is induced during egg deposition itself (Klomp *et al.*, 1980).

Suzuki *et al.* (1984) found that inexperienced female *T. chilonis* ovipositing into previously parasitized *P. xuthus* hosts do not alter the number of male or female eggs deposited, and the sex ratio is unchanged. The eggs of *P. xuthus* are much larger than those of *E. kuehniella* and *S. cerealella*. Superparasitism of *P. xuthus* does not usually result in mortality of all the superparasitizing larvae. For singly parasitized hosts, female egg survival was about 99% and male egg survival was 85% (Suzuki *et al.*, 1984). In doubly parasitized hosts female egg survival was reduced to 68% and male egg survival was 80%, resulting in a slight reduction of the female bias of the sex ratio. Under these conditions superparasitism probably confers considerable benefit to the superparasitizing wasp when the number of unparasitized hosts is small, since the offspring of the superparasitizing wasp have a significant likelihood of survival (Bakker *et al.*, 1985).

Dijken & Waage (1987) distinguish between two types of super-parasitism. Self-superparasitism, which occurs when a female oviposits into a host which it has itself previously parasitized, and conspecific-superparasitism, which occurs when a parasitoid attacks a host previously parasitized by a conspecific other than itself. Dijken & Waage (1987) suggested that if a parasitoid can identify hosts parasitized by itself and hosts parasitized by other conspecifics, and can determine or remember the number of eggs allocated to a previously self-parasitized host, then the parasitoid may be able to initially allocate a submaximal number of eggs to an unparasitized host. If the female does not encounter additional hosts, she may subsequently adjust egg numbers by self-superparasitism to achieve an optimal distribution of its egg complement among the available hosts. Such a strategy could be used by *Trichogramma* to adjust progeny allocation to patchily distributed or clustered hosts. However, if the wasps cannot discriminate between self- and conspecific-parasitized eggs, the situation is more complex (Alphen *et al.*, 1987; Dijken & Waage, 1987). Under these conditions if the parasitoid cannot determine the number of eggs present in a host or remember previous oviposition sites, it should avoid superparasitism if unparasitized hosts are available. The wasps may use elapsed time between ovipositions as a substitute for remembering oviposition sites. An actively foraging wasp encountering a parasitized host after a prolonged interval of searching probably is not encountering a host parasitized by itself (Dijken & Waage, 1987). In consequence, superparasitism of widely separated eggs may be more frequent than superparasitism of host aggregations (however, see Chapter 12 for additional discussion). The use of such a strategy is supported by the observations of Klomp *et al.* (1980) that the proportion of experienced wasps accepting previously parasitized hosts is correlated with the interval between host encounters.

In a series of experiments, Dijken & Waage (1987) tested the ability of *T. evanescens* to detect and regulate self- and conspecific-superparasitism of *M. brassicae* eggs. *M. brassicae* eggs are smaller than those of *P. xuthus*, but are still capable of supporting the complete development of 4–6 larvae (Schmidt & Pak, 1991). Unlike superparasitized *E. kuehniella* and *S. cerealella* hosts, superparasitized *M. brassicae* often produce some survivors. On unparasitized *M. brassicae* hosts the wasps avoid self-superparasitism (Dijken & Waage, 1987). On previously parasitized hosts the distribution of attacks is not significantly different from random and there is no evidence that the wasps

can distinguish between their own clutches and those laid by other conspecifics on the basis of individual-specific markers. Like Salt (1934) and Klomp *et al.* (1980), Dijken & Waage (1987) found that larger clutches are laid during attacks on unparasitized hosts than during attacks on previously parasitized hosts. They also detected a tendency for *T. evanescens* to lay smaller clutches in hosts containing large clutches laid by another wasp. Like Suzuki *et al.* (1984), Dijken & Waage (1987) also found that the probability that male eggs will be laid did not differ for wasps attacking parasitized and unparasitized hosts. Furthermore, although local male competition models suggest that sex ratios should differ for self- and conspecific-superparasitism, Dijken & Waage (1987) did not detect significant differences. In general, they found that superparasitizing *T. evanescens* cannot identify hosts parasitized by themselves, and lay the same reduced number of eggs during self- and conspecific-superparasitism, concluding 'there is no evidence that *T. evanescens* makes a different allocation of sex and progeny under self- and conspecific-superparasitism' (Dijken & Waage, 1987). Wajnberg *et al.* (1989), using smaller *E. kuehniella* hosts and higher parasitoid to host ratios did find that female *T. maidis* (= *T. brassicae*) will attack previously parasitized hosts more often during conspecific-superparasitism. Additional studies are required to determine the mechanism underlying this ability to discriminate between conspecific- and self-superparasitism. Waage & Lane (1984) found that progeny allocation and sex ratios are unaffected by encounters with other conspecific females attacking the same patch of hosts. Instead, the tendency of female wasps to lay male eggs early in the oviposition sequence leads indirectly to an increase in the proportion of male progeny, since the wasps avoid previously parasitized hosts and lay fewer eggs (Waage & Lane, 1984) (see also Chapter 12 for a modelization of this process). However, these results do not preclude the use of elapsed time between ovipositions as a measure of the likelihood of self- vs. conspecific-superparasitism by the wasps. Indeed the results of Klomp *et al.* (1980) suggest that the wasps may employ such a strategy as a simple method of 'estimating' spatial displacement from an earlier oviposition site (cf. Dijken & Waage, 1987).

Multiparasitism

Multiparasitism occurs when a parasitoid oviposits into an egg previously parasitized by a different parasitoid species. *Trichogramma* species, like most other parasitoids studied (Bakker *et al.*, 1985), does not avoid parasitizing hosts parasitized by other parasitoid species. Strand & Vinson (1983) have described facultative hyperparasitism of *T. heliothidis* larvae by *T. pretiosum* Riley in the eggs of *H. virescens* (F.) (Noctuidae). It is likely that *Trichogramma* species can often consume other parasitoid larvae as readily as the host embryo (cf. Klomp & Teerink, 1978). The extent to which *Trichogramma* species multiparasitize hosts attacked by other members of the same genus has not been investigated in detail.

Host volume

Behavioural responses to host size are widespread among Hymenoptera parasitoids (Vinson, 1985). Many parasitoids use host size as a cue for host acceptance (Arthur, 1967; Vinson, 1976; Strand & Vinson, 1983). Since larvae are confined to a single host during parasitoid development, host volume can play a critical role in nutrition (Flanders, 1935; Klomp & Teerink, 1967). Clutch size adjustment to host volume by gregarious parasitoids has been reported to occur in several families (Vinson, 1976, 1985), and the regulation of sex ratio with host volume has been described for many species (Clausen, 1939; Flanders, 1946, 1965; Waage, 1982, 1986; Werren, 1984). In general, species that use measures of host volume to set clutch size or progeny sex ratio also use host size as a component of host recognition behaviour. Host size is a critical factor in host acceptance by female *Trichogramma* species, and is also a major factor regulating clutch size and sex ratio adjustment. In addition, the time spent examining the host is correlated with host diameter (Salt, 1935, 1958; Klomp & Teerink, 1962, 1967; Taylor & Stern, 1971; Schmidt & Smith, 1985a, 1987a).

Female *Trichogramma* species will attack a wide variety of objects of appropriate size and shape regardless of surface odour (Holloway, 1912; Salt, 1935; Marchal, 1936; Klomp & Teerink, 1962). Hosts ranging in size from 0.3 mm (0.014 mm^3) to 3 mm in diameter (14 mm^3) are parasitized (Schmidt & Smith, 1987c). Most *Trichogramma* species prefer hosts of intermediate or large diameter (0.8–1.8 mm) that can support several larvae to smaller diameter hosts (0.3–0.7 mm) that support only one or two developing larvae (Salt, 1935; Taylor & Stern, 1971; Boldt *et al.*, 1973). Oviposition into large hosts reduces the likelihood that smaller hosts will be accepted during subsequent encounters (Taylor & Stern, 1971). However, in some host-specific strains of *Trichogramma*, size preference is not simply correlated with host size. Instead, a reduced range of intermediate host diameters may be accepted for oviposition (De Jong & Pak, 1984). Flanders (1935) suggested that these thresholds for host acceptance may vary with the size of the parasitoid. This effect has been confirmed by Schmidt & Smith (1987a), who found that host acceptance is based on a relative measure of host size. The effects of host size on acceptance are also often obscured by other differences between host species. In particular, the effects of size can be confounded by the effects of chemical factors such as host odour, internal composition and developmental stage.

Small hosts such as the eggs of *E. kuehniella* (0.5 × 0.3 × 0.3 mm) are allocated only one or two eggs. Larger hosts such as *M. sexta* (diameter

1.36–1.45 mm) may be allocated as many as 40 developing larvae, although 18–25 progeny are more usually allocated to hosts of this volume (Klomp & Teerink, 1962, 1967; Schmidt & Smith, 1985a). In general, host species of similar volume are allocated similar numbers of progeny, and interspecific differences play a less important role than the absolute volume of the host (Klomp & Teerink, 1967; Taylor & Stern, 1971; Boldt *et al.*, 1973). Response to host size can occur independently of species differences. For example, Salt (1934, 1935) found that larger eggs of *E. kuehniella* tended to contain two rather than one *T. evanescens* larvae, and were superparasitized more frequently than smaller hosts.

Many studies have found that parasitoid body size, fecundity and longevity depend upon larval feeding (Flanders, 1935; Klomp & Teerink, 1967; Legner, 1969; Southard *et al.*, 1982; Beckage & Riddiford, 1983; Charnov & Skinner, 1985). For gregarious parasitoids such as *Trichogramma* attacking hosts of similar volume, allocation of large clutches results in progeny of reduced size and fecundity (Klomp & Teerink, 1967, 1978; Wylie, 1967) and there is a general tendency for smaller broods to produce progeny of greater fitness. Clutch size adjustment by gregarious parasitoids represents a compromise between increasing the number of offspring produced and enhancing the fitness of each offspring (Iwasa *et al.*, 1984; Parker & Courtney, 1984; Charnov & Skinner, 1985; Waage, 1986).

Since progeny survival and fecundity can both be influenced by clutch size, simply allocating more progeny to a host may not result in a corresponding increase in overall maternal fitness. For all gregarious parasitoids there is an upper limit for clutch size above which none of the larvae obtain sufficient nutrients for survival (Klomp & Teerink, 1967). A lower limit to clutch size has also been found for parasitoids attacking hosts that are larger than themselves (Clausen, 1940). Inadequate feeding by insufficient *Trichogramma* species larvae can lead to increased mortality due to the decomposition of host residues and deterioration of the physical environment of the larvae (Fisher, 1971; Hoffman *et al.*, 1975).

In general, the relationship between the number of eggs deposited and host volume between these upper and lower bounds for clutch size is non-linear. Small hosts are allocated a single egg and produce small adults limited by the nutrient and volume available for development. However, above a certain host volume, the size of adult *T. minutum* is no longer restricted by host volume and depends entirely on the number of eggs allocated to the host. Small hosts such as *E. kuehniella* are allocated a disproportionately large number of eggs per unit of host volume (Schmidt & Smith, 1987c). Barrett & Schmidt (1991) determined that *T. minutum* reared on *S. cerealella* contain an average of 1.0 μg of amino acid. The total amino acid content of *S. cerealella* eggs, including the chorion, averaged 2.2 μg. Allowing for the chorion and metabolic losses, there is only sufficient protein available in *S. cerealella* eggs to support development of a single *T. minutum* larva. In contrast, *M. sexta* hosts produce much larger *T. minutum* adults containing 8.2 µg of amino acid. Based on a mean of 94 µg of amino acid per *M. sexta* host, between 10 and 12 *T. minutum* adults would be anticipated from each host, a prediction that corresponds well with observed clutch sizes for this host (Schmidt & Smith, 1985a, 1987c). However, in theory, *M. sexta* hosts could produce between 80 and 90 progeny, based on the size of *T. minutum* reared from *S. cerealella*. This discrepancy reflects the non-linearity of the wasp's response to host size (Barrett & Schmidt, 1991).

Several different methods for the adjustment of clutch size have been described. In some gregarious species 'scramble' competition occurs, in which no fixed amount of host nutrient is allocated to each developing parasitoid (Lenteren & DeBach, 1981). The offspring that survive to emergence are those that obtained enough food; the remainder starve. Scramble competition does not require the ovipositing wasp to ascertain host volume and the adjustment of clutch size occurs after oviposition. In other gregarious species, including *Trichogramma* species, clutch size is adjusted to host volume during oviposition, and an adequate amount of host volume and nutrient is allocated to each larva (Lenteren & DeBach, 1981).

Klomp & Teerink (1962, 1967) and Suzuki *et al.* (1984) found that the number of eggs deposited usually equals the number of adults that subsequently emerge, demonstrating that scramble competition between larvae does not usually regulate clutch size. Only if the same host is repeatedly parasitized are supernumerary larvae eliminated by starvation or physical attack (Klomp & Teerink, 1978). Instead, progeny allocation is set according to a measurement of host volume made during the examination walk, before the wasp begins drilling into the host (Klomp & Teerink, 1967). When a *T. embryophagum* female about to oviposit into a small host is removed and replaced on a larger host, the wasp lays the number of eggs into the larger host appropriate to the original smaller host (Klomp & Teerink, 1962).

Schmidt & Smith (1985a) presented *T. minutum* females with spherical *M. sexta* hosts that were either resting fully exposed on the substrate or were partially embedded in holes in the substrate. More progeny were allocated to each of the fully exposed hosts than to the partially embedded hosts, indicating that some factor dependent upon host height above the substrate can be used to discriminate between hosts of the same internal volume and surface curvature. Furthermore, the use of the same host species for both treatments rules out the use of chemical, textural or nutritional differences. The ability to discriminate between fully exposed and partially embedded hosts is retained when oviposition occurs in complete darkness, showing that visual cues are not required.

Several parameters of the examination, including turn number and frequency, and the duration of the initial transit across the host surface, correlate with exposed host volume and surface area. Of these, only the duration of the wasp's initial transit was found to significantly affect the

number of eggs allocated to a host. Interruption of the initial transit by blocking the wasp's first path across the host surface resulted in fewer eggs being laid compared to the number oviposited by wasps that were allowed to complete their initial transit (Schmidt & Smith, 1987c). Both large and small wasps have the same walking speed during host examination (Schmidt & Smith, 1987a, 1989), and although small wasps take more steps than large wasps, the duration of the initial transit is independent of wasp size. This constancy of walking speed provides *Trichogramma* females with the means to determine absolute volume. When large and small *T. minutum* females are presented with *M. sexta* hosts of the same exposed volume and diameter, the wasps allocate the same number of progeny regardless of their own body size (Schmidt & Smith, 1987c). This ability to measure absolute rather than relative volumes ensures that the progeny of different sizes of wasp are allocated a constant optimal volume of host, and are exposed to similar levels of larval competition.

In addition to examination time, host diameter also sets an upper limit for clutch size. When hosts are mounted on the tips of very thin wires, the wasps do not come into contact with the substrate, and initial transit length is of indeterminate length. Such hosts are allocated only slightly more eggs than hosts of the same size resting on the substrate (Schmidt & Smith, 1987a). This upper limit to clutch size is set relative to wasp body length: small wasps lay more eggs than larger wasps on these 'point-mounted' hosts (Schmidt & Smith, 1987a). Smaller wasps also take longer to examine hosts of the same diameter (Schmidt & Smith, 1987a).

A possible mechanism for host diameter measurement, using the angular relationships between the positions of the body parts of the wasp has been proposed by Schmidt & Smith (1986). Most body angles and the height of the wasp above the host surface do not change with relative host diameter. Only the scapal–head and flagellar–scapal angles of the antennae are significantly correlated with relative host diameter and the maximum angle between the antennae and head may be monitored to detect host curvature (Schmidt & Smith, 1986). The bases of the antennal scapes have hairplates which articulate against the head when the antennae are depressed and provide information about the extent of antennal displacement (Schmidt & Smith, 1986). This sensory mechanism can only detect relative host diameter, a prediction confirmed by the observed behaviour of *Trichogramma* (Schmidt & Smith, 1986, 1987a). The same mechanism may also be responsible for host size acceptance. Flanders (1935) reported that small hosts that are rejected by large wasps are accepted by smaller wasps, demonstrating that host recognition is dependent upon relative host size. Ablation experiments, structurally modifying the hairplates on the antennae, are needed to confirm the proposed model.

The mechanism used by *T. dendrolimi* Matsumura to set clutch size functions independently of ambient temperature (Schmidt & Pak, 1991).

Although walking speed is temperature dependent, and initial transits last longer at lower temperatures, clutch size is unaffected. Furthermore, wasps can discriminate between fully exposed and partially exposed hosts at both 13 and 26°C. Both clutch size and the relationship between initial transit duration and temperature are compensated for changes in temperature and its effects on metabolism. Wasps at 13°C took longer to complete their initial transits and to lay each of their eggs than wasps at 26°C (Schmidt & Pak, 1991). However, at both temperatures, the duration of the initial transits showed the same linear correlation with oviposition duration, and clutch size was unaffected. Additional experiments demonstrated that the measurement of initial transit duration must vary according to ambient temperature, since initial transits of the same duration made at different temperatures differ in their effects (Schmidt & Pak, 1991). An initial transit at 13°C is equivalent in its effect on clutch size to an initial transit of shorter duration at a higher temperature. Short interval timing by *T. dendrolimi* is not temperature compensated and the endogenous mechanism used to measure elapsed time during initial transits proceeds more slowly at low temperatures (Schmidt & Pak, 1991).

The evolutionary development of the measuring behaviours in *Trichogramma* remains uncertain. The observation that host curvature measurement is a factor in three behavioural components (i.e. acceptance, examination time and the upper limit to clutch size), while initial transit affects only clutch size regulation, suggests that curvature measurement is the earlier mechanism. The use of initial transit may have arisen later, primarily as an improved method for adjusting progeny allocation to clustered hosts. Clarification of the relationships between these mechanisms must await more detailed studies of comparable behaviours in other parasitoids.

Responses to host density

Use of initial transit time to measure volume leads to appropriate responses to hosts which occur in clusters. As host density increases, it becomes more probable that a wasp will find sufficient hosts to oviposit its entire egg complement. Under these conditions, it is advantageous for the wasp to reduce the number of eggs laid in each host in order to produce larger progeny of greater fecundity and reproductive fitness (Klomp & Teerink, 1967; Waage & Godfray, 1984; Charnov & Skinner, 1985). A reduction in clutch size per host has been reported for *Trichogramma* species attacking clusters (i.e. groups in which the hosts are in direct contact with their neighbours) (Hirose *et al.*, 1976; Schmidt & Smith, 1985b). When clustered hosts are examined, the mean initial transit time is shortened due to the reduction of the surface area accessible to the wasps (Schmidt & Smith, 1985b). As a result, clutch sizes are smaller for clustered hosts, and hosts in the centres of clusters receive fewer

eggs than those on the perimeter (Schmidt & Smith, 1985b). Other host density dependent factors, including chemical cue concentrations, may be involved in these responses, and the effects of the duration of wasp exposure must also be considered (Kfir, 1981; Waage & Godfray, 1984; Vinson, 1985). Dijken & Waage (1987) have suggested that under some conditions of host density, superparasitism may also be a mechanism for adjusting progeny allocation to the local availability of hosts. However, Schmidt & Smith (1987b) found no evidence that *T. minutum* adjusts clutch size by super-parasitism. When presented with either single hosts or groups of hosts, the maximal number of progeny was allocated to the single hosts during the first oviposition. The number of progeny allocated to the single hosts during a single oviposition by inexperienced wasps always equalled or exceeded the number allocated to hosts occurring in groups and exposed to multiple oviposition attempts.

Schmidt & Smith (1987b) found that the spacing between hosts not in direct contact with each other affected progeny allocation. Clutch size was directly proportional to the separation between hosts, indicating that the wasps may reduce clutch sizes when the temporal or spatial interval between host encounters is small. Waage & Ng (1984) demonstrated that clutch sizes were reduced with successive ovipositions when *T. evanescens* encountered a patch of hosts, an effect that may also be dependent on the interval between host encounters (however, see Chapter 12).

Temperature effects

In the field, *Trichogramma* must cope with both seasonal and daily fluctuations in temperature (Quednau, 1957; Calvin *et al.*, 1984; Pak & van Heiningen, 1985). In addition, conditions may vary between the base and canopy of vegetation, and between the upper and lower surfaces of leaves. In general, the wasps are active over a temperature range between 10 and 35°C (Quednau, 1957; Pak & van Heiningen, 1985). Although these transient temperature changes affect the searching and locomotory activity of the wasps, they do not influence the adjustment of clutch size to host volume (Schmidt & Pak, 1991). However, it is unclear how larger scale, persistent or seasonal temperature conditions will influence the lifetime reproductive output of female *Trichogramma*. Studies by Pak & van Heiningen (1985) suggest that *Trichogramma* species employ a variety of strategies to maintain parasitization levels over the range of temperatures at which they remain active. It may be that increased longevity at constant low temperatures compensates for reduced rates of searching and host encounter (Boldt, 1974; Calvin *et al.*, 1984). The wasps may also be less fecund at lower temperatures (Quednau, 1957; Russo & Voegelé, 1982), and require fewer hosts to accommodate their supply of eggs. Calvin *et al.* (1984) found that developmental times and sex ratios of *T. pretiosum* were unaffected by crowding levels over most of the active temperature range. Thus, the deleterious effects of both under- and overcrowding of larvae may be similar at different temperatures. The observation that clutch size is temperature compensated (Schmidt & Pak, 1991) provides some evidence that the reproductive parameters governing the adjustment of clutch size are also independent of temperature.

Internal contents

Changes in internal chemical cues have been correlated with host acceptance and clutch size regulation. Ovipositing *Trichogramma* are sensitive to variations in host content arising from host embryonic development and the effects of ageing and deterioration due to storage (Marston & Ertle, 1969; Pak, 1986).

Chemosensilla on the ovipositor play a major role in detecting host suitability and factors associated with ageing (LeRalec & Wajnberg, 1990). *Trichogramma* species females will oviposit into encapsulated mixtures of various cations, tissue culture media and artificial diets (Rajendram, 1978b; Wu & Qin, 1982; Nettles *et al.*, 1983). Nettles *et al.* (1983, 1985) have demonstrated that solutions containing a concentration of $MgSO_4$ and KCl comparable to host haemolymph induce high rates of oviposition into wax capsules (see Chapter 4). Glucose, protein hydrosylates and diet components inhibit the stimulatory effects of K^+ and Mg^{2+} (Nettles *et al.*, 1983). It remains to be shown whether the effects of KCl and $MgSO_4$ are comparable to those of the stimulatory agents responsible for egg laying into natural hosts and the biological significance of this response to cations remains unclear.

Wu & Qin (1982) and Lu (1979) found that leucine, isoleucine, phenylalanine and histidine, alone or in mixtures, evoked significant oviposition at concentrations between 20 and 160 mg $(10\,ml)^{-1}$. Phenylalanine and leucine were the most potent and histidine was the least effective. In their studies wasps oviposited into encapsulated mixtures and continued to adjust clutch size in response to the volume of the capsule, demonstrating that the normal control of oviposition was not bypassed by these stimulants. All four effective amino acids are essential nutrients and may be limiting nutritional factors for larval development (Barrett & Schmidt, 1991). Barrett & Schmidt (1991) determined that free isoleucine, leucine and phenylalanine occur at concentrations ranging between 10 and 40 mg $(100\,ml)^{-1}$ in the eggs of *M. sexta*. Histidine is present at higher concentrations $(50–100\,mg\,(100\,ml)^{-1})$. These concentrations would be sufficient to induce oviposition in the absence of other factors (Lu, 1979; Wu & Qin, 1982). However, they are at the lower range of efficacy, suggesting that they may provide the wasp with some information about host nutritional quality and the extent of amino acid

depletion in the host (Barrett & Schmidt, 1991).

Host age can affect host examination. Examination of older hosts may be prolonged, and in many cases older hosts are not rejected until the chorion is penetrated by the ovipositor (Marston & Ertle, 1969). Hirose (1982) reported that although fresh *P. xuthus* eggs are readily accepted by *T. papilionis*, older eggs are rejected only after repeated drilling of the chorion and probing with the ovipositor.

In nearly all cases, young hosts are accepted more readily than older hosts (Pak, 1986, 1988). Different patterns of changing acceptance with host age arise depending upon the rate at which acceptability declines. Marston & Ertle (1969) found that inhibition of acceptance coincided with blastokinesis or rotation of the host embryo (see also Benoit & Voegelé, 1979). In many instances sclerotization of the head capsule of the embryonic host marks the end of host acceptability (Lewis & Redlinger, 1969), although Benoit & Voegelé (1979) found that the eggs of the European corn borer (*Ostrinia nubilalis* (Hübner) (Pyralidae)) were accepted for oviposition by *T. evanescens* females until immediately prior to hatching. *Trichogramma* species females often insert their ovipositors into older hosts without ovipositing. These hosts often cease development and collapse, probably as a result of factors injected by the wasps (Eidmann, 1934; Benoit & Voegelé, 1979). Hosts in which development is arrested by freezing or irradiation, or which are unfertilized, are usually accepted for oviposition (Eidmann, 1934; Houseweart *et al.*, 1982).

Less acceptable, older hosts are usually allocated fewer progeny than more frequently accepted younger hosts (Lewis & Redlinger, 1969; Marston & Ertle, 1969; Taylor & Stern, 1971). However, a reverse trend has also been reported: Kennel-Heckel (1963) found that progeny allocation by *T. embryophagum* increased for older eggs of *Bupalus pinarum* L. (Geometridae).

The effects of host age on progeny sex ratios are closely linked with the effects on clutch size. In general, hosts allocated more progeny are allocated more female progeny. As a result, younger hosts are likely to produce a greater proportion of female offspring than older hosts. Juliano (1982) and Navarajan (1979) have reported results consistent with this mechanism of sex ratio adjustment. However, Stern & Bowen (1963) found no effects of host age on sex ratio, and Taylor & Stern (1971) found that proportionally more female offspring emerged from older hosts. Since Taylor & Stern (1971) did not observe ovipositions, their results may reflect differential mortality.

Adjustment of Sex Ratio

Most *Trichogramma* species, like other Hymenoptera, regulate the sex ratio of their offspring by arrhenotoky: fertilized zygotes produce females, unfertilized eggs become males (Suzuki *et al.*, 1984). A few *Trichogramma* species are thelytokous and produce only female progeny (Nobuchi, 1961; Flanders, 1965; Suzuki & Hiehata, 1985; see Chapter 1). The adjustment of sex ratio to host size has been reported for numerous parasitoid species and has been the subject of several reviews (Clausen, 1939; Flanders, 1965; Hamilton, 1967; Charnov, 1982; Green *et al.*, 1982; Jones, 1982; Waage & Godfray, 1984; Werren, 1984; Waage, 1986). In large hosts, mated *Trichogramma* usually produced highly skewed sex ratios, with female offspring out-numbering male offspring (Suzuki *et al.*, 1984; Waage & Ng, 1984; Waage, 1986).

The female-biased sex ratios seen in the clutches of most gregarious parasitoids were first explained by Hamilton (1967) in terms of the high degree of inbreeding that occurs between siblings emerging from the same host. A gregarious parasitoid needs only to allocate sufficient males to each host to fertilize all the females in the same host, resulting in a highly female-biased sex ratio (Hamilton, 1967; Charnov & Skinner, 1985; see Chapter 12). If the hosts are likely to be parasitized by more than one female, then the condition of strict inbreeding no longer holds (Hamilton, 1967; Suzuki & Iwasa, 1980; Werren, 1984). In this situation, the reproductive fitness of the superparasitizing female is increased if at least one of her sons survives to mate with the remaining female offspring, whether or not they are siblings. Analyses of this process by Suzuki & Iwasa (1980), Charnov (1982), Waage & Godfray (1984), Werren (1984), and Charnov & Skinner (1985) have generally suggested that the reproductive fitness of the superparasitizing female will be increased if she allocates proportionally more male offspring to the previously parasitized host.

The effects of this process (local mate competition) and sibling inbreeding are evident in the correlations between host size, clutch size and sex ratio observed for various *Trichogramma* species (Nobuchi, 1961; Suzuki *et al.*, 1984; Waage & Lane, 1984; Waage & Ng, 1984; Suzuki & Hiehata, 1985). In hosts that support only one offspring, the smallest acceptable hosts are more frequently allocated male offspring, whereas female progeny predominate in larger hosts. Since female fitness requires a greater proportional investment of host nutrient and developmental space than male fitness, wasps that overproduce female progeny on large hosts and overproduce males on small hosts increase their overall reproductive success (Salt, 1934; Flanders, 1935; Charnov *et al.*, 1981; Charnov & Skinner, 1985; Opp & Luck, 1986). In larger hosts supporting the development of at least two offspring or on patches of closely spaced hosts, selection favours the allocation of only sufficient males to mate with all the female progeny (Suzuki & Iwasa, 1980; Green *et al.*, 1982; Waage & Godfray, 1984). For most *Trichogramma* species, as clutch size or patch size increases sex ratios are increasingly female-biased (Nobuchi, 1961; Taylor & Stern, 1971; Boldt *et al.*, 1973; Waage & Ng, 1984). When presented with either fully exposed or partially embedded *M. sexta* hosts, *T. minutum* allocate the same average number of male offspring

to hosts of either type. However, fully exposed hosts are allocated significantly more female offspring, resulting in a more female-biased clutch in the hosts with larger exposed volumes (Schmidt & Smith, 1985a).

Several mechanisms for the adjustment of sex ratio by parasitoid Hymenoptera have been identified (Flanders, 1946, 1965; Wylie, 1973). Sex ratios can be adjusted by scramble competition, in which one sex has a higher incidence of mortality under conditions of larval crowding (Clausen, 1939; Flanders, 1965). This may be a factor in cases of superparasitism by some *Trichogramma* species. Suzuki *et al.* (1984) found that female *T. chilonis* egg survival was 98.7% in single parasitized *P. xuthus* hosts, but only 68% in doubly parasitized hosts. In contrast, male egg survival was only marginally worse in superparasitized hosts, resulting in sex ratios that are slightly more male-biased in superparasitized hosts (see Waage & Ng, 1984). These results differ from the observations of Dijken & Waage (1987), who found that, when *T. evanescens* superparasitize *M. brassicae* hosts, female offspring are more likely to survive than male offspring. However, Suzuki *et al.* (1984) did find that in singly parasitized hosts, male mortality was higher than female mortality.

Although local mate competition theory predicts that sex ratios should differ for superparasitism, Suzuki *et al.* (1984) found that inexperienced *T. chilonis* ovipositing into previously parasitized eggs do not alter the number of male or female eggs laid into the host, and that sex ratios are unchanged. In contrast, Alphen *et al.* (1987) found that sex ratios did differ among singly parasitized and superparasitized *M. brassicae* hosts. Inexperienced *T. evanescens* allocated more progeny to unparasitized hosts than to previously parasitized hosts (see Salt, 1934). On unparasitized hosts, the first egg laid was invariably female, whereas the first egg laid into a parasitized host was a male in 27% of the observed ovipositions (Alphen *et al.*, 1987). As a result, the female-biased sex ratio observed for unparasitized hosts was slightly reduced on superparasitized hosts. Dijken & Waage (1987), also using *T. evanescens* and *M. brassicae*, found that fewer eggs were allocated to previously parasitized hosts. However, they found no clear differences in sex allocation among unparasitized and previously parasitized hosts. There was also no evidence that *T. evanescens* changes its sex and progeny allocation based on the recognition of individual-specific markers (Dijken & Waage, 1987).

In *Trichogramma* species, differential mortality of male and female larvae is probably only a significant factor in the case of superparasitism. More frequently the number of adults emerging from a singly parasitized host is equal to the number of eggs laid into the host (Klomp *et al.*, 1980; Suzuki *et al.*, 1984), and the progeny sex ratio is determined directly by the ovipositing female. Suzuki *et al.* (1984) determined that *T. chilonis* lay an unfertilized egg as one of the first two eggs oviposited, followed by additional female eggs, with male eggs interspersed at approximately eight egg intervals. Similar results were obtained for *T. evanescens* (Waage & Lane, 1984; Waage & Ng, 1984) and for *T. brassicae* (Wajnberg, 1993; see Chapter 12). Such a fixed pattern of fertilization would result in the appropriate 1 : 1 sex ratio in small hosts allocated only one or two offspring, and the increasingly female-biased sex ratios observed in larger hosts allocated more offspring (Schmidt & Smith, 1985a). The rate of egg laying and host encounters can similarly affect the frequency of fertilization (Suzuki *et al.*, 1984). Cumulative sex ratios decrease with patch size when closely spaced hosts are parasitized (Waage & Ng, 1984). Male progeny are allocated to the first or second hosts parasitized, and at intervals of several female eggs thereafter (Waage & Ng, 1984; Waage, 1986; Wajnberg, 1993; see Chapter 12).

Inherited Factors Affecting Host Acceptance and Progeny Allocation

Although many cues release fixed patterns of behaviour, there is increasing evidence for the plasticity of the recognition response. Thresholds for acceptance may be set by previous experience and encounters with hosts, resulting in selectivity and the expression of preferences. The time required by *T. maidis* (= *T. brassicae*) to recognize and parasitize a host may decrease as the parasitoid learns to handle hosts more efficiently (Wajnberg, 1989). Chassain & Boulétreau (1987) and Chassain *et al.* (1988) have shown that differences in responses by *T. maidis* to host distribution are inherited. In addition, host examination shows significant variability within individuals, at least part of which is genetically determined (Wajnberg, 1989). Numerous recent studies have demonstrated that strains and species of *Trichogramma* differ significantly in oviposition behaviour (Brand *et al.*, 1984; De Jong & Pak, 1984; Pak & van Heiningen, 1985; Pak 1986, 1988). The results of these studies illustrate the need for caution when attempting to make general statements about *Trichogramma* behaviour.

      Although Pak (1988) found that strains of *T. maidis* Pintureau & Voegelé (= *T. brassicae* Bezdenko), *T. buesi* Voegelé (= *T. brassicae* Voegelé) and *T. evanescens* were all able to discriminate among *M. brassicae*, *P. brassicae* and *P. rapae* L. hosts of different ages, the stage of the examination process at which rejection occurred differed among strains. Some strains rejected the hosts before drilling into the chorion, whereas other strains probed the host internally prior to rejection. The results obtained by Pak (1988) also demonstrated the ability to detect previously parasitized hosts. Salt (1937) found that female *T. evanescens* used separate external and internal markers to detect parasitized hosts. Ables *et al.* (1981) found that female *T. pretiosum* appear to use only an external marker. In contrast, Klomp *et al.* (1980) reported that female *T. embryophagum* rely almost entirely on an internal marker. Pak (1988) found that *T. brassicae* discriminated between parasitized and unparasitized hosts on the basis of internal examination with the

ovipositor, whereas *T. maidis* and *T. evanescens* did not. Subsequent experiments by Wajnberg *et al.* (1989) demonstrated that the intensity of superparasitism is a heritable trait. In addition, the variability of clutch size distribution and mean clutch size per available host are also genetically determined, but by independent mechanisms (see Chapter 12).

Equally striking are the differences in host acceptance among strains. Pak (1988) and De Jong & Pak (1984) found that, whereas *M. brassicae* eggs are readily accepted by all tested strains, most species also showed a preference for the eggs of *M. brassicae* over those of *P. brassicae* and *P. rapae*. However, in no choice tests, some strains accepted *Pieris* eggs as readily as *M. brassicae* eggs (Pak, 1988). Some strains rejected *Pieris* eggs even when these were the only hosts available. Strains that rejected *Pieris* eggs were repelled by the accessory gland secretions coating the eggs, and rejected yellow glass bead models more frequently than transparent or white models (De Jong & Pak, 1984). Furthermore, larval survival in *Pieris* hosts was significantly lower for strains that rejected these eggs than for strains that accepted them more readily (Pak, 1988).

Schmidt (1991) found that the mean clutch size of a Romanian strain of *T. dendrolimi* was significantly greater than that of a strain coming from former Czechoslovakia. This effect was largely the result of the higher percentage of *M. brassicae* hosts allocated a single egg by females of the strain originating from former Czechoslovakia. Hybrids of the two strains were fertile, and produced clutches of intermediate size. However, the distribution of clutch sizes allocated by the hybrids resembled that of the Romanian strain, with only a few hosts allocated a single egg.

The ability to discriminate between hosts differing in exposed volume is, at least in part, a heritable behavioural trait in *Trichogramma* species (Schmidt, 1991). *Trichogramma* females can be selected for constancy of clutch size in response to either embedded or surface-mounted hosts. In addition, clutch size itself is an inherited trait (see Chapter 12).

Conclusion

Although some strains of *Trichogramma* show indications of host preference, most are able to opportunistically recognize and utilize a wide variety of host species. An important component in the evolution of this polyphagy has been the use of general physical and tactile cues such as curvature and a correspondingly reduced reliance on species-specific chemical cues. The use of chemical cues by *Trichogramma* is largely reduced to the evaluation of host developmental and nutritional suitability and the avoidance of repellent or previously parasitized hosts. The use of these chemical cues places relatively few restrictions on the exploitation of potential hosts by *Trichogramma* females. Instead, by coupling direct physical measures of distance, time and curvature to the determination of clutch size and progeny sex ratios, these wasps have achieved the necessary plasticity to deal with the complex problems of assessing host availability and distribution.

In an insect as small as *Trichogramma*, sensory and neuronal processing resources are limited and the wasps must employ relatively simple problem-solving strategies. In this context, it is important to note the frequently encountered correlations between specific time intervals and *Trichogramma* behaviour. The duration of the initial transit is used as a simple measure of host volume and establishes the number of eggs allocated. The same mechanism can be used to adjust clutch size when the wasp encounters clustered hosts. There is also evidence that the time interval between successive host encounters is used as a measure of local host density. In at least some species, the interval between encounters with previously parasitized hosts also regulates the occurrence of superparasitism. Finally, the time interval between successive ovipositions has a significant affect on the production of male and female eggs. These examples suggest that the ability of *Trichogramma* to detect the passage of time between specific events may be a central neurophysiological mechanism underlying many of the behavioural responses reported for these parasitoids.

References

Ables, J.R., Vinson, S.B. & Ellis, J.S. (1981) Host description by *Chelonus insularis* (Hym.: Braconidae), *Telenomus heliothidis* (Hym.: Scelionidae), and *Trichogramma pretiosum* (Hym.: Trichogrammatidae). *Entomophaga* 26, 149–156.

Alphen, J.J.M. van, van Dijken, M.J. & Waage, J.K. (1987) A functional approach to superparasitism: host discrimination needs not be learnt. *Netherlands Journal of Zoology* 37, 167–179.

Arthur, A.P. (1967) Influence of position and size of host shelter on host-searching by *Itoplectis conquisitor* (Hymenoptera: Ichneumonidae). *Canadian Entomologist* 99, 877–886.

Bakker, K., van Alphen, J.J.M., van Batenburg, F.H.D., van der Hoeven, N., Nell, H.W., van Strien-van Liempt,W.T.F.H. & Turlings, T.C.J. (1985) The function of host discrimination and superparasitization in parasitoids. *Oecologia* 67, 572–576.

Barrett, M. & Schmidt, J.M. (1991) A comparison between the amino acid composition of an egg parasitoid wasp and some of its hosts. *Entomologia Experimentalis et Applicata* 59, 29–41.

Beckage, N.E. & Riddiford, L.M. (1983) Growth and development of the endoparasitic wasp *Apanteles congregatus*: dependence on host nutrient status and parasite load. *Physiological Entomology* 8, 231–241.

Benoit, M. & Voegelé, J. (1979) Choix de l'hôte et comportement trophique des larves de *Trichogramma evanescens* (Hym.: Trichogrammatidae) en fonction du développement embryonnaire de *Ephestia kuehniella* et *Ostrinia nubilalis* (Lep.: Pyralidae). *Entomophaga* 24, 199–207.

Boldt, P.E. (1974) Temperature, humidity and host: effect on rate of search of

Trichogramma evanescens and *T. minutum* auctt. (not Riley, 1871). *Annals of the Entomological Society of America* 67, 706–708.

Boldt, P.E., Marston, N. & Dickerson, W.A. (1973) Differential parasitism of several species of lepidopteran eggs by two species of *Trichogramma*. *Environmental Entomology* 2, 1121–1122.

Brand, A.M., van Dijken, M.J., Kole, M. & van Lenteren, J.C. (1984) Host-age and host-species selection of three strains of *Trichogramma evanescens* Westwood, an egg parasite of several lepidopteran species. *Mededelingen van de Faculteit Landbouwwetenschappen Rijksuniversiteit Gent* 1984, 839–847.

Burbutis, P.P., Morse, B.W., Morris, D. & Benzon, G. (1983) *Trichogramma nubilale* (Hymenoptera: Trichogrammatidae): progeny distribution and superparasitism in European corn borer (Lepidoptera: Pyralidae). *Environmental Entomology* 12, 1587–1589.

Calvin, D.D., Knapp, M.C., Welch, S.M., Poston, F.L. & Elzinga, R.J. (1984) Impact of environmental factors on *Trichogramma pretiosum* reared on Southwestern corn borer eggs. *Environmental Entomology* 13, 774–780.

Charnov, E.L. (1982) *The Theory of Sex Allocation*. Princeton University Press, Princeton, New Jersey, 355 pp.

Charnov, E.L. & Skinner, S.W. (1985) Complementary approaches to the understanding of parasitoid oviposition decisions. *Environmental Entomology* 14, 383–391.

Charnov, E.L., Los-den Hartogh, R.L., Jones, W.T. & Van den Assem, J. (1981) Sex ratio evolution in a variable environment. *Nature* 289, 27–33.

Chassain, C. & Boulétreau, M. (1987) Genetic variability in the egg-laying behaviour of *Trichogramma maidis*. *Entomophaga* 32, 149–157.

Chassain, C., Boulétreau, M. & Fouillet, P. (1988) Host exploitation by parasitoids: local variations in foraging behaviour among populations of *Trichogramma* species. *Entomologia Experimentalis et Applicata* 48, 195–202.

Clausen, C.P. (1939) The effect of host size upon the sex ratio of hymenopterous parasites and its relation to methods of rearing and colonization. *Journal of the New York Entomological Society* 47, 1–9.

Clausen, C.P. (1940) *Entomophagous Insects*. McGraw-Hill, New York, 688 pp.

De Jong, E.J. & Pak, G.A. (1984) Factors determining differential host-egg recognition of two host species by different *Trichogramma* spp. *Mededelingen van de Faculteit Landbouwwetenschappen Rijksuniversiteit Gent* 49, 815–825.

Dijken, M.J. van & Waage, J.K. (1987) Self and conspecific superparasitism by the egg parasitoid *Trichogramma evanescens*. *Entomologia Experimentalis et Applicata* 43, 183–192.

Dijken M.J. van, Kole, M., van Lenteren, J.C. & Brand, A.M. (1986) Host-preference studies with several strains of *Trichogramma evanescens* Westwood (Hym.: Trichogrammatidae) for *Mamestra brassicae*, *Pieris brassicae* and *Pieris rapae*. *Zeitschrift für Angewandte Entomologie* 101, 64–85.

Edwards, R.L. (1954) The host-finding and oviposition behaviour of *Mormoniella vitripennis* (Walker), a parasite of muscoid flies. *Behaviour* 7, 88–112.

Eidmann, H. (1934) Zur Kenntnis der Eiparasiten der Forleule insbesondere über die Entwicklung und Ökologie von *Trichogramma minutum* Riley. *Mittellungen Forstwirschaft und Forstwissenschaft* 5, 56–77.

Fisher, R.C. (1971) Aspects of the physiology of endoparasitic Hymenoptera. *Biological*

Reviews, Cambridge Philosophical Society 46, 243–278.

Flanders, S.E. (1935) Host influence on the prolificacy and size of *Trichogramma*. *Pan-Pacific Entomology* 11, 175–177.

Flanders, S.E. (1937) Habitat selection of *Trichogramma*. *Annals of the Entomological Society of America* 30, 208–210.

Flanders, S.E. (1946) Control of sex and sex-limited polymorphism in the Hymenoptera. *The Quarterly Review of Biology* 21, 135–143.

Flanders, S.E. (1965) On the sexuality and sex ratios of hymenopterous populations. *American Naturalist* 99, 489–494.

Green, R.F., Gordh, G. & Hawkins, B.A. (1982) Precise sex ratios in highly inbred parasitic wasps. *American Naturalist* 120, 653–665.

Hamilton, W.D. (1967) Extraordinary sex ratios. *Science* 156, 477–488.

Hase, A. (1925) Beiträge zur Lebensgeschichte der Schlupfwespe *Trichogramma evanescens* Westwood. *Arbeiten aus der Biologischen Reichsanstadt für Land- und Fortswirtschaft* 14, 171–224.

Hirose, H. (1982) Recognition time as a factor affecting the parasitism by *Trichogramma*. In: Voegelé, J. (ed.) *Les Trichogrammes. Les Colloques de l'INRA* 9, 111–115.

Hirose, Y., Kimoto, H. & Hiehata, K. (1976) The effect of host aggregation on parasitism by *Trichogramma papilionis* Nagarkatti (Hym.: Trichogrammatidae), an egg parasitoid of *Papilio xuthus* L. (Lepid.: Papilionidae). *Applied Entomology and Zoology* 11, 116–125.

Hoffman, J.D., Ignoffo, C.M. & Dickerson, W.A. (1975) *In vitro* rearing of the endoparasitic wasp *Trichogramma pretiosum. Annals of the Entomological Society of America* 68, 335–336.

Holloway, T.E. (1912) An experiment on the oviposition of a hymenopterous egg parasite. *Entomology News* 23, 329–330.

Houseweart, M.W., Southard, S.G. & Jennings, D.T. (1982) Availability and acceptability of spruce budworm eggs to parasitism by the egg parasitoid, *Trichogramma minutum* (Hymenoptera: Trichogrammatidae). *Canadian Entomologist* 114, 657–666.

Iwasa, Y., Suzuki, Y. & Hiroyuki, M. (1984) Theory of oviposition strategy of parasitoids. I. Effects of mortality and limited egg number. *Theoretical Population Biology* 26, 205–227.

Jackson, D.J. (1966) Observations on the biology of *Caraphractus cinctus* Walker (Hymenoptera: Mymaridae), a parasitoid of the eggs of Dytiscidae (Coleoptera). III. The adult life and sex ratio. *Transactions of the Royal Entomology Society, London* 118, 23–49.

Jones, R.L., Lewis, W.J., Beroza, M., Bierl, B.A. & Sparks, A.N. (1973) Host-seeking stimulants (kairomones) for the egg-parasite *Trichogramma evanescens. Environmental Entomology* 2, 593–596.

Jones, W.T. (1982) Sex ratio and host size in a parasitoid wasp. *Behavioral Ecology and Sociobiology* 10, 207–210.

Juliano, S.A. (1982) Influence of host age on host acceptability and suitability for a species of *Trichogramma* (Hymenoptera: Trichogrammatidae) attacking aquatic Diptera. *Canadian Entomologist* 114, 713–720.

Keller, M.A. (1987) Influence of leaf surfaces on movements by the hymenopterous parasitoid *Trichogramma exiguum. Entomologia Experimentalis et Applicata* 43, 55–59.

Kennel-Heckel, W. (1963) Experimentell-ökologische Untersuchungen an *Trichogramma embryophagum* Hartig (Chalc. Hym.) sowie am Ei des Kiefernspanners *Bupalus piniarius* L. (Geom. Lep.). *Zeitschrift für Angewandte Entomologie* 52, 142–184.

Kfir, R. (1981) Effect of hosts and parasite density on the egg parasite *Trichogramma pretiosum* (Hymenoptera: Trichogrammatidae). *Entomophaga* 26, 445–452.

Klomp, H. & Teerink, B.J. (1962) Host selection and number of eggs per oviposition in the egg parasite *Trichogramma embryophagum* Htg. *Nature* 195, 1020–1021.

Klomp, H. & Teerink, B.J. (1967) The significance of oviposition rates in the egg parasite, *Trichogramma embryophagum* Htg. *Archives Neerlandaises de Zoologie* 17, 350–375.

Klomp, H. & Teerink, B.J. (1978) The elimination of supernumerary larvae of the gregarious egg-parasitoid *Trichogramma embryophagum* (Hym.: Trichogrammatidae) in eggs of the host *Ephestia kuehniella* (Lep.: Pyralidae). *Entomophaga* 23, 153–159.

Klomp, H., Teerink, B.J. & Ma, W.-C. (1980) Discrimination between parasitized and unparasitized hosts in the egg parasite *Trichogramma embryophagum* (Hym., Trichogrammatidae): a matter of learning and forgetting. *Netherlands Journal of Zoology* 30, 254–277.

Laing, J. (1938) Host-finding by insect parasites. II. The chance of *Trichogramma evanescens* finding its hosts. *Journal of Experimental Biology* 15, 281–302.

Legner, E.F. (1969) Adult emergence interval and reproduction in parasitic Hymenoptera influences by host size and density. *Annals of the Entomological Society of America* 62, 220–226.

Lenteren, J.C. van & DeBach, P. (1981) Host discrimination in three ectoparasites (*Aphytis coheni, A. lingnanensis,* and *A. melinus*) of the oleander scale (*Aspidiotus nerii*). *Netherlands Journal of Zoology* 31, 504–532.

LeRalec, A. & Wajnberg, E. (1990) Sensory receptors of the ovipositor of *Trichogramma maidis* (Hym.: Trichogrammatidae). *Entomophaga* 35, 293–299.

Lewis, W.J. & Redlinger, L.M. (1969) Suitability of eggs of the almond moth, *Cadra cautella,* of various ages for parasitism by *Trichogramma evanescens. Annals of the Entomological Society of America* 62, 1482–1484.

Lewis, W.J., Jones, R.L., Nordlund, D.A. & Gross, H.R. Jr. (1975) Kairomones and their use for management of entomophagous insects. II. Mechanisms causing increase in rate of parasitization by *Trichogramma* spp. *Journal of Chemical Ecology* 1, 349–360.

Lu, W. (1979) Ovipositional behaviour of Trichogrammatid wasps. *Acta Entomologica Sinica* 22, 361–363.

Maier, K. (1960) Verhaltensstudien bei Eiparasiten der Gattung *Trichogramma* (Hym., Chalcididae). *Mittelungen aus der Biologischen Reichsanstalt für Land und Forstwirtschaft, Berlin-Dahlem* 100, 3–10.

Marchal, P. (1936) Les Trichogrammes. *Annales Epiphyties et de Phytogénétique, Paris* 2, 448–550.

Marston, N. & Ertle, L.R. (1969) Host age and parasitism by *Trichogramma minutum* (Hymenoptera: Trichogrammatidae). *Annals of the Entomological Society of America* 62, 1476–1482.

Morrison, G., Lewis, W.J. & Nordlund, D.A. (1980) Spatial differences in *Heliothis zea* egg density and the intensity of parasitism by *Trichogramma* spp.: an experimental analysis. *Environmental Entomology* 9, 79–85.

Navarajan, A.V. (1979) Influence of host age on parasitism by *Trichogramma australicum* Gir. and *T. japonicum* Ash. (Trichogrammatidae: Hymenoptera). *Zeitschrift für Angewandte Entomologie* 87, 277–281.

Nettles, W.C. Jr., Morrison, R.K., Xie, Z.N., Ball, D., Shenkir, C.A. & Vinson, S.B. (1983) Effect of cations, anions and salt concentrations on oviposition by *Trichogramma pretiosum* Riley in wax eggs. *Entomologia Experimentalis et Applicata* 33, 283–289.

Nettles, W.C. Jr., Morrison, R.K., Xie, Z.N., Ball, D., Shenkir, C.A. & Vinson, S.B. (1985) Effect of artificial diet media, glucose, protein hydrolyzates, and other factors on oviposition in wax eggs by *Trichogramma pretiosum*. *Entomologia Experimentalis et Applicata* 38, 121–129.

Nobuchi, H. (1961) Insect natural enemies of *Dendrolimus spectabilis*: egg parasitoids. *Shinrinshokubo News* 10, 179–182.

Noldus, L.P.J.J. & van Lenteren, J.C. (1985) Kairomones for the egg parasite *Trichogramma evanescens* Westwood. II. Effect of contact chemicals produced by two of its hosts, *Pieris brassicae* L. and *Pieris rapae* L. *Journal of Chemical Ecology* 11, 793–800.

Nordlund, D.A., Lewis, W.J., Todd, J.W. & Chalfant, R.B. (1977) Kairomones and their use for management of entomophagous insects: VII. The involvement of various stimuli in the differential response of *Trichogramma pretiosum* Riley to two suitable hosts. *Journal of Chemical Ecology* 3, 513–518.

Nordlund, D.A., Strand, M.R., Lewis, W.J. & Vinson, S.B. (1987) Role of kairomones from host accessory gland secretion in host recognition by *Telenomus remus* and *Trichogramma pretiosum*, with partial characterization. *Entomologia Experimentalis et Applicata* 44, 37–43.

Opp, S.B. & Luck, R.F. (1986) Effects of host size on selected fitness components of *Aphytis melinus* and *A. lingnanensis* (Hymenoptera: Aphelinidae). *Annals of the Entomological Society of America* 79, 700–704.

Pak, G.A. (1986) Behavioural variations among strains of *Trichogramma* spp.: a review of the literature on host-age selection. *Journal of Applied Entomology* 101, 55–64.

Pak, G.A. (1988) Selection of *Trichogramma* for inundative biological control. PhD Thesis, Wageningen Agricultural University, The Netherlands, 224 pp.

Pak, G.A. & van Heiningen, T.G. (1985) Behavioural variations among strains of *Trichogramma* spp.: adaptability to field-temperature conditions. *Entomologia Experimentalis et Applicata* 38, 3–13.

Parker, G.A. & Courtney, S.P. (1984) Models of clutch size in insect oviposition. *Theoretical Population Biology* 26, 27–48.

Quednau, W. (1955) Über einige neue *Trichogramma*-Wirte und ihre Stellung im Wirt-Parasit-Verhaltnis. Ein Beitrag zur Analyse des Parasitismus bei Schlupfwespen. *Nachrichten blatt des Deutschen Pflanzenschutzdienstes Systematik, Braunschweig* 7, 145–148.

Quednau, W. (1957) Über den Einfluss von Temperatur und Luftfeuchtigkeit auf den Eiparasiten *Trichogramma cacoeciae* Marchal (Eine biometrische Studie). *Mitteilungen aus der Biologischen Bundesanstalt für Land und Forstwirtschaft, Berlin-Dahlem* 90, 5–63.

Quednau, W. (1960) Über die Identität der *Trichogramma*-Arten und einiger ihrer Ökotypen (Hymenoptera, Chalcidoidea, Trichogrammatidae). *Mitteilungen aus der*

Biologischen Bundesanstalt für Land und Forstwirtschaft, Berlin-Dahlem 100, 11–50.

Rajendram, G.F. (1978a) Oviposition behavior of Trichogramma californicum on artificial substrates. Annals of the Entomological Society of America 71, 92–94.

Rajendram, G.F. (1978b) Some factors affecting oviposition of Trichogramma californicum (Hymenoptera: Trichogrammatidae) in artificial media. Canadian Entomologist 110, 345–352.

Russo, J. & Voegelé, J. (1982) Influence de la température sur quatre espèces de Trichogrammes (Hym., Trichogrammatidae) parasites de la pyrale du maïs, Ostrinia nubilalis Hbn. (Lep., Pyralidae). II. Reproduction et survie. Agronomie 2, 517–524.

Sachtleben, H. (1929) Die Forleule. Monographien zum Pflanzenschutz. Springer Verlag, Berlin.

Salt, G. (1934) Experimental studies in insect parasitism. II. Superparasitism. Proceedings of the Royal Society, London 114, 455–476.

Salt, G. (1935) Experimental studies in insect parasitism. III. Host selection. Proceedings of the Royal Society, London 117, 413–435.

Salt, G. (1937) The sense used by Trichogramma to distinguish between parasitized and unparasitized hosts. Proceedings of the Royal Society, London 122, 57–75.

Salt, G. (1958) Parasite behaviour and the control of insect pests. Endeavour July, 145–148.

Schmidt, J.M. (1991) Inheritance of clutch size regulation by Trichogramma species. In: Bigler, F. (ed.) Proceedings of the 5th International Workshop on Quality Control of Mass-reared Arthropods. Wageningen, The Netherlands, pp. 26–36.

Schmidt, J.M. & Pak, G.A. (1991) The effect of temperature on progeny allocation and short interval timing in a parasitoid wasp. Physiological Entomology 16, 345–354.

Schmidt, J.M. & Smith, J.J.B. (1985a) Host volume measurement by the parasitoid wasp Trichogramma minutum: the roles of curvature and surface area. Entomologia Experimentalis et Applicata 39, 213–221.

Schmidt, J.M. & Smith, J.J.B. (1985b) The mechanism by which the parasitoid wasp Trichogramma minutum responds to host clusters. Entomologia Experimentalis et Applicata 39, 287–294.

Schmidt, J.M. & Smith, J.J.B. (1986) Correlations between body angles and substrate curvature in the parasitoid wasp Trichogramma minutum: a possible mechanism of host radius measurement. Journal of Experimental Biology 125, 271–285.

Schmidt, J.M. & Smith, J.J.B. (1987a) Measurement of host curvature by the parasitoid wasp Trichogramma minutum, and its effect on host examination and progeny allocation. Journal of Experimental Biology 129, 151–164.

Schmidt, J.M. & Smith, J.J.B. (1987b) The effect of host spacing on the clutch size and parasitization rate of Trichogramma minutum Riley. Entomologia Experimentalis et Applicata 43, 125–131.

Schmidt, J.M. & Smith, J.J.B. (1987c) The measurement of exposed host volume by the parasitoid wasp Trichogramma minutum and the effects of wasp size. Canadian Journal of Zoology 65, 2837–2845.

Schmidt, J.M. & Smith, J.J.B. (1989) Host examination walk and oviposition site selection of Trichogramma minutum: studies on spherical hosts. Journal of Insect Behavior 2, 143–171.

Southard, S.G., Houseweart, M.W., Jennings, D.T. & Halteman, W.A. (1982) Size differences of laboratory reared and wild populations of *Trichogramma minutum* (Hymenoptera: Trichogrammatidae). *Canadian Entomologist* 114, 693–698.

Stern, V.M. & Bowen, W. (1963) Ecological studies of *Trichogramma semifumatum*, with notes on *Apanteles medicaginis* and their suppression of *Colias eurytherme* in southern California. *Annals of the Entomological Society of America* 56, 358–372.

Strand, M.R. & Vinson, S.B. (1983) Factors affecting host recognition and acceptance in the egg parasitoid *Telenomus heliothidis* (Hymenoptera: Scelionidae). *Environmental Entomology* 12, 1114–1119.

Suzuki, Y. & Hiehata, K. (1985) Mating systems and sex ratios in the egg parasitoids, *Trichogramma dendrolimi* and *T. papilionis* (Hymenoptera: Trichogrammatidae). *Animal Behaviour* 33, 1223–1227.

Suzuki, Y. & Iwasa, Y. (1980) A sex ratio theory of gregarious parasitoids. *Research on Population Ecology* 22, 366–382.

Suzuki, Y., Tsuji, H. & Sasakawa, M. (1984) Sex allocation and effects of superparasitism on secondary sex ratios in the gregarious parasitoid, *Trichogramma chilonis* (Hymenoptera: Trichogrammatidae). *Animal Behaviour* 32, 478–484.

Taylor, T.A. & Stern, V.M. (1971) Host preference studies with the egg parasite *Trichogramma semifumatum* (Hymenoptera: Trichogrammatidae). *Annals of the Entomological Society of America* 64, 1381–1390.

Vinson, S.B. (1976) Host selection by insect parasitoids. *Annual Reviews of Entomology* 21, 109–133.

Vinson, S.B. (1985) The behaviour of parasitoids. In: Kerkut, G.A. & Gilbert, L.I. (eds) *Comprehensive Insect Physiology, Biochemistry and Pharmacology*. Pergamon Press, Oxford, vol. 9, pp. 417–469.

Waage, J.K. (1982) Sib-mating and sex ratio strategies in scelionid wasps. *Ecological Entomology* 7, 103–112.

Waage, J.K. (1986) Family planning in parasitoids: adaptive patterns of progeny and sex allocation. In: Waage, J.K. & Greathead, D.J. (eds) *Insect Parasitoids*. Academic Press, London, pp. 63–95.

Waage, J.K. & Godfray, H.C.J. (1984) Reproductive strategies and population ecology of insect parasitoids. In: Sibly, R.M. & Smith, R.H. (eds) *Behavioural Ecology: the Population Consequences of Adaptive Behaviour*. Blackwell Scientific Publishers, Oxford, pp. 449–470.

Waage, J.K. & Lane, J.A. (1984) The reproductive strategy of a parasitic wasp. II. Sex allocation and local mate competition in *Trichogramma evanescens*. *Journal of Animal Ecology* 53, 417–426.

Waage, J.K. & Ng, S.M. (1984) The reproductive strategy of a parasitic wasp. I. Optimal progeny and sex allocation in *Trichogramma evanescens*. *Journal of Animal Ecology* 53, 401–415.

Wajnberg, E. (1989) Analysis of variations of handling-time in *Trichogramma maidis*. *Entomophaga* 34, 397–407.

Wajnberg, E. (1993) Genetic variation in sex allocation in a parasitic wasp. Variation in sex pattern within sequences of oviposition. *Entomologia Experimentalis et Applicata* 69, 221–229.

Wajnberg, E., Pizzol, J. & Babault, M. (1989) Genetic variation in progeny allocation in *Trichogramma maidis*. *Entomologia Experimentalis et Applicata* 53, 177–187.

Walter, S. (1983) Zur Biologie und Ökologie von Eiparasiten aus der Gattung

Trichogramma Westwood (Hym., Chalc.). Teil 1: Untersuchungen in ausge-
wählten Forstbiozönosen der DDR. *Zoologishes Jahrbuch. Systematik* 110,
271–299.

Werren, J.H. (1984) A model for sex ratio selection in parasitic wasps: local mate
competition and host quality effects. *Netherlands Journal of Zoology* 34, 81–96.

Wu, Z.X. & Qin, J. (1982) Oviposition response of *Trichogramma dendrolimi* to the
chemical contents of artificial eggs. *Acta Entomologia Sinica* 25, 363–372.

Wylie, H.G. (1967) Some effects of host size on *Nasonia vitripennis* and *Muscidifurax
raptor* (Hymenoptera: Pteromalidae). *Canadian Entomologist* 99, 742–748.

Wylie, H.G. (1973) Control of egg fertilization by *Nasonia vitripennis* (Hymenoptera:
Pteromalidae) when laying on parasitized house fly (*Musca domestica*: Dipt.,
Muscidae) pupae. *Canadian Entomologist* 105, 709–718.

Physiological Interactions between Egg Parasitoids and Their Hosts 10

S. Bradleigh Vinson
*Department of Entomology, Texas A&M University, College Station,
Texas 77843-2475, USA*

Abstract

The physiological interaction of oophages with their host involves both the location and utilization of insect eggs. These eggs represent a stationary, finite, changing and short-lived resource. As a finite resource the size of the egg is of major importance, but there may be limits to the small size of both the host and the parasitoid. Evidence indicates that the quality of the resource (egg) also decreases with time. Further, the short period of availability of the host presents additional challenges to the parasitoid in terms of both location and utilization. These challenges and the ways that oophages have evolved to meet them are discussed.

Introduction

The literature concerning the physiological interactions between parasitoids and their hosts and the importance of these interactions in determining host suitability for parasitoid development has expanded rapidly over the last decade (Beckage, 1985; Lawrence, 1986, 1990; Stoltz, 1986; Vinson, 1990a; Coudron, 1991). The impetus driving these studies has been the development of techniques that may be applied to the *in vitro* mass-production of beneficial insects (Greany *et al.*, 1984; Greany, 1986; Thompson, 1986; Vinson, 1986; Nettles, 1990; see Chapter 4), and to efforts that describe the mechanisms by which parasitoids influence the physiology of their host which may be applied to pest management (Jones, 1986; Vinson, 1988a). However, the majority of studies have utilized larval parasitoids, even though the egg parasitoids,

particularly *Trichogramma*, have been used extensively in biological control (DeBach & Rosen, 1991).

Parasitoids may be grouped as solitary or gregarious, endo- or ectophagous, or idio- or koinobionts, and may parasitize egg, larval, pupal or adult life forms. This discussion will be restricted to the egg parasitoids or oophages. The terms idiobiont and koinobiont refer to the ecological condition of the parasitized host, the latter referring to hosts that continue to develop after parasitism. Idiobionts in contrast feed as larvae on a host whose development has either been arrested by the female parent prior to oviposition (Askew & Shaw, 1986) or is in a resting or quiescent stage (Halselbarth, 1979). By definition all egg parasitoids are idiobionts with the exception of the egg-larval species which are excluded from this discussion.

However, unless eggs are in diapause, they are neither resting or quiescent, but undergo rapid physiological changes once embryogenesis is initiated. The mechanisms by which oophages deal with these physiological changes have received much less attention than corresponding studies for the larval parasitoids (Fisher, 1971; Vinson & Iwantsch, 1980a,b; Beckage, 1985). Mellini (1986) discussed the importance of egg age at the time of parasitization and discussed various physiological interactions between oophages and their host. Strand (1986) emphasized egg parasitoids in his review of the consequence of host–parasitoid interactions on reproductive strategies. He included the impact of the host–parasitoid physiological interaction on host discrimination, competition, and the resulting effect on ovipositional decisions. This review will focus on the physiological interaction of the egg (host) with the oophage.

The Organisms

Oophages

Insect eggs are utilized as a resource by 16 families of parasitic Hymenoptera (Clausen, 1940) with three families, i.e. the Mymaridae, Scelionidae and Trichogrammatidae, being exclusively endo-oophagous. A few ecto-oophagous species occur, such as *Podagryon mantis* (Torymidae) which attacks mantid eggs within an ootheca and *Habrocytus lixi* (Pteromalidae) which attacks cerambycid and curculionid eggs laid within the pith of plant stems (Bin *et al.*, 1988), and are restricted to groups of eggs within an enclosure.

Host specificity of oophages varies greatly with some families restricted to a particular group of hosts such as the predaceous Ichneumonidae that feed exclusively on spider (Arachnids) eggs that occur within a silken egg-sack (Carlson, 1979). The Evaniidae are restricted to Blattodea and the Chrysididae to Phasmadea (Askew, 1971). In contrast, trichogrammatids have been reported attacking host species occurring in ten orders (Burks, 1979).

Similarly, some hosts are attacked by a limited number of oophage species. For example, the eggs of Psocoptera or Tysanoptera are only known to be attacked by Mymaridae or Trichogrammatidae, respectively. In contrast, the eggs of some Lepidoptera or Coleoptera are attacked by species from seven oophage families (Krombein *et al.*, 1979).

The host

Eggs are initially an enlarged cell packed with storage forms of all of the nutrients needed to initiate embryogenesis. The cell is enclosed in a protective shell referred to as a chorion. The chorion consists primarily of layers of a protein–lipid–cement complex, varying in thickness and complexity, that regulates air and water exchange (Margaritis, 1985). Insect eggs lack a cellular defence reaction (Salt, 1970) which is a major factor in protecting the larval stages from invading parasitoids (Vinson, 1990b).

In contrast, hormones, which have a major impact on insect development (Hoffmann & Lagueux, 1985), are known to occur in eggs and to play a major role in egg diapause (Yamashita, 1983) and embryogenesis (Bergot *et al.*, 1981; Scalia & Morgan, 1982; Gharib & de Reggi, 1983; Gharib *et al.*, 1983). These hormones appear to be stored in the egg in a bound form which is released as the embryo develops (Gharib *et al.*, 1983; Hoffmann & Lagueux, 1985). What effect, if any, these compounds have on the oophage–host relationship is unknown. Hormones are involved in the physiological inter-relationship between larval parasitoids and their host (Dover *et al.*, 1988; Coudron *et al.*, 1990; Tanaka & Vinson, 1991).

Although eggs are a finite stationary resource, when not in diapause they are rapidly changing and short lived. These changes involve the conversion of stored nutrients into structural and metabolic components of the embryo. As embryogenesis proceeds, the egg, which originally consisted primarily of yolk surrounded by a thin cytoplasmic layer, becomes more complex. Depending on the species, embryos may develop to gastrulation within a day or two at which time the organization and composition of the internal contents of the egg are very different. As development proceeds, any invader is exposed to an increasingly complex environment. Further, as embryonic tissues mature and cell numbers increase, there is increasing competition for remaining stored resources.

Physiological Interactions

The interaction between an oophage and its host determines both host and

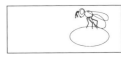

parasitoid specificity, which are important considerations in biological control. This specificity is the result of complex physiological interactions between the parasitoid and host that not only involve the direct physiological interaction between the developing larvae within its host, but also the impact of physiological changes in the host that influence the physiologically based behaviour of the parasitoid involved in host selection. For these reasons, the physiological interactions have been divided into three categories: behavioural, constrained and regulatory. This requires consideration of the changes that occur in eggs as a resource over time, many of which can be considered as a form of defence.

One factor influencing the resource is age (Vinson & Barbosa, 1987). This is particularly important for oophages because the host undergoes extensive developmental changes during a relatively short period (Agrell & Lundquist, 1973; Sander *et al.*, 1985). Several authors (Lewis & Redlinger, 1969; Taylor & Stern, 1971; Leibee *et al.*, 1979) have reported that the susceptibility of eggs to successful parasitism decreases as they age, although Pak *et al.* (1986), working with species of *Trichogramma*, assumed that the nutritional resource and energy an egg contained remained constant throughout embryonic development. Ruberson *et al.* (1987) utilizing *Edovum puttleri* challenged this assumption and showed that this species not only oviposited less often in older eggs, but, in such a case, oophage mortality was greater and body size decreased. These results support a hypothesis that host egg quality as a resource declines with age. As suggested by Ruberson *et al.* (1987), various nutrients may become incorporated into forms that reduce their accessibility to developing parasitoid larvae. Nettles (1990) suggested that parasitoids may have evolved a dependency on certain specific nutrients. If these nutrients were utilized early in the host's embryonic development, they could also limit the host's susceptibility to parasitism over time.

Detecting changes that occur in eggs over time may be important, as suggested by Strand (1986). For example, a hypothetical host with a 60 h embryonic developmental period could only be successfully parasitized by species requiring 35 h for its own embryonic development if they oviposited before the host was 25 h old (Strand, 1986). Otherwise, the host would hatch before the parasitoid could hatch and feed. Thus, one strategy of the developing embryo to reduce its vulnerability to parasitism is rapid development. Other strategies include the sequestration of toxins to which the host species has evolved a tolerance. For example, the cardinolides found in the eggs of *Paekilocerus bufanius* (Orthoptera) which feed on Asclepeadaceae (Eww *et al.*, 1967), cyanogenic glycosides in eggs of *Zygaena* spp. (Lepidoptera) (Jones *et al.*, 1962), or cantharidin present in the eggs of myeloid beetles (Carrel *et al.*, 1975) may reduce the suitability of the egg, at least for generalist oophages. Although these defences are a constraint, the specialist oophage may evolve to avoid, sequester or detoxify the toxin.

Behavioural

Early host location appears to be important. Some parasitoids have evolved to respond to sex pheromones (Buleza & Mikheev, 1979; Aldrich *et al.*, 1984) or defensive secretions (Mattiacci *et al.*, 1993) of adults that orient and attract the oophage to habitats where oviposition is imminent or is taking place. This ensures the location of young eggs.

Another approach is for the female parasitoid to hitch a ride on a gravid female host, a phenomena referred to as phoresy, only leaving to oviposit her eggs as host eggs are oviposited (Clausen, 1976). Phoresy is most common among the scelionids and trichogrammatids that attack egg masses protected by a hardened covering or oviposited in various substrates in widely scattered locations which do not provide consistent cues to their location (Clausen, 1976). Phoresy thus allows the parasitoid access to multiple clutches of eggs which would otherwise be difficult to locate and also provides access to eggs prior to the protection of age becoming effective.

Egg defence is often provided by the ovipositing female covering her eggs with materials that conceal their location, or deter or inhibit the attack of parasitoids. The eggs of some species are covered by a jelly-like material referred to as spermaline (Hodson & Weinman, 1945; Gower, 1967) that prevents oviposition. Many insects glue their eggs to the substrate or force them into plant tissue while others cover their eggs with hairs or scales (Roonwal, 1954; Levine & Chandler, 1976) some of which are toxic (Clements, 1951). These defensive activities result in cues that are exploited by the parasitoid for host location and recognition (Vinson, 1988b). However, these factors associated with the ovipositing female decrease with time once eggs are deposited. Thus, responding to such cues increases the likelihood that only young hosts are located.

Another evolutionary approach to the utilization of rapidly changing host resources is to locate the host and through some mechanism only accept young eggs. The factors involved are concerned with host recognition which includes size, shape, texture and contact kairomones (Klomp & Teerink, 1962; Strand & Vinson, 1982, 1983a,b; Schmidt & Smith, 1986, 1987; Vinson & Piper, 1986; Bin *et al.*, 1993). As a host ages, these factors change and can be considered defensive. For example, Elsey (1972) reported that the eggs of the hawk-moth, *Manduca sexta*, was protected from stylet penetration after 19 h by the newly formed serosal membrane of the embryo. Leibee *et al.* (1979) suggested that the inability of *Patossom lamcerei*, a mymarid, to attack older eggs successfully was due to the hardened chorion of older eggs. However, instead of being a host defence, these changes may signal to the ovipositing female that the resources quality has decreased to an unacceptable level. For example, the eggs of *Heliothis virescens* are spherical when oviposited, but become more conical and less acceptable as a shape for *Telenomus heliothidis* oviposition (Strand & Vinson, 1983b). The kairomones may also change. C. Rosi (unpubl.) found that the

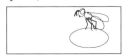

kairomone activity of older *Nezara viridula* egg extracts decreased with host age. All these changes reduce the acceptance of older eggs and ensure that only the younger eggs are attacked.

Constraints

Eggs occur in a variety of shapes that range from flattened or spindle-shaped to round spheres. They may occur singly or in groups, and they may be exposed, covered, or embedded in a substrate (Hinton, 1981). The shape of the host may influence the development of the parasitoid, for example *Trissolcus basalis* (scelionid) is barrel-shaped developing from a barrel-shaped host and *Telenomus tabanivorus* (scelionid) is long and thin developing from an elongated egg. If all other things were equal, whether these species could develop within the opposite shaped host is unknown. The occurrence of groups of eggs covered or embedded in a substrate has certainly been important to the evolution of ecto-oophagy and predation.

Eggs consist of a finite pool of energy and other required components necessary for life (Calow, 1983), which can vary due to size and age. Insect eggs occur in a variety of sizes ranging from over 6 mm in some Orthoptera to less than 0.5 mm in some Thysanoptera (Hinton, 1981), although the eggs of many parasitic species are an order of magnitude smaller. As a containerized resource, the size of the egg has a major influence on the evolution of oophagous Hymenoptera. Insect eggs are generally small and some of their oophages, such as mymarids and trichogrammatids, are the smallest insects known. For example, adults of the mymarid genus *Alaptus* measure less than 0.22 mm in length (Sweetman, 1963). Because the smallest insect species are not smaller than 0.2 mm in length, we may assume that eggs smaller than those in which *Alaptus* develop may not be susceptible to oophagous Hymenoptera. However, smaller egg sizes are almost always confined to species having no storage forms of resources. Such eggs depend upon incubation within a resource-rich media, such as a host. Thus, there may also be a limit to small egg size for free-living forms.

Egg size can vary within and between species (Hinton, 1981) and dictates the development of the insect's maximum size which can be translated into increased survival potential and fecundity (Capinera, 1979). Thus, female parasitoids emerging from small hosts, which do not provide optimal resources, are smaller, have fewer eggs and have a shorter life span than conspecific females emerging from larger hosts (Flanders, 1935; Salt, 1940; Klomp & Teerink, 1967; Marston & Ertle, 1973; Bigler *et al.*, 1987; Hoffmann *et al.*, 1988; Bai *et al.*, 1992). There is also an increased tendency to oviposit males in a smaller host (Walter, 1983), or to decrease their tendency to oviposit (Flanders, 1935; Klomp & Teerink, 1967; Salt, 1940; Strand & Vinson, 1982; Bin *et al.*, 1993).

As the size of the host egg increases, the solitary species either increase in size, fail to use the entire resource and die (Strand & Vinson, 1985), or decrease their tendency to oviposit (Strand & Vinson, 1983a; Bin *et al.*, 1993). Gregarious species have the added option of increasing the number of progeny allocated per host (Pu *et al.*, 1956; Waage & Ng, 1984; Schmidt & Smith, 1985, 1987; Bai *et al.*, 1992); however, for a given size of host egg as the number of parasitoids increases, the average size of the parasitoid decreases (Barber, 1937; Iyatomi, 1958; Klomp & Teerink, 1962, 1967; Suzuki *et al.*, 1984; Waage & Ng, 1984; Bigler *et al.*, 1987). This adjustment to the resource can also be extended to the parasitoids' response to host age. As observed by Ruberson *et al.* (1987), the size of a developing parasitoid declines when older host eggs are accepted.

Regulation

Instead of ensuring that only young hosts are located or accepted, or evolving to be more adaptable and conform to the resource, developing parasitoids sometimes seem to be able to increase the period of vulnerability of the host. As suggested by Strand (1986), selective forces directed towards the evolution of the ability to arrest host development should be strongest on parasitoids attacking rapidly developing hosts. To examine this hypothesis, Strand (1986) compared the duration of host embryogenesis to that proportion of host embryogenesis which is susceptible to parasitism using data on non-diapausing hosts. He used 60% or greater progeny survival as an indication of the maximum age that was successfully parasitized and found that a significant negative correlation existed between the duration of host development and the percentage of the embryonic developmental period that was susceptible to parasitism. As predicted, more rapidly developing hosts tended to be susceptible to parasitism over a greater percentage of their developmental period (Strand, 1986). Thus, for parasitoids attacking a resource that is only available for a short time, host embryonic arrestment appears to be a sound strategy.

Many early workers (Clausen, 1940; Salt, 1968) suggested that indirect stress induced by larval feeding damage was responsible for host development disruption, but as hypothesized by Strand (1986) feeding damage would not provide sufficient time for the parasitoid larvae to develop, if well-developed embryonic hosts were attacked.

Although it is often difficult to separate host pathologies that are due to indirect stress from those attributable to the parasitoid (Thompson, 1983), there is considerable evidence to show that larval parasitoids regulate the physiology of their host to their advantage (Vinson & Iwantsch, 1980b). Changes in host growth, development and immune response have been attributed to symbiotic viruses possessed by certain braconids, ichneumonids

and chalcids (Stoltz & Vinson, 1979; Rizki & Rizki, 1990; Flemming, 1992; Krell, 1992) or to venoms and other secretions (Piek *et al.*, 1974; Beard, 1978; Leluk *et al.*, 1989; Coudron *et al.*, 1990; Jones & Coudron, 1993). However, the physiological arrestment of hosts by oophages has only been investigated in two species.

The scelionid *Telenomus heliothidis* is a solitary oophage of several species of *Heliothis* (Strand & Vinson, 1983a) and can successfully parasitize *H. virescens* eggs as old as 63 h (9 h before the host embryo matured and hatched or 9 h before the parasitoid egg would be expected to hatch) (Strand *et al.*, 1986). In parasitized hosts prior to hatching of the parasitoid larvae, the internal structural integrity of the host egg was lost and by 24 h extensive degeneration of host tissue was observed (Strand, 1986). These changes in host tissues were found to be due to two factors. By interrupting oviposition just before an egg was deposited, Strand *et al.* (1983) reported that the host egg did not develop. The results suggested that the female injected a developmental arrestment factor which was determined to be released by cells located in the distal region of the common oviduct (Strand *et al.*, 1986).

Host tissue degeneration did not begin until the parasitoid larvae hatched. Similar observations led both Balduf (1926) and Sahad (1982) to attribute host tissue dissolution by scelionids and mymarids to the larvae. However, larvae are not a likely candidate for the effect, at least in the scelionid (Strand, 1986). Strand *et al.* (1985) reported that host tissue necrosis began shortly after teratocytes are released as the embryo hatches. Teratocytes, which are cells that constitute the embryonic membranes of the embryo, are primarily found in braconids, but also occur in scelionids (Volkoff *et al.*, 1992; Dahlman & Vinson, 1993). These teratocytes appear to release enzymes into the egg that result in decomposition (Strand *et al.*, 1988). Thus, *T. heliothidis* larvae basically develop on dead and partially digested host tissue.

Trichogramma pretiosum also parasitizes *H. virescens* eggs and can success-fully attack host eggs to within 2 h of hatching (Strand, 1986). Whether eggs are placed in the vitellophages, which is typical, or in the embryo, tissue degeneration begins immediately. By the time the parasitoid hatches, the host embryonic tissue is totally dissociated due to a poison gland factor injected during oviposition (Strand, 1986). Although Voegelé *et al.* (1974a) suggested that *Trichogramma brasiliensis* released one teratocyte, the observed cell is more likely a polar body (Strand, 1986). As a result of the poison gland factor(s), the hatching *Trichogramma* larvae also feed on predigested host material on which they rapidly grow.

A number of authors have demonstrated that oophages can successfully develop on host eggs that have been frozen or heated (Singh, 1969; Lewis & Young, 1972; Egwuatu & Taylor, 1977; Hu & Xu, 1988; Ma, 1988; Morrison, 1988), sterilized by chemicals (Young & Hamm, 1967; Lewis & Young, 1972), or sterilized by irradiation (Brénière, 1965; Barton & Stehr, 1970;

Bjegovic, 1972; Voegelé *et al.*, 1974b; Egwuatu & Taylor, 1977; Brower, 1982). These results demonstrate that many oophages, certainly the scelonids and trichogrammatids, basically feed in a container consisting of a nutrient broth, egg viability being unimportant. Recognition that oophages feed on a nutritional broth has stimulated attempts to rear these insects in artificial conditions (see Chapter 4 for a detailed review).

When reared on an artificial medium, several authors have observed that larval development proceeds, but the larvae fail to pupate (Xie *et al.*, 1986b; Takasu & Yagi, 1992). Nutritional factors may be important and Nettles (1990), based partly on the results of Irie *et al.* (1987), suggested that parasitoids may require some specific nutritional factors that they have lost the ability to synthesize, such factors being present in the hosts on which they develop. However, such a possibility does not mean that the parasitoids are specialists due to nutritional specialization. As suggested by House (1977), it is likely that parasitic species will require the same basic nutritional and accessory growth factors as other insects. By the same token, most insects probably contain and supply these same basic nutritional and accessary growth factors.

The rearing of parasitoids on factitious hosts (Fedde *et al.*, 1982) supports this view. However, rearing of a parasitoid on a factitious host is often not successful. But this lack of success may be due to non-nutritional factors (Strand & Vinson, 1985; Volkoff *et al.*, 1992). As suggested by Lee *et al.* (1988), the inability to rear some *Trichogramma* in some host species may be due to pH or water content rather than a nutritional problem.

There are many other factors that play an important role, such as excess diet (Strand & Vinson, 1985; Xie *et al.*, 1986a; Strand *et al.*, 1988; Masutti *et al.*, 1992). Masutti *et al.* (1992) were able to rear *Ooencyrtus pitzocampae* to the adult stage without insect haemolymph, but found that success depended on the careful manipulation of the last larval stage which died when their connection to their egg stalk broke. If any fluid remained around the larvae, it drowned. The third instar of *Edovum puttleri* were found to line the inside shell of their host egg with a salivary secretion and to open holes in the chorion with their oxidant mandibles (Colazza & Bin, 1992). These holes, which can number over a dozen per host, may aid in the exchange of air and regulate humidity (Colazza & Bin, 1992). Thus, proper timing with respect to access to air or the regulation of humidity are as important as the nutritional factors.

From the available evidence, the physiological interrelationship between developing oophages and their host appears less complex than those of the larval parasitoids which have been investigated (for example, see Pennacchio et al., 1994). Although the concept of host regulation can be extended to the Scelionidae where the venom stops embryogenesis, the dissolution of host tissues forming a broth within a container on which the parasitoid develops goes beyond the concept of host regulation. Thus, understanding the

physiological interaction between oophages and their hosts is not so much a study of the physiological interrelationship between both systems, but more an understanding of the physiological adaptations that the developing oophages have evolved to overcome the host's defences and to allow for their development within small, constrained, changing and fleeting resources.

Acknowledgements

Approved as TA 31162 by the Director of the Texas Agricultural Experiment Station. The author wishes to thank Dr Howard Williams for his help and suggestions and Ms Debbie Sennett for her persistence in typing.

References

Agrell, I.P.S. & Lundquist, A.M. (1973) Physiological and biochemical changes during insect development. In: Rockstein, M. (ed.) *The Physiology of Insecta.* Academic Press, New York, vol. 1, pp. 159–247.

Aldrich, J.R., Kochansky, J.P. & Abrams, C.B. (1984) Attractant for a beneficial insect and its parasitoids: pheromone of the predatory spined soldier bug, *Podisus maculiventris* (Hemiptera: Pentatomidae). *Environmental Entomology* 13, 1031–1036.

Askew, R.R. (1971) *Parasitic Insects.* American Elsevier Press, New York, 316 pp.

Askew, R.R. & Shaw, M.R. (1986) Parasitoid communities: their size, structure, and development. In: Waage, J.K. & Greathead, D. (eds) *Insect Parasitoids.* Academic Press, London, pp. 224–264.

Bai, B., Luck, R.F., Forster, L., Stephens, B. & Jansen, J.A.M. (1992) The effect of host size on quality attributes of the egg parasitoid, *Trichogramma pretiosum. Entomologia Experimentalis et Applicata* 64, 37–48.

Balduf, W.U. (1926) *Telenomus cosmopeplae* Gahan, an egg parasite of *Cosmopepla bimaculata* Thomas. *Journal of Economic Entomology* 19, 829–841.

Barber, G.W. (1937) Variation in populations and in size of adults of *Trichogramma minutum* Riley emerging from eggs of *Heliothis obsoleta* Fabr. *Annals of the Entomological Society of America* 30, 263–268.

Barton, L.C. & Stehr, F.W. (1970) Normal development of *Anaphes flavipes* in cereal leaf beetle eggs killed with X-radiation, and potential field use. *Journal of Economic Entomology* 63, 128–130.

Beard, R.L. (1978) Venoms of Braconidae. In: Bettini, S. (ed.) *Handbuch der experimentellen Pharmakologie.* Springer-Verlag, Berlin, vol. 48, pp. 773–800.

Beckage, N.E. (1985) Endocrine interactions between endoparasitic insects and their hosts. *Annual Review of Entomology* 30, 371–413.

Bergot, B.J., Baker, F.C., Cerf, G., Janieson, G. & Schooley, D.A. (1981) Qualitative and quantitative aspects of juvenile hormone titers in developing embryos of several insect species: discovery of a new JH-like substance extracted from eggs of *Manduca sexta.* In: Pratt, G.H. & Brookes, G.T. (eds) *Juvenile Hormone Biochemistry.* Elsevier Press, New York, pp. 33–45.

Bigler, F., Meyer, A. & Bosshart, S. (1987) Quality assessment in *Trichogramma maidis* Pintureau et Voegelé reared from eggs of the factitious hosts *Ephestia kuehniella* Zell. and *Sitotroga cerealella* (Oliver). *Journal of Applied Entomology* 104, 340–353.

Bin, F., Strand, M.R. & Vinson, S.B. (1988) Host relationships and associations in oophagous parasitic hymenoptera. In: Voegelé, J., Waage, J.K. & van Lenteren, J.C. (eds) Trichogramma *and Other Egg Parasitoids 2nd International Symposium. Les Colloques de l'INRA* 43, 153–154.

Bin, F., Vinson, S.B., Strand, M.R., Colazza, S. & Jones, W.A.Jr. (1993) Source of an egg kairomone for *Trissolcus basalis*: a parasitoid of *Nezara viridula. Physiological Entomology* 18, 7–15.

Bjegovic, P. (1972) Reproduction of *Ooencyrtus kuwanae* Howard in the killed gypsy moth eggs with radiation. *Zastita Bilja* 23, 3–6.

Brénière, J. (1965) Les Trichogrammes parasites de *Proceras sacchariphagus* Boj. borer de la canne à sucre à Madagascar. II. Etude biologique de *Trichogramma australicum* Gir. *Entomophaga* 10, 99–117.

Brower, J.H. (1982) Parasitization of irradiated eggs and eggs from irradiated adults of the indian meal moth (Lepidoptera: Pyralidae) by *Trichogramma pretiosum* (Hymenoptera: Trichogrammatidae). *Journal of Economic Entomology* 75, 939–944.

Buleza, V.B. & Mikheev, A.V. (1979) On the interactions of *Trissolcus grandis* and *T. simoni*, egg parasites of *Eurygaster integriceps. Zoology Zhurnal* 58, 54–60.

Burks, B.D. (1979) Trichogrammatidae. In: Krombein, K.V., Hurd, P.H., Smith, D.R. & Burks, B.D. (eds) *Catalog of Hymenoptera in America North of Mexico*. Smithsonian Institution Press, Washington, DC, pp. 1033–1041.

Calow, P.C. (1983) Energetics of reproduction and its evolutionary implications. *Linnian Society Biological Journal* 20, 153–165.

Capinera, J.L. (1979) Qualitative variation in plants and insects. Effects of propagule size on ecological plasticity. *American Naturalist* 114, 350–361.

Carlson, R.W. (1979) Ichneumonidae. In: Krombein, K.V., Hurd, P.H., Smith, D.R. & Burks, B.D. (eds) *Catalog of Hymenoptera in America North of Mexico*. Smithsonian Institution Press, Washington, DC, pp. 315–742.

Carrel, J.E., Thompson, W. & MacLaughlin, M. (1975) Parental transmission of a defensive chemical (cantharidin) in blister beetles. *America Society of Zoology* 13, 1258.

Clausen, C.P. (1940) *Entomophagous Insects*. Hafner Press, New York, 688 pp.

Clausen, C.P. (1976) Phoresy among entomophagous insects. *Annual Review of Entomology* 21, 343–368.

Clements, A.N. (1951) On the urticating properties of adult Lymantriidae. *Proceedings of the Royal Entomological Society, London* 26, 104–108.

Colazza, S. & Bin, F. (1992) Introduction of the oophage *Edovum puttleri* Griss (Hymenoptera: Eulophidae) in Italy for the biological control of Colorado potato beetle. *Redia* 75, 203–225.

Coudron, T.A. (1991) Host-regulating factors associated with parasitic hymenoptera. In: Hedin, P.A. (ed.) *Naturally Occurring Pest Bioregulators*. American Chemical Society Symposium, Series No. 449, pp. 41–65.

Coudron, T.A., Kelley, T.J. & Puttler, B. (1990) Developmental responses of *Trichoplusia ni* (Lepidoptera: Noctuidae) to parasitism by the ectoparasite *Euplectrus*

plathypenae (Hymenoptera: Eulophidae). *Archives of Insect Biochemistry and Physiology* 13, 83–94.

Dahlman, D.L. & Vinson, S.B. (1993) Teratocytes: developmental and biochemical characteristics. In: Beckage, W.E., Thompson, S.N. & Federici, B.A. (eds) *Parasites and Pathogens of Insects*. Academic Press, New York, vol. 1, pp. 145–165.

DeBach, P. & Rosen, D. (1991) *Biological Control by Natural Enemies*. Cambridge University Press, Cambridge, 440 pp.

Dover, B.A., Davies, D.H. & Vinson, S.B. (1988) Degeneration of last instar *Heliothis virescens* prothoracic glands by *Campoletis sonorensis* polydnavirus. *Journal of Invertebrate Pathology* 51, 80–91.

Egwuatu, R.I. & Taylor, T.A. (1977) Development of *Gryon gnidus* (Nixon) (Hymenoptera: Scelionidae) in eggs of *Acanthomia tomentosicollis* (Stal) (Hemiptera: Coreidae) killed either by gamma irradiation or by freezing. *Bulletin of Entomological Research* 67, 31–33.

Elsey, K.D. (1972) Defenses of eggs of *Manduca sexta* against predation by *Jalysus spinosus*. *Annals of the Entomological Society of America* 65, 896–897.

Eww, J. von, Fishelson, L., Parsons, J.A., Reichstein, T. & Rothschild, M. (1967) Cardenolides (heart poisons) in a grasshopper feeding on milkweeds. *Nature* 214, 35–39.

Fedde, V.H., Fedde, G.F. & Drooz, A.T. (1982) Factitious hosts in insect parasitoid rearings. *Entomophaga* 27, 379–386.

Fisher, R.C. (1971) Aspects of the physiology of endoparasitic Hymenoptera. *Biological Reviews* 46, 243–278.

Flanders, S.E. (1935) Host influence on the prolificacy and size of *Trichogramma*. *Pan-Pacific Entomologist* 11, 175–177.

Flemming, J.G.W. (1992) Polydnaviruses: mutualists and pathogens. *Annual Review of Entomology* 37, 401–425.

Gharib, B. & de Reggi, M. (1983) Changes in ecdysteroid and juvenile hormone levels in developing eggs of *Bombyx mori*. *Journal of Insect Physiology* 29, 871–876.

Gharib, B., de Reggi, M., Connat, J. & Chaix, J. (1983) Ecdysteroid and juvenile hormone changes in *Bombyx mori* eggs related to the initiation of diapause. *Federation of the European Biochemistry Society* 160, 119–123.

Gower, A.M. (1967) A study of *Limnephilus lunatus* Curtis (Trichoptera: Limnephilidae) with reference to its life cycle in watercress beds. *Transactions of the Royal Entomology Society, London* 119, 283–302.

Greany, P.D. (1986) *In vitro* culture of hymenopterous larval parasitoids. *Journal of Insect Physiology* 32, 409–420.

Greany, P.D., Vinson, S.B. & Lewis, W.J. (1984) Insect parasitoids: finding new opportunities for biological control. *Bioscience* 34, 690–696.

Halselbarth, E. (1979) Zur Parasitierung der Puppen von Forleule (*Panolis flammea* Schiff.), Kiefernspanner (*Bupalus piniarius* L.) and Heidelbeerspanner (*Boarmia bistortana* Goezc) in bayerischen Kiefernwäldern. *Zeischrift für Angewandte Entomologie* 87, 186–202; 311–333.

Hinton, H.E. (1981) *Biology of Insect Eggs*. Pergamon Press, New York, vols 1–3, 1125 pp.

Hodson, A.C. & Weinman, C.J. (1945) Factors affecting recovery from diapause and hatching of eggs of the forest tent caterpillar *Malacosoma disstria* Hbn. *Technical Bulletin, Minnesota Agricultural Experiment Station* 170, 1–31.

Hoffmann, J.A. & Lagueux, M. (1985) Endocrine aspects of embryonic development in insects. In: Kerkut, G.A. & Gilbert, L.I. (eds) *Comprehensive Insect Physiology, Biochemistry, and Pharmacology.* Pergamon Press, New York, vol. 1, pp. 435–460.

Hoffmann, C.L., Luck, R.F. & Oatman, E.R. (1988) A comparison of longevity and fecundity of adult *Trichogramma platneri* (Hymenoptera: Trichogrammatidae) reared from the eggs of the cabbage looper and the angiomas grain moth, with and without access to honey. *Journal of Economic Entomology* 81, 1307–1312.

House, H.L. (1977) Nutrition of natural enemies. In: Ridgway, R.L. & Vinson, S.B. (eds) *Biological Control by Augmentation of Natural Enemies.* Plenum Press, New York, pp. 151–182.

Hu, Z. & Xu, Q. (1988) Studies on frozen storage of eggs of rice moth and oak silkworm. In: Voegelé, J., Waage, J.K. & van Lenteren, J.C. (eds) Trichogramma *and Other Egg Parasitoids. 2nd International Symposium. Les Colloques de l'INRA* 43, 327–338.

Irie, K., Xie, Z.N., Nettles, W.C., Morrison, R.K., Chen, A.C., Holman, G.M. & Vinson, S.B. (1987) The partial purification of a *Trichogramma pretiosum* pupation factor from hemolymph of *Manduca sexta. Insect Biochemistry* 17, 269–275.

Iyatomi, K. (1958) Effect of superparasitism on reproduction of *Trichogramma japonicum* Ashmead. In *Proceedings of the 10th International Congress on Entomology, Montréal, Canada, August 17–25, 1956.* Mortimer Press, Ottawa, pp. 897–900.

Jones, D. (1986) Use of parasite regulation of host endocrinology to enhance the potential of biological control. *Entomophaga* 31, 153–157.

Jones, D. & Coudron, T. (1993) Venoms of parasitic Hymenoptera as investigatory tools. In: Beckage, W.E., Thompson, S.N. & Federici, B.A. (eds) *Parasites and Pathogens of Insects.* Academic Press, New York, vol. 1, pp. 227–249.

Jones, D.A., Parsons, J. & Rothschild, M. (1962) Release of hydrocyanic acid from crushed tissues of all stages in the life-cycle of species of Zygaeninae (Lepidoptera). *Nature* 193, 52–53.

Klomp, H. & Teerink, B.J. (1962) Host selection and number of eggs per oviposition in the egg parasite *Trichogramma embryophagum* Htg. *Nature* 195, 1020–1021.

Klomp, H. & Teerink, B.J. (1967) The significance of oviposition rates in the egg parasite, *Trichogramma embryophagum* Htg. *Archives Neerland Zoology* 17, 350–375.

Krell, P.J. (1992) Polydnaviridae. In: Adams, J.R. & Bonami, J.R. (eds) *Atlas of Invertebrate Viruses.* CRC Press, Boca Raton, Florida, pp. 321–338.

Krombein, K.V., Hurd, P.H., Smith, D.R. & Burks, B.D. (eds) (1979) *Catalog of Hymenoptera in America North of Mexico.* Smithsonian Institution Press, Washington, DC, vol. 1, 1198 pp.

Lawrence, P.O. (1986) Host–parasite hormonal interactions: an overview. *Journal of Insect Physiology* 32, 295–298.

Lawrence, P.O. (1990) The biochemical and physiological effects of insect hosts on the development and ecology of their insect parasites: an overview. *Archives of Insect Biochemistry and Physiology* 13, 217–228.

Lee, K.Q., Jiang, F.L. & Guo, J.J. (1988) Preliminary study on the reproductive behavior of the parasitic wasps (*Trichogramma*). In: Voegelé, J., Waage, J.K. & van Lenteren, J.C. (eds) Trichogramma *and Other Egg Parasitoids. 2nd International Symposium. Les Colloques de l'INRA* 43, 215–219.

Leibee, G.L., Pass, B.C. & Yeargan, K.V. (1979) Developmental rates of *Patasson lameerei* (Hym.: Mymaridae) and the effect of host egg age on parasitism. *Entomophaga* 24, 345–348.

Leluk, J., Schmidt, J. & Jones, D. (1989) Comparative studies on the protein composition of Hymenopteran venom reservoirs. *Toxicon* 27, 105–114.

Levine, E. & Chandler, L. (1976) Biology of *Bellura gortynoides* (Lepidoptera: Noctuidae), a yellow water lily borer, in Indiana. *Annals of the Entomological Society of America* 69, 405–414.

Lewis, W.J. & Redlinger, L.M. (1969) Suitability of eggs of the almond moth, *Cadra cautela*, of various ages for parasitism by *Trichogramma evanescens*. *Annals of the Entomology Society of America* 62, 1482–1484.

Lewis, W.J. & Young, J.R. (1972) Parasitism by *Trichogramma evanescens* of eggs from tepa-sterilized and normal *Heliothis zea*. *Journal of Economic Entomology* 65, 705–708.

Ma, H.Y. (1988) Studies on long-term storage of hosts for propagating *Trichogramma*. In: Voegelé, J., Waage, J.K. & van Lenteren, J.C. (eds) Trichogramma *and Other Egg Parasitoids. 2nd International Symposium. Les Colloques de l'INRA* 43, 369–371.

Margaritis, L.H. (1985) Structure and physiology of the eggshell. In: Kerkut, G.A. & Gilbert, L.I. (eds) *Comprehensive Insect Physiology, Biochemistry, and Pharmacology*. Pergamon Press, New York, pp. 153–230.

Marston, E. & Ertle, L.R. (1973) Host influence of the bionomics of *Trichogramma minutum*. *Annals of the Entomological Society of America* 66, 1155–1162.

Masutti, L., Battisti, A., Milani, N. & Zanata, M. (1992) First success in the *in vitro* rearing of *Ooencyrtus pitzocampae* (Mercet) (Hym: Encyrtidae). Preliminary note. *Redia* 75, 222–232.

Mattiacci, L., Vinson, S.B., Williams, H.J., Aldrich, J.R. & Bin, F. (1993) A long-range attractant kairomone for egg parasitoid *Trissolcus basalis*, isolated from defensive secretion of its host, *Nezara viridula*. *Journal of Chemical Ecology* 19, 1165–1179.

Mellini, E. (1986) Importanza dell'eta dell'novo, al momento della parassitizzazions, per la biologia degli Imenotteri oofagi. *Bollettino dell'Istituto di Entomologia 'Guido Grandi' della Universita di Bologna* 41, 1–21

Morrison, R.K. (1988) Methods for the long-term storage and utilization of eggs of *Setolioga cerealella* (Olivers) for production of *Trichogramma pretiosum* Riley. In: Voegelé, J., Waage, J.K. & van Lenteren, J.C. (eds) Trichogramma *and Other Egg Parasitoids. 2nd International Symposium. Les Colloques de l'INRA* 43, 373–377.

Nettles, W.C. Jr (1990) *In vitro* rearing of parasitoids: role of host factors in nutrition. *Archives of Insect Biochemistry and Physiology* 13, 167–175.

Pak, G.A., Buis, H.C.E.M., Heck, I.C.C. & Hermans, M.L.G. (1986) Behavioral variations among strains of *Trichogramma* spp.: host-age selection. *Entomologia Experimentalis et Applicata* 40, 247–258.

Pennacchio, F., Vinson, S.B., Tremblay, E. & Tanaka, T. (1994) Biochemical and developmental alterations of *Heliothis virescens* (F.) (Lepidoptera: Noctuidae) larvae induced by the endophagous parasitoid *Cardiochiles nigriceps* (Vierick) (Hymenoptera: Braconidae). *Archives of Insect Biochemistry and Physiology* (in press).

Piek, T., Spanjer, W., Njio, K.D., Veenendaal, R.L. & Mantel, P. (1974) Paralysis caused by the venom of the wasp, *Microbracon gelechiae*. *Journal of Insect Physiology* 20, 2307–2319.

Pu, C., Tehai, T., Dheng, L.C., Hung, S.C. & Yu-shih, M. (1956) On the rearing of *Trichogramma evanescens* Westw. and its utilization for control of sugar cane borers. *Acta Entomology Sinica* 6, 1–36.

Rizki, R.M. & Rizki, T.M. (1990) Parasitoid virus-like particles destroy *Drosophila* cellular immunity. *Proceedings National Academy of Science of the USA* 87, 8388–8392.

Roonwal, M.L. (1954) Structure of the egg-masses and their hairs in some species of *Lymantria* of importance to forestry (Insecta: Lepidoptera: Lymantriidae). *Indian Forest Records, Entomology* 8, 265–276.

Ruberson, J.R., Tauber, M.J. & Tauber, C.A. (1987) Biotypes of *Edovum puttleri* (Hymenoptera: Eulophidae) responses to developing eggs of the Colorado potato beetle (Coleoptera: Chrysomelidae). *Annals of the Entomological Society of America* 80, 451–455.

Sahad, K.A. (1982) Biology and morphology of *Gonatocerus* spp. (Hymenoptera, Mymaridae), an egg parasitoid of the green rice leafhopper, *Nephotettix cincticeps* Uhler (Homoptera: Deltocephalidae) I. Biology. *Kontyu, Tokyo* 50, 246–260.

Salt, G. (1940) Experimental studies in insect parasitism. VII. The effects of different hosts on the parasite *Trichogramma evanescens* Westw. (Hym. Chalcidoidea). *Proceedings of the Royal Entomological Society, London* 15, 81–95.

Salt, G. (1968) The resistance of insect parasitoids to the defense reactions of their hosts. *Biological Reviews* 43, 200–223.

Salt, G. (1970) *The Cellular Defence Reaction of Insects.* Cambridge University Press, Cambridge, 118 pp.

Sander, K., Gutzeit, H.O. & Jäckle, H. (1985) Insect embryogenesis: morphology, physiology, genetical and molecular aspects. In: Kerkut, G.A. & Gilbert, L.I. (eds) *Comprehensive Insect Physiology, Biochemistry, and Pharmacology.* Pergamon Press, New York, pp. 319–385.

Scalia, S. & Morgan, E.D. (1982) A reinvestigation on the ecdysteroids during embryogenesis in the desert locust *Schistocerca gregaria. Journal of Insect Physiology* 28, 647–654.

Schmidt, J.M. & Smith, J.J.B. (1985) The mechanism by which the parasitoid wasp *Trichogramma minutum* responds to host clusters. *Entomologia Experimentalis et Applicata* 39, 287–294.

Schmidt, J.M. & Smith, J.J.B. (1986) Correlations between body angles and substrate curvature in the parasitoid wasp *Trichogramma minutum*: a possible mechanism of host radius measurement. *Journal of Experimental Biology* 125, 271–285.

Schmidt, J.M. & Smith, J.J.B. (1987) The measurement of exposed host volume by the parasitoid wasp *Trichogramma minutum* and the effects of wasp size. *Canadian Journal of Zoology* 65, 2837–2845.

Singh, R.P. (1969) A simple technique for rendering host eggs inviable for the laboratory rearing of *Trichogramma* spp. *Indian Journal of Entomology* 31, 83–84.

Stoltz, D.B. (1986) Interactions between parasitoid-derived products and host insects: an overview. *Journal of Insect Physiology* 32, 347–350.

Stoltz, D.B. & Vinson, S.B. (1979) Viruses and parasitism in insects. *Advances in Virus Research* 24, 125–171.

Strand, M.R. (1986) The physiological interactions of parasitoids with their hosts and their influence on reproductive strategies. In: Waage, J.K. & Greathead, D. (eds) *Insect Parasitoids.* Academic Press, London, pp. 97–136.

Strand, M.R. & Vinson, S.B. (1982) Source and characterization of an egg recognition kairomone of *Telenomus heliothidis*, a parasitoid of *Heliothis virescens*. *Physiological Entomology* 7, 83–90.

Strand, M.R. & Vinson, S.B. (1983a) Factors affecting host recognition and acceptance in the egg parasitoid *Telenomus heliothidis* (Hymenoptera: Scelionidae). *Environmental Entomology* 12, 1114–1119.

Strand, M.R. & Vinson, S.B. (1983b) Analysis of an egg recognition kairomone of *Telenomus heliothidis* (Hymenoptera: Scelionidae). *Journal of Chemical Ecology* 9, 423–432.

Strand, M.R. & Vinson, S.B. (1985) *In vitro* culture of *Trichogramma pretiosum* on an artificial medium. *Entomologia Experimentalis et Applicata* 39, 203–209.

Strand, M.R., Ratner, S. & Vinson, S.B. (1983) Maternally induced host regulation by the egg parasitoid *Telenomus heliothidis*. *Physiological Entomology* 8, 469–475.

Strand, M.R., Quarles, J.M., Meola, S.M. & Vinson, S.B. (1985) Cultivation of teratocytes of the egg parasitoid *Telenomus heliothidis* (Hymenoptera: Scelionidae). In vitro *Cellular and Developmental Biology* 21, 361–367.

Strand, M.R., Meola, S.M. & Vinson, S.B. (1986) Correlating pathological symptoms in *Heliothis virescens* eggs with development of the parasitoid *Telenomus heliothidis*. *Journal of Insect Physiology* 32, 389–402.

Strand, M.R., Vinson, S.B., Nettles, W.C. & Xie, Z.N. (1988) *In vitro* culture of the egg parasitoid *Telenomus heliothidis*. *Entomologia Experimentalis et Applicata* 46, 71–78.

Suzuki, Y., Tsuji, H. & Sasakawa, M. (1984) Sex allocation and effects of superparasitism on secondary sex ratios in the gregarious parasitoid, *Trichogramma chilonis* (Hymenoptera: Trichogrammatidae). *Animal Behavior* 32, 478–484.

Sweetman, H.L. (1963) *Principles of Biological Control*. Wm. C. Brown Co., Dubridge, Iowa, 560 pp.

Takasu, K. & Yagi, S. (1992) *In vitro* rearing of the egg parasitoid, *Ooencyrtus nezarae* Ishii (Hymenoptera: Encyrtidae). *Applied Entomology and Zoology* 27, 171–173.

Tanaka, T. & Vinson, S.B. (1991) Depression of prothoracic gland activity of *Heliothis virescens* by venom and calyx fluids from the parasitoid *Cardiochiles nigriceps*. *Journal of Insect Physiology* 37, 139–144.

Taylor, T.A. & Stern, V.M. (1971) Host-preference studies with the egg parasite *Trichogramma semifumatum* (Hymenoptera: Trichogrammatidae). *Annals of the Entomological Society of America* 64, 1381–1390.

Thompson, S.N. (1983) Biochemical and physiological effects of metazoan endoparasites on their host species. *Comparative Biochemistry and Physiology* 74, 183–211.

Thompson, S.N. (1986) Nutrition and *in vitro* culture of insect parasitoids. *Annual Review of Entomology* 31, 197–220.

Vinson, S.B. (1986) The role of behavioral chemicals for biological control. In: Franz, J.M. (ed.) *Biological Plant and Health Protection*. Fortschritte Zoologie 32. G. Fischer Verlag, Stuttgart, pp. 75–87

Vinson, S.B. (1988a) Physiological studies of parasitoids reveal new approaches to the biological control of insect pests. *IST Atlas Science. Animal and Plant Science* 1988, 25–32.

Vinson, S.B. (1988b) Comparison of host characteristics that elicit host recognition behavior of parasitoid hymenoptera. In: Gupta, V. (ed.) *Advances in Parasitic*

Hymenoptera Research. E. J. Brill Publishers, Amsterdam, pp. 285–291.

Vinson, S.B. (1990a) Physiological interactions between the host genus *Heliothis* and its guild of parasitoids. *Archives of Insect Biochemistry and Physiology* 13, 63–81.

Vinson, S.B. (1990b) How parasitoids deal with the immune system of their host: an overview. *Archives of Insect Biochemistry and Physiology* 13, 3–27.

Vinson, S.B. & Barbosa, P. (1987) Interrelationship of nutritional ecology of parasitoids. In: Stansky, F.Jr. & Rodriquez, J.G. (eds) *Nutritional Ecology of Insects, Mites and Spiders.* John Wiley & Sons, New York, pp. 673–693.

Vinson, S.B. & Iwantsch, G.F. (1980a) Host suitability for insect parasitoids. *Annual Review of Entomology* 25, 397–419.

Vinson, S.B. & Iwantsch, G.F. (1980b) Host regulation by insect parasitoids. *Quarterly Review of Biology* 55, 143–165.

Vinson, S.B. & Piper, G.L. (1986) Source and characterization of host recognition kairomones of *Tetrastichus hagenowii,* a parasitoid of cockroach eggs. *Physiological Entomology* 11, 459–468.

Voegelé, J., Brun, P. & Daumal, J. (1974a) Modalités de la prise de possession et de l'élimination de l'hote chez le parasite embryonnaire *Trichogramma brasiliensis.* *Annales de la Societé Entomologique de France* 10, 757–762.

Voegelé, J., Daumal, J., Brun, P. & Onillon, J. (1974b) Action du traitement au froid et aux ultraviolets de l'oeuf d'*Ephestia kuehniella* (Pyralidae) sur le taux de multiplication de *Trichogramma evanescens* et *T. brasiliensis* (Hymenoptera: Trichogrammatidae). *Entomophaga* 19, 341–348.

Volkoff, A.N., Vinson, S.B., Wu, Z.W. & Nettles, W.C. Jr (1992) *In vitro* rearing of *Trissolcus basalis* (Hymenoptera: Scelionidae) an egg parasitoid of *Nezara viridula* (Heteroptera: Pentatomidae). *Entomophaga* 37, 141–148.

Waage, J.K. & Ng, S.M. (1984) The reproductive strategy of a parasitic wasp. I. Optimal progeny and sex allocation in *Trichogramma evanescens. Journal of Animal Ecology* 53, 401–415.

Walter, G.N. (1983) Differences in host relationships between male and female heteronomous parasitoids: a review of host location, oviposition, and pre-imaginal physiology and morphology. *Journal of the Entomological Society of South Africa* 46, 261–282.

Xie, Z.N., Nettles, W.C. Jr., Morrison, R.K., Irie, K. & Vinson, S.B. (1986a) Three methods for the *in vitro* culture of *Trichogramma pretiosum* Riley. *Entomological Science* 21, 133–138.

Xie, Z.N., Nettles, W.C. Jr., Morrison, R.K., Irie, K. & Vinson, S.B. (1986b) Effect of ovipositional stimulants and diets on the growth and development of *Trichogramma pretiosum in vitro. Entomologia Experimentalis et Applicata* 42, 119–124.

Yamashita, O. (1983) Egg diapause. In: Downer, R.G.H. & Laufer, H. (eds) *Invertebrate Endocrinology, vol. 1: Endocrinology of Insects.* Allan R. Liss, New York, pp. 337–342.

Young, J.R. & Hamm, J.J. (1967) Reproduction of *Trichogramma fasciatum* in eggs from tepa-sterilized fall armyworms. *Journal of Economic Entomology* 60, 723–724.

Overwintering Strategies of Egg Parasitoids

11

Guy Boivin

Research Station, Agriculture Canada, 430 Boul. Gouin,
Saint-Jean-sur-Richelieu, Québec, Canada J3B 3E6

Abstract

Most egg parasitoids overwinter as dormant immatures inside their host egg. For these species, winter survival is complicated by the fact that they have little opportunity to choose their overwintering sites. Therefore, their survival is almost entirely based on physiological adaptation. The dormant parasitoids are either in quiescence, where development is halted during the adverse conditions, or diapause, a more profound interruption of development in response to a cue normally preceding adverse conditions. These cues are generally photoperiod, temperature or a combination of both. The conditions inducing both quiescence and diapause are of interest for egg parasitoids because the control of dormancy could increase the efficiency of mass-rearing by increasing the fecundity of females or storing mass-produced insects until needed. To survive at low temperatures during winter, the dormant parasitoids need special adaptations. They can avoid the coldest locations either by selecting their overwintering hosts or by increasing their cold hardiness. Increases in cold hardiness have been related in egg parasitoids to increases in cryoprotectants that prevent ice formation in the parasitized egg. Such cryoprotectant synthesis also responds to photoperiodic and temperature cues. The ability to overwinter successfully is an important characteristic of egg parasitoids that influences their value as biological control agents. The selection of exotic parasitoids for introduction must be based on their capacity to overwinter and the presence of suitable overwintering hosts.

Introduction

In temperate and northern climates and in alpine areas, the ability of arthropods to survive at low temperatures in winter is one of the major factors influencing both their intergenerational population dynamics and their geographical distribution. Severe winters, characterized either by low temperatures or wide variation of temperatures, cause high mortality, resulting in greatly reduced spring populations and modifying distribution within a habitat. In agroecosystems, the combined effect of winter mortality of both pests and their natural enemies influences the potential for crop damage.

Egg parasitoids face particular problems of winter survival. Most of these parasitoids overwinter as immature stages in their host eggs and the possibility of behavioural adaptations to avoid extreme temperatures is limited to the choice of a suitable host by females ovipositing in the autumn. Specific habitat selection or migration to protected habitats, behaviour widely used by overwintering insects, is not available to these species. The majority of egg parasitoids parasitize eggs oviposited above the soil level where no protection against the cold is available, apart from snow cover and vegetation. They must rely on physiological adaptations to survive the cold season and be synchronized with a highly specific and ephemeral resource, their host (Tauber *et al.*, 1983). Interseasonal synchronization is especially important for parasitoids that are species-specific, require a host at a particular stage, and prefer a certain developmental age of the egg for oviposition (Pak *et al.*, 1986, Reznik & Umarova, 1990). In addition, most egg parasitoids are short-lived and therefore active for only a brief period, leaving little room for imprecision in the synchronization of their life cycle with that of their hosts.

Overwintering adaptations comprise dormancy, when the abiotic conditions are too difficult or when a resource is scarce, regulation of developmental rates, use of alternative hosts and the ability to switch to a different trophic level (Tauber *et al.*, 1983). Some parasitoids can control their developmental rates, such as the braconid *Chelonus curvimaculatus* that oviposits in the eggs of several lepidopterans and develops in the egg, larva and pupa of its host. It can modify its developmental rate according to the life cycle of the available host (Broodryk, 1969). The overwintering capabilities of egg parasitoids vary greatly at both the interspecific and intraspecific levels. In Trichogrammatidae, within the genus *Trichogramma*, some species are not adapted to cold conditions (*T. achaeae*, *T. pintoi*, *T. perkinsi* and *T. semifumatum*), others experience considerable winter mortality (*T. embryophagum* and *T. daumalae*) and others are relatively cold hardy (*T. maidis* (= *T. brassicae*), *T. oleae*, *T. evanescens*, *T. dendrolimi*, *T. rhenana*, *T. buesi*, *T. principium* and *T. semblidis*) (Voegelé *et al.*, 1988).

The most common physiological adaptation used by insects to survive winter is dormancy that comprises quiescence and diapause (Danks, 1987). Quiescence is an immediate direct response to a limiting factor, such as low temperature, while diapause is a more profound interruption of development that routes the metabolic programme of the organism away from direct developmental pathways and normally precedes the advent of adverse conditions (Danks, 1987). Egg parasitoids use both tactics against low temperature.

Dormancy in Egg Parasitoids

Much of the research done on the dormancy of egg parasitoids has been on the genus *Trichogramma* because of its value in biological control. Control of quiescence or diapause could increase the efficiency of mass-rearing and field use. Possible advantages include an increase of the fecundity of the F_1, the ability to store a full year of mass-production in diapause, greater ease of use in the field, the possibility of quality control on the emergent population, conditioning of the produced parasitoids for commercial release, greater ease of distribution of the conditioned parasitoids, greater genetic stability of the rearing because a stock population can be maintained in diapause and then used to start rearing batches, and finally the possibility of better synchronization between the pest and the parasitoid (Voegelé *et al.*, 1986). Some research results are already being applied in the commercial production of egg parasitoids, as in *T. maidis* that is put in diapause by controlling the photoperiod and temperature and then maintained at $-6°C$ for a year (Voegelé, 1976). Although most egg parasitoids do enter dormancy, some species are continuously active throughout the year, especially in the southern part of their distribution as in the mymarid *Anagrus giraulti*, a parasitoid of the eggs of leafhoppers in California (Meyerdirk & Moratorio, 1987), and *Anaphes ovijentatus* a parasitoid of the eggs of *Lygus hesperus* in the southwestern United States (Jackson, 1987).

Diapause in egg parasitoids

Diapause is a cessation of activity that is actively induced and happens at a generally species-specific point in the insect life cycle. The onset of diapause is decided by environmental factors, called token stimuli, that are not adverse *per se* but predict unfavourable conditions (Saunders, 1982). Such conditioning is essential for diapause induction and survival through winter as for *Trichogramma nubilale* in Delaware (Burbutis *et al.*, 1976). There are three types of diapause. Parapause is an obligatory diapause whose induction is entirely determined genetically and not subject to environmental modifications and its termination is controlled by an environmental factor, either photoperiod or temperature. This type of diapause is generally found in univoltine insects. Eudiapause is a facultative diapause induced by photoperiod and terminated only after a rather long period of low temperatures.

Oligopause is also a facultative diapause induced by seasonal changes in either photoperiod or temperature and terminated by a reversal of the inducing factors (Saunders, 1982).

Diapause enables insects to avoid periods when living conditions are unacceptable because of low or high temperatures, scarcity of resources or, as in the case of parasitoids, low host population density. In the diapause, the insects entered diapause as eggs (10%), larvae (82%), pupae (4%) or adults (4%) (Danks, 1987). In egg parasitoids, diapause generally occurs inside the host egg either in the mature larval or prepupal stage, as in the eupelmids *Anastatus disparis* (Sullivan *et al.*, 1977) and *Mesocomys pulchriceps* (van den Berg, 1971), the encyrtid *Ooencyrtus pityocampae* (Battisti *et al.*, 1990), the mymarid *Caraphractus cinctus* (Jackson, 1963) and most *Trichogramma* species (Voegelé, 1976; Zaslavski & Umarova, 1982, 1990; Voegelé *et al.*, 1986; Sorokina & Maslennikova, 1988) or as pupae as the trichogrammatids *Aphelinoidea plutella* and *Abbella subflava* (Henderson, 1955). Some species of egg parasitoids overwinter as adults but in quiescence. The mymarid *Anagrus epos* parasitizes the eggs of leafhoppers in British Columbia and the time of oviposition by the autumn generation determines when the eggs will hatch in the spring. Eggs laid early in autumn develop to the pupal stage and emerge in May while those laid later do not start developing until the spring and do not produce adults until June (McKenzie & Beirne, 1972).

Many insects use photoperiod as a cue for timing the induction of diapause. At a given location, the photoperiod is a reliable and precise indication of the period of the year and, therefore, the seasonal biology of several temperate insect species is timed so that two or more generations are produced annually, and the last generation uses this cue to prepare for winter. Four types of diapause induction curves reflect the percentage incidence of diapause for a given photoperiod (Beck, 1980). In Type I, also known as a long-day response, long daylength favours continuous development while short daylength triggers diapause (Fig. 11.1). At a low ambient temperature, the curve will be shifted to the right, the insect then entering diapause at a longer photoperiod. In the Type II response, also called short-day response, insects enter diapause at long daylength but develop without interruption at a short photoperiod. The Type III response has no critical daylength as in responses of Types I and II. In this short-day–long-day response, no diapause is observed at either short or long daylength and only intermediate photo-periods will induce a complete diapause. Finally, the Type IV response is rare among insect species. This response is characterized by the absence of diapause only in a restricted range of rather long photoperiods (Fig. 11.1). When daylength is shorter or longer than this critical value, the insect enters diapause (Beck, 1980).

Most insects rely on abiotic diapause-inducing signals and, for para-sitoids, photoperiod is the most common diapause-inducing stimulus (Tauber *et al.*, 1986). Photoperiodic influence on diapause has been demonstrated in

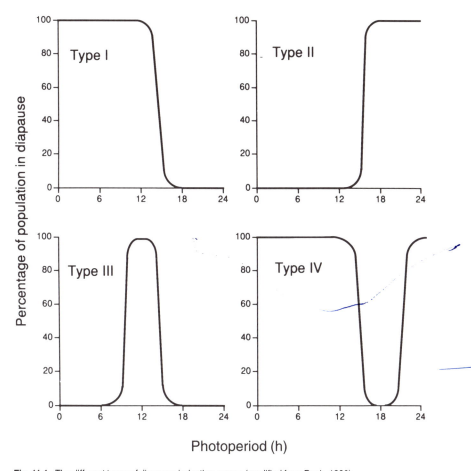

Fig. 11.1. The different types of diapause induction curves (modified from Beck, 1980).

about 325 insect species, among them 30 hymenopterous parasitoids (Saunders, 1982). In egg parasitoids, several species are also known to respond to either temperature or other signals. Both temperature and photoperiod can influence diapause induction when they act on different sensitive stages of an insect. In *Trichogramma evanescens*, photoperiod acts mostly on the adults while temperature is the most important factor for the developing larva and, together, these two factors determine the incidence of diapause (Zaslavsky & Umarova, 1982). Even within this genus, there is considerable interspecific and intraspecific variation in the expression of diapause (Voegelé *et al.*, 1988). Dormancy ranges from an intense diapause controlled by photoperiod to a less intense diapause controlled mainly by

temperature (Tauber *et al.*, 1983), and some species and strains of *Tricho-gramma* do not enter diapause (Oatman & Platner; 1972; Keller, 1986; Voegelé *et al.*, 1986). Finally, *T. rhenana* and *T. semblidis* are in quiescence in October and November and enter a true diapause only after that period (Voegelé *et al.*, 1986).

Response to photoperiod

Two types of photoperiodic responses have been shown in parasitoids, Type I and Type IV (Tauber *et al.*, 1986; Brown & Phillips, 1990). Type I response, where the insect enters diapause under short photoperiod, has been observed in *C. cinctus*, a species that parasitizes eggs of Dytiscidae. Diapause occurs when the fully grown larva is exposed to less than 12 h of light (Jackson, 1963). Such a response was also found in *M. pulchriceps*, an egg parasitoid of emperor moths that enters diapause at photoperiods of less than 12.5 h (van den Berg, 1971), *Ooencyrtus ennomus* with a critical photoperiod of 14–15 h (Anderson & Kaya, 1974) and *T. evanescens* (Bonnemaison, 1972; Zaslavski & Umarova, 1982) where, in addition, a strong maternal effect was noted.

Ooencyrtus ennomus overwinters as a late-stage larva in diapause within its host egg, the elm spanworm, *Ennomos subsignarius* (Anderson & Kaya, 1975). Diapause is induced at short and at long photoperiods, but development continues when the photoperiod is between 16 and 22 h day^{-1} (Anderson & Kaya, 1974). This is the only example of a Type IV response known outside the Lepidoptera. The diapause is thus facultative and is induced by photoperiod in the maternal generation (see later).

A response that appears to be a Type III response has recently been described in *T. pintoi* but in this case the photoperiod experienced by the parental generation influenced the progeny generation. At long and at short photoperiods only about 30% of the individuals entered diapause whereas up to 80% of the progeny generation from parents exposed to between 8 and 14 h of light entered diapause (Zaslavski & Umarova, 1990). In some species photoperiod has no effect on diapause induction, as in the scelionid *Telenomus californicus* that does not enter diapause under either short (12L : 12D) or long (18L : 6D) photoperiod while kept at 21°C (Ryan *et al.*, 1981).

The light threshold for response is low: *C. cinctus* responds to a light intensity of 0.32 lx. When light above this threshold is perceived for more than 12 h, the larvae will not enter diapause (Jackson, 1963). For comparison, the human eye can detect intensities as low as 0.000001 lx (Danks, 1987). The light intensity of a full moon is 0.25 lx and the threshold of fully exposed insects is unlikely to be much lower than this to avoid erratic responses during full moon. Lower thresholds are to be expected for insects living in habitats where light intensity is low.

Response to temperature

Low temperature seems to be the major factor that induces diapause in *Trichogramma* spp. although this response is modified through the so-called maternal influence (see later). Species like *T. evanescens*, *T. embryophagum*, *T. pintoi* and *T. principium* enter diapause at temperatures below 15°C, whereas *T. semblidis* responds at a lower temperature (12.5°C) (Bonnemaison, 1972; Zaslavski & Umarova, 1982, 1990). Similar results were obtained by Sorokina & Maslennikova (1988) who found that diapause occurred at 10°C in *T. pintoi*, *T. evanescens*, *T. embryophagum*, *T. cacoeciae*, *T. ingricum* and *T. aurosum*. The major role of temperature in diapause induction was also demonstrated for *T. maidis*, *T. nagarkattiae*, *T. exiguum* and *T. achaeae* (Voegelé *et al.*, 1986). When the developing larvae of *T. maidis* were exposed to temperatures close to their lower temperature limit for development, they entered diapause (Voegelé *et al.*, 1986).

Temperature can also hasten diapause termination, even if that diapause was induced mostly by photoperiod, as for *O. ennomus* (Anderson & Kaya, 1975) where chilling terminates diapause, for *C. cinctus* where frost was able to break the diapause (Jackson, 1963) and for *O. pityocampae* where the winter diapause of the mature larvae could be interrupted by heating them at 28–30°C (Battisti *et al.*, 1990). Low temperatures also terminated diapause in late autumn in *Aphelinoidea plutella* and *Abbella subflava*, egg parasitoids of the beet leafhopper *Circulifer tenellus* in Idaho, the rest of the winter being spent in quiescence (Henderson, 1955).

Response to other factors

Diapause can also respond to other factors such as humidity or host conditions. Moisture was found to break diapause in *M. pulchriceps* (van den Berg, 1971). Neither ultraviolet radiation, age of host at oviposition time or storage at low temperature influenced the diapause duration in *T. maidis* (Pizzol & Voegelé, 1987).

Several interactions characterize the relationship between parasitoids and hosts before and during diapause. Some larval parasitoids can regulate host diapause as in the braconid *Cotesia koebelei*, a larval parasitoid of the butterfly *Euphydryas editha*, that controls the number of pre-diapause feeding instars of its host. Parasitized larvae are more likely to pass through an extra feeding instar before entering diapause (Moore, 1989). Such modification of host biology may enable the parasitoid to better synchronize its life cycle with the host species. It is not known if egg parasitoids can influence directly the diapause of their host eggs.

In most insect species, the onset of egg diapause is determined by the photoperiod experienced maternally (Saunders, 1982). Egg diapause can occur at any stage of embryogenesis. Some species enter diapause as fully

formed larvae within the egg shell while others enter diapause at an earlier stage of development. Some parasitoids rely exclusively on the host for a diapause induction cue but in others the induction of diapause is totally independent of the host (Tauber et al., 1986). In some egg parasitoids, the host condition influences the induction of diapause as in T. cacoeciae where diapause is induced by some factor in the diapause eggs of its host Archips rosanus (Keller, 1986). In T. evanescens, diapause is induced at different temperatures and photoperiods when different lepidopteran eggs are used as hosts (Bonnemaison, 1972). The propensity to enter diapause in Trichogramma also seems to be influenced by the superparasitism of host eggs. When Arctia caja or Dendrolimus pini are used as hosts, high levels of superparasitism (8.7 larvae per host) inhibits diapause at 20°C and 16 h light while 100% of the parasitoids enter diapause under these conditions when superparasitism is low (Bonnemaison, 1972).

Some species of egg parasitoids need a diapausing host egg in order to enter diapause, as in Anastalus sp. that overwinters in the diapause eggs of the Indian gypsy moth, Lymantria obfuscata (Singh et al., 1987), and in Acmopolynema hervali that enters diapause in the diapause eggs of its host, the sugarcane froghopper, Mahanarva posticata (Barbosa et al., 1979). T. minutum cannot overwinter in the eggs of the spruce budworm, Choristoneura fumiferana, in Maine because this host does not overwinter in the egg stage (Jennings & Houseweart, 1983). When the host species does overwinter as eggs, Trichogramma spp. can overwinter successfully, as for T. evanescens in the eggs of the imported cabbageworm, Artogeia rapae, in Missouri (Parker & Pinnell, 1971). C. cinctus overwinters in diapause in the eggs of the dytiscid Agabus bipustulatus but these host eggs do not overwinter in a diapause stage but rather in quiescence (Jackson, 1963). Consequently the seasonal diapause of C. cinctus cannot be conditioned by a host diapause.

The nutrition of adult parasitoids affects their response to photoperiod and eventually diapause induction in their progeny. Host deprivation in species that feed on the host, or unavailability of suitable plants for species feeding on nectar, as in the Mymaridae, increases the proportion of diapause offspring. Starvation in sarcophagid larvae can also increase the proportion of diapause pupae following an increase in development time (Saunders, 1982). In egg parasitoids, if overcrowding from superparasitism results in insufficient food supply, an increase in diapause could also occur, because prolonged development would enable the larva to detect a greater number of inductive cycles before the end of its sensitive period. Such an effect would be present in species that are photoperiod-sensitive during larval development. This research area remains to be studied.

In some phytophagous species, nutrition affects the propensity of eggs to enter diapause. Red spider mite, Panonychus ulmi, females lay non-diapause eggs in long daylength when reared on young apple foliage. Under the same photoperiod, females transferred to old foliage produced 68% diapausing eggs

(Lees, 1953). For species of egg parasitoids relying on host egg diapause to induce their own diapause, nutrition of the host adult could influence the proportion of their offspring that enter diapause. Because the nutritional quality of most plant species is expected to decrease in the autumn such a cue could be reliable.

The maternal effect

In some families of Hymenoptera, including the Trichogrammatidae and the encyrtid *O. ennomus* (Anderson & Kaya, 1974), the induction of diapause is triggered by conditions occurring over two generations. This maternal effect can either dominate completely the induction of diapause or merely modify the threshold responses of the daughter generation (Zaslavski, 1988). In most species, however, no such maternal effect is found, as in *C. cinctus* (Jackson, 1963).

Maternal effect on the induction of diapause has been demonstrated in *T. evanescens, T. embryophagum, T. aurosum, T. pintoi* and *T. cacoeciae* but not in *T. ingricum* (Sorokina & Maslennikova, 1988). In *T. evanescens* and *T. pintoi*, short days in the parental generation increase the probability of diapause in the daughter generation (Zaslavski, 1988). The developing larvae respond to different temperature thresholds depending on the photoperiodic regimen that their mothers experienced. Particular conditions during larval development are necessary so that the maternal effect is expressed. Higher temperatures prevent diapause even in the progeny of females reared under short photoperiods, and at least a certain percentage of the progeny of females reared under long photoperiod will enter diapause if those larvae experience low temperatures (Zaslavski, 1988). The temperature experienced by the developing larvae of *Trichogramma* is the decisive factor in the induction of diapause. In *T. pintoi* the temperature threshold of the progeny of females that develop under short photoperiod is about 2–3°C higher than the progeny of long-photoperiod females (Zaslavski & Umarova, 1990).

The temperature experienced by the females also modifies the induction of diapause. Two types of reactions are possible. When females are reared at high temperature, their progeny may show less propensity to enter diapause compared to the progeny of females reared at low temperature, as in *T. pintoi* (Zaslavski, 1988), whereas the reverse is also possible, as shown for *T. evanescens* (Zaslavski & Umarova, 1982). *O. ennomus* has a facultative diapause that is induced by photoperiod during the parental generation and temperature during the daughter generation. The number of annual generations is largely determined by the photoperiods to which the adults are exposed (Anderson & Kaya, 1974).

The maternal influence may also be detected over two generations. When the daughter (F_1) generation develops in conditions not favourable to diapause induction, the maternal effect can still influence diapause induction

in the progeny of that generation (F_2). Such a relationship was shown in *T. evanescens* (Zaslavski & Umarova, 1982). When diapause is terminated, the resulting females generally are no longer responsive to the conditions that induced diapause, as in *T. maidis* (Voegelé *et al.*, 1986) and *C. cinctus* (Jackson, 1963).

An annual rhythmicity in the propensity to enter photoperiodically induced diapause has also been reported for *T. evanescens* (Zaslavski & Umarova, 1982). Such an annual rhythm involves reinstating the maternal influence on photoperiodic induction of diapause. However, these results on the annual rhythmicity have not been confirmed and the possibility of inadvertent selection has not been ruled out (Tauber *et al.*, 1986). When an insect terminates its diapause, it is generally not sensitive to the token stimuli that first induced its diapause. In some species, this insensitive period may last for several generations (Danks, 1987). Such circannual rhythms could be controlled through photoperiodic or geomagnetic forces (Danks, 1987).

Quiescence in egg parasitoids

In quiescence, the dormancy is directly imposed by adverse conditions and recovery occurs soon after these restrictions are removed (Saunders, 1982). Some insects overwinter in a quiescent state while others enter quiescence after their period of diapause is terminated. In both cases, the insect resumes activity when the conditions become favourable again. When dormancy ends in a period of quiescence, the insect will be synchronized with the seasonal changes but not necessarily with the seasonal development of its host. If an egg parasitoid must delay development to coincide with the oviposition period of its spring host, diapause is probably essential.

Although quiescence has been demonstrated for several species, for others quiescence is assumed because diapause has not been positively identified. For these species, more research is likely to detect diapause if the right cues are tested. Several species of *Trichogramma* are reported to overwinter in a state of quiescence, such as *T. minutum* as a larva in the eggs of the variegated cutworm *Peridromia saucia* (Parker & Pinnell, 1971; Jennings & Houseweart, 1983), *T. nubilale* as a pupa in the eggs of *Ostrinia nubilalis* (Curl & Burbutis, 1977), *T. pretiosum* in the eggs of *Heliothis virescens* (López & Morrison, 1980a), and *T. exiguum* in eggs of different species of Noctuidae but also as adults in North Carolina (Keller, 1986). *T. daumalae* and *T. dendrolimi* spend all winter in quiescence while *T. rhenana* and *T. semblidis* enter quiescence in autumn but switch to diapause during winter (Voegelé *et al.*, 1988). *Anaphes diana* appears to overwinter as larvae in quiescence in Delaware (Dysart, 1990) and as larvae and adults in the Mediterranean region (Aeschlimann, 1977). Many species that overwinter as adults do so in a state of quiescence. *Ooencyrtus kuwanai* (Griffiths & Sullivan, 1978) and the

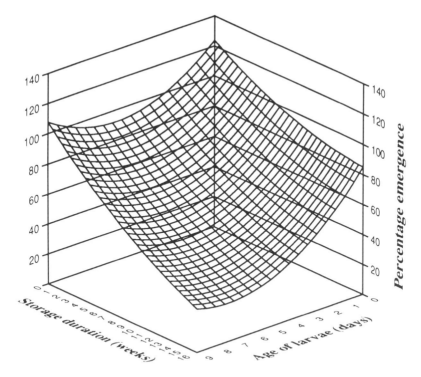

Fig. 11.2. Predicted surface response for emergence of *Anaphes* n. sp. subjected to storage at 2°C as larva.

scelionids *T. californicus* (Torgersen & Ryan, 1981), *Ceratobaeus masneri*, *C. clubionus* (Austin, 1984) and *Trissolcus biproruli*, a parasitoid of Pentatomidae eggs (James, 1988), overwinter as quiescent adults.

In a quiescent state, the overwintering insect generally is not sensitive to cues that do not directly constrain its development. For example, when in quiescence *T. nubilale* does not respond to photoperiod (Curl & Burbutis, 1977).

Quiescence can be used to store egg parasitoids at low temperature for mass-production. Peterson (1931) was among the first to note that *T. minutum* could survive up to 6 months at 4°C when eggs of the common bagworm *Thyridopteryx ephemeraeformis* were used as a host. He also found that the developmental stage of the parasitoids when they were placed at low temperature influenced their survival. Survival of up to 93% was obtained with *T. pretiosum* when last-day pupae were stored at 16.7°C for 6 days and then at 15°C for another 6 days (Stinner *et al.*, 1974). Species like *T. chilonis*, *T. achaeae* and *Trichogrammatoidea eldanae* can survive up to 49 days at 10°C with emergence still above 30%, while *T. japonicum* had less than 10% survival under

the same conditions (Jalali & Singh, 1992). The adult parasitoids can also be stored at low temperature. *Telenomus remus* females can be stored at 5°C for a week without significant losses of fecundity (Gautam, 1986) and the developing larvae can survive up to 2 weeks at 5°C (Kumar *et al.*, 1984). Adults of *A. ovijentatus* can survive for 4 weeks at 1.7°C (Jackson, 1986).

When different developing stages of *Anaphes* sp. were placed at 2°C for various lengths of time, both age of larvae and duration of cold exposure influenced the emergence rate (unpubl.). Emergence gradually decreases as the duration of cold storage increases but the age of the larva also has a major influence, as shown on a model based on a surface response analysis (Fig. 11.2). Young larvae, either first or second instar, appear to survive cold exposure much better.

Mechanisms of Cold Resistance

To survive throughout winter, either in diapause or quiescence, egg parasitoids had to develop adaptations of behaviour to avoid cold temperatures or of physiology to make them less susceptible to damage caused by cold. Physical damage to the cells is generally caused by the freezing of intercellular fluids, although biochemical and physiological damage also results from more or less prolonged exposure to low temperatures above the freezing point of the insect.

Cold resistance should be an important characteristic in the development of a biological control programme. Lack of success in several biological control programmes has been attributed to high mortality of natural enemies due to climatic extremes (Bale, 1991) but surprisingly, although diapause has been studied extensively, relatively little attention has been given to cold resistance of egg parasitoids.

Cold avoidance

An obvious way to avoid cold is to migrate to locations where winter temperatures can be tolerated, as for the monarch butterfly (*Danaus plexippus*) that migrates annually from northern North America to Mexico and Central America (Walker, 1980). However, the small size of egg parasitoids precludes this option. Another option is to avoid cold either by selecting host species that overwinter in protected habitats or, within a species, by parasitizing eggs in the most sheltered microhabitats. Again, little information is available for egg parasitoids.

Ooencyrtus kuvanae, an egg parasitoid of the gypsy moth, *Lymantria dispar*, overwinters as an adult in Pennsylvania and it changes its distribution according to the season. In summer, the parasitoid abundance is positively correlated with the height of the egg mass but in autumn this relationship becomes negative, the parasitoid activity switching to the lower portion of the bole (Brown & Cameron, 1982). This modification of behaviour enables the species to experience less extreme temperatures because of the insulating property of snow cover. However, other species do not use cold-avoidance behaviour and overwinter in the same habitats they use throughout the summer. *Anagrus epos*, an egg parasitoid of the white apple leafhopper *Typhlocyba pomaria* in Michigan, overwinters in the same location as its host. This species showed no preference to parasitize overwintering eggs of its host in orchard areas protected from cold (Seyedoleslami & Croft, 1980). *Telenomus nitidulus* overwinters as adults in The Netherlands in the bark of poplars where its host the satin moth, *Leucoma salicis*, develops (Grijpma, 1984).

When no particular cold-avoidance behaviour is detected and both the egg parasitoid and its host eggs have about the same cold hardiness, parasitism is expected to be constant throughout the host's geographic distribution. Another egg parasitoid of the gypsy moth, *A. disparis*, attacks eggs both near ground level on rocks and logs but also in more exposed positions along the trunk of trees and the underside of branches (Sullivan *et al.*, 1977). When snow protection is available, survival of both species is good at soil level but parasitized and unparasitized eggs exposed to temperatures below −28°C failed to emerge. Such low temperature is experienced every winter in southern Quebec and Ontario resulting in a low survival rate of the parasitoid and a low efficacy early in the season. If *A. disparis* were to parasitize proportionally more of the eggs located near soil level, its efficiency in the northern distribution of its host would be greater.

Cold hardiness

Cold hardiness is the capacity to resist damage caused by exposure to low temperatures. This damage may result either from the transformation of the internal fluids into ice or from prolonged exposure to low temperatures that are above the freezing point of the insect. The resistance to this type of damage relies on a series of biochemical and physiological adaptations.

Insects can be either freezing-tolerant if they survive the formation of intra-body ice or freezing-intolerant if they do not survive such ice formation. As expected, the cold hardiness strategy of each type is different. Freezing-tolerant insects have developed adaptations that favour the creation of ice crystals at relatively high subfreezing temperatures. Ice formation at very low temperatures is more damaging because of more rapid crystallization that both causes physical damage to the cells and rapidly creates a large osmotic pressure that may collapse the cells. Freezing-tolerant insects typically freeze at about −5°C to −8°C. Adaptations to accelerate ice formation include the presence of ice-nucleating agents in the digestive

system or in the haemolymph of the insect. These ice-nucleating agents mimic the surface of an ice crystal and therefore facilitate the creation of the first ice nucleus.

Freezing-intolerant insects must delay ice formation as long as possible. They avoid ice formation by eliminating or concealing ice-nucleating agents, generally by not feeding in the period immediately before winter, by dehydrating, thereby increasing their solute concentrations, or by synthesizing cryoprotectants, low-molecular-weight antifreezes such as polyols and sugars (glycerol, sorbitol, mannitol, ethylene glycol, ribitol, erythritol, inositol, fructose, trehalose, glucose) or thermal hysteresis proteins. These adaptations enable insects to overwinter in a supercooled state where cold-induced damage is kept to a minimum.

The capacity of an insect to supercool is measured by its supercooling point, the temperature at which spontaneous ice formation is observed. These cryoprotectants, by their colligative action, depress the melting and supercooling points of the insects (Lee, 1991). Supercooling points are determined by gradually cooling an insect, generally at the rate of $1°C$ min^{-1}, while measuring its temperature. The freezing of body fluid is detected by a small increase in temperature that corresponds to the release of latent heat. Supercooling points as low as $-60°C$ have been recorded for Hymenoptera (Ring & Tesar, 1981).

For both freezing-intolerant and freezing-tolerant species, continuous exposure to low temperature, even above the supercooling point, causes damage. Such damage could be due to modifications in transmembrane diffusion and transport, disruption of the higher orders of protein structure and of enzyme pathways, changes in dielectric permittivity and ionic activities and rate effects on diffusion processes, active and carrier-mediated transport and flux through metabolic pathways (Storey & Storey, 1988). Generally, the effects of damage are cumulative and an increasing proportion of the population is affected according to the duration of the exposure and the temperature experienced. In some species, damage occurs rapidly, as for the scelionid *Platytelenomus hylas*, an egg parasitoid of *Sesamia cretica* in Egypt, adults of which can survive for only about 5 h when exposed to $-1°C$ (Hafez *et al.*, 1977).

For several of the species of egg parasitoids that overwinter as adults, such as *T. exiguum* in North Carolina (Keller, 1986), *T. californicus* in Oregon (Torgersen & Ryan, 1981), *A. diana* in the Mediterranean region (Aeschlimann, 1977), and the scelionids *Trissolcus biproruli* (James, 1988) and *Ceratobaeus* sp., an egg parasitoid of spiders (Austin, 1984), in Australia, no data on cold hardiness is available. The level of low temperatures experienced by these species varies greatly and resistance to both freezing and cumulative cold damage is certainly involved. Some information is available for *O. kuwanai* that overwinters as an adult in Maine and New Jersey. Unacclimated males have a supercooling point of $-10.6°C$ and acclimation of these males at 0°C for 4 or 20 days did not result in a significant decrease of the supercooling point. Unacclimated females have a very similar supercooling point, −10.8°C, but acclimation at 0°C for about 20 days decreased the supercooling point to −14.5°C. However, cumulative cold damage at 0°C caused over 70% mortality in these females (Griffiths & Sullivan, 1978). These results indicate that this species is unable to survive in the northern part of the distribution of its host, the gypsy moth. *T. nitidulus* placed outside in vials during winter survived temperatures of −11.1°C and more than 75% of the individuals survived when fed (Grijpma, 1984).

The majority of egg parasitoids overwinter as immature stages in host eggs and, although they cannot avoid cold behaviourally except for the selection of autumn hosts that are located in sheltered habitats, they can benefit from the intrinsic cold hardiness of host eggs. The supercooling points of insect eggs are generally lower than those of adults (Sømme, 1982) probably because of their small size, because they are enclosed in a more or less rigid chorion, and for biochemical reasons, several cryoprotectants being the building blocks of developing embryos. Crystallization of water is a probabilistic event (Lee, 1991); small volumes of water freeze at relatively low temperatures and insect eggs react the same way. Therefore, even without any specific biochemical or physiological adaptations, the immature egg parasitoid has some protection against freezing within its host egg. In addition, when the host eggs are inserted in plant material, as is the case with curculionid eggs, the plant itself offers protection, through its insulating property, and by keeping pressure on the chorion. Low supercooling points have been observed in the eggs of the orthopteran *Locusta migratoria* (−30°C), different species of psocids (−27°C to −37°C) (Block, 1982), the curculionid *Listronotus oregonensis* (−22°C to −25°C) (Hance & Boivin, 1993) and the eggs of a geometrid moth *Epirrita autumnata* in Scandinavia (Tenow & Nilssen, 1990). In some species, such as *L. migratoria*, this low supercooling point is due to the dehydration of the eggs during the dry season. Some species have eggs that are susceptible even to moderately cold temperatures, such as the notodontid *Clostera inclusa*, the eggs of which are killed if exposed at 5°C for a month (Drooz & Solomon, 1980).

The unparasitized eggs of the gypsy moth have a supercooling point of −29.7°C. When those eggs are parasitized by *A. disparis*, a species that overwinters as a mature larva inside its host eggs, they have a supercooling point of −28.8°C and conditioning of the eggs does not change this supercooling point (Sullivan *et al.*, 1977). These data indicate that *A. disparis* does not modify the cold hardiness of its host. *T. evanescens* is also reported to withstand temperatures down to −33°C to −37°C (Maslennikova, 1959, cited in Jackson, 1963). The unparasitized eggs of *L. oregonensis* have supercooling points that increased from −24.9°C to −22.1°C as eggs matured. When these eggs are parasitized by *Anaphes* n.sp., their supercooling point remains stable at −22.9°C in spite of the parasitoid embryogenesis, that

creates new nucleating points and thus should have increased the super-cooling point. Cold hardiness of this parasitoid is due to a multicomponent system based on glycerol and fructose (Hance & Boivin, 1993). Concentrations of 1.6% of fresh weight for glycerol and 2.9% for fructose were measured. Glycerol is the most common low-molecular-weight cryoprotectant in insects and the overwintering larval stage of *Bracon cephi* produces 5 M concentrations of glycerol, an amount equal to 25% of its body weight (Salt, 1961). A multicomponent system could achieve a high level of cold hardiness without the possible toxic effect of a very high level of a single cryoprotectant (Zachariassen, 1985).

The measurement and interpretation of cryoprotectant concentrations in egg parasitoids pose special problems. Immature egg parasitoids must not only protect themselves from freezing but also protect their immediate habitat, the host egg. Freezing of the host vitellus is probably fatal for the developing parasitoid larva. Although protecting their host egg from freezing implies a relatively large investment in cryoprotectant synthesis, it also enables the larvae to produce concentrations that could be lethal if present in their bodies but acceptable when present in their surroundings. In addition, this investment would not be lost because the pathways for synthesis and catabolism of cryoprotectants like glycerol in insect fat body are part of lipid metabolism (Storey & Storey, 1991). Separate analysis of the cryoprotectant contents of the parasitoid larva and of the host egg would give interesting insights into the cold hardiness strategy of the egg parasitoid, but would necessitate delicate dissection and sensitive analytical techniques.

Cumulative cold damage affects survival of egg parasitoids that are exposed to temperatures above their supercooling points. Typically such damage is studied by exposing insects to increasingly low temperatures and measuring their mortality after different exposure periods. *A. disparis* had less than 20% mortality when exposed at 0°C for 10 months or −18°C for 4 days followed by 2–3 months at 0°C (Sullivan *et al.*, 1977). For *T. pretiosum*, the duration of exposure resulting in 50% mortality (LT_{50}) decreased with temperature and varied with the age of the immature parasitoid (López & Morrisson, 1980b). LT_{50} ranged from 15 days for 6-day-old larvae at 0°C to less than 1 h for 8-day-old larva at −20°C. The mortality of 3-day-old larvae of *Anaphes* n. sp., exposed to temperatures of 0°C to −25°C, varied according to the duration of exposure. At −25°C no larvae survived, even after only 1 h exposure. However, the supercooling point of the developing larvae of *Anaphes* n. sp. is −23°C (Hance & Boivin, 1993). At 0°C, more than 50% of the larvae survived after 120 days (L. Traoré & G. Boivin, unpubl.) and LT_{50} gradually decreased with temperature (Fig. 11.3).

When cold hardiness data are available for both the minimum temperature that can be survived (supercooling point) and the gradual increase in mortality with cold exposure (cumulative cold damage), it is possible to integrate them in a model of overwintering survival. The input for such a model is temperature data

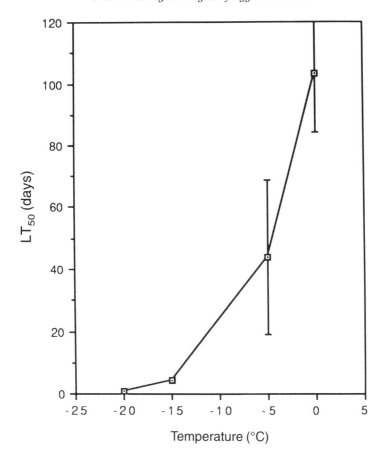

Fig. 11.3. Relationship between the time necessary to cause 50% mortality in larva of *Anaphes* n. sp. and the temperature.

from a specific habitat and its output is a prediction of percentage survival for a given population. Such a model can predict the population level of egg parasitoids to be expected in the spring. These overwintered populations can have an important effect on the population dynamics of the first generations of their host. The cold hardiness of a species often limits its northern distribution. In southern Quebec, minimum air temperatures are often below the super-cooling point of *Anaphes* n. sp. Therefore, survival during the coldest nights probably is possible only in sheltered habitats. The individuals overwintering in poorly protected habitats would not survive and, in spring, a given habitat would be colonized by individuals that survived in the few sheltered micro-habitats. Because of shifting snow banks it is generally impossible in the autumn

to predict where the most sheltered habitats will be during the coldest nights. A parasitoid would thus gain an advantage by dispersing as much as possible in the autumn. Such bet-hedging in a seasonally unpredictable environment is observed in *Anaphes sordidatus*, a gregarious egg parasitoid of *L. oregonensis*. As the season progresses, the average number of parasitoids emerging per host egg increases from 1 to 2.3 (unpubl.). This increase in superparasitism could result partly from a numerical response, because the number of parasitoids available per host egg increases also during the season (Boivin, 1993). Similar results are reported for a pteromalid, *Pteromalus puparum*, a parasitoid of lepidopterous pupae. A late-season increase in superparasitism, although it resulted in smaller adults with a higher mortality, could be favourable because it increased their dispersion at the end of the season (Takagi, 1987). In that case, this dispersal was related to the decrease in suitable host availability.

Evolutionary and Practical Implications

Several important ecological characteristics of insects are influenced by their winter survival strategies and one of the most obvious is their geographical distribution. In the northernmost part of their distribution, survival is a function of these strategies and of the availability of suitable overwintering sites. Distinctive patterns of distribution and abundance can be related to extreme winter conditions, as in the geometrid moth *Epirrita autumnata* in northern Scandinavia. This insect is an important defoliator of mountain birch forests in this area and major outbreaks occurred in 1955 and 1964–1965 (Tenow & Nilssen, 1990). However, forests in the lowermost parts of the mountain valleys remained unattacked while the surrounding areas were completely defoliated. This pattern occurred because cold air accumulated in the valleys and made them especially cold. *E. autumnata* overwinters as eggs on the branches and stems of the mountain birch and these eggs have a mean supercooling point of $-36°C$. Such a low temperature was reached in the valley, killing the overwintering eggs and therefore protecting the mountain birch stands from defoliation (Tenow & Nilssen, 1990). Similar impact on egg parasitoid populations could occur but field data on winter mortality are not available.

The many photoperiodic responses observed in diapause induction are adaptive, genetically determined characteristics that explain the distribution of egg parasitoids and their exploitation of a great diversity of habitats and niches (Beck, 1980). The diapause induction cues and the cold hardiness adaptations can act to form reproductively isolated populations and therefore are implicated in the evolutionary process of speciation. Because egg parasitoids must depend almost exclusively on physiological adaptations to survive winter, one can expect those characteristics to be strongly selected for. As a consequence, cold hardiness probably is a significant factor in the speciation process of egg parasitoids in temperate climates.

Significant differences between the cold hardiness of a host and its egg parasitoids can create a 'climatic' refuge for that host, at least early in the season before the parasitoids have had time to disperse. If a secondary host is necessary for an egg parasitoid to overwinter, differences in the distribution of the principal and secondary hosts would affect the impact of the egg parasitoid on both host species.

Several examples are known where winter survival strategies of egg parasitoids had a marked effect on their efficacy to control their host. When *T. minutum* was mass-released against the spruce budworm in Ontario, the absence of overwintering in this area was explained by the harsh winter and the lack of ecological diversity in the forests that probably reduced the availability of alternative autumn hosts (Smith *et al.*, 1987). The southwestern corn borer, *Diatraea grandiosella*, has been expanding its range since the beginning of this century from its native Mexico to most of the southern United States (Overholt & Smith, 1990). In its northern expansion, the geographic isolation and the adaptations gained to colder climates have reduced the variety of its natural enemies and limited the effectiveness of those that followed it, such as *T. pretiosum* (Rodriguez-del-Bosque *et al.*, 1989). When *T. atopovirilia* was introduced to the Texas High Plains from Mexico to control *D. grandiosella*, it was unable to overwinter and thus failed to establish itself (Overholt & Smith, 1990). The introduction of the eulophid *Edovum puttleri*, an egg parasitoid of the Colorado potato beetle, *Leptinotarsa decemlineata*, from Colombia to North America and Europe was also unsuccessful because this parasitoid had no dormancy and therefore was unable to overwinter either in New York state (Obrycki *et al.*, 1985) or in central Italy (Pucci & Dominici, 1988).

Other introduced exotic egg parasitoids have become established, although no formal cold hardiness studies were available for some of them. In France, the introduction of *T. evanescens* in Alsace resulted in the establishment of this species, probably because it was able to enter dormancy in winter (Stengel *et al.*, 1977). In western Canada, *Anaphes* (*Patasson*) *luna* was recovered from the eggs of the alfalfa weevil, *Hypera postica*, in 1978 probably following its releases in the United States from 1910–1933 (Schaber, 1981). The long interval between the original release and its recovery in Alberta could be due to the gradual selection of strains able to resist the harsh winters of the Canadian prairies.

The presence of an alternative host is sometimes essential for the overwintering of egg parasitoids as it is for the egg parasitoids of the tarnished plant bug that overwinters as adults. *Polynema pratensiphagum*, a Mymaridae, has to overwinter in the eggs of another mirid species, *Adelphocoris lineolatus*, in Quebec (Al-Ghamdi *et al.*, 1993). The northern distribution of egg parasitoids will therefore be much influenced by the distribution of their alternative hosts, a distribution that can be very different from the distribution of another host that is targeted in a biological control programme. *A. epos* is

highly effective in reducing populations of the grape leafhopper, *Erythroneura elegantula*, in California vineyards. However, this leafhopper does not over-winter as eggs and *A. epos* has to overwinter in other leafhopper species in different habitats (Doutt & Nakata, 1965; Kido *et al.*, 1984). The effectiveness of *A. epos* therefore depends on the presence of refuges where it can overwinter and the distance of these refuges from the vineyards (Kido *et al.*, 1984). A similar situation is known for *Anagrus incarnatus* that parasitizes the eggs of several species of rice planthoppers in Japan. None of these planthoppers overwinters in the egg stage in Japan and eggs of other species of Delphacidae were parasitized in the autumn by overwintering *A. incarnatus* (Chantarasa-Ard, 1984).

If the overwintering strategies of egg parasitoids were better known, this information could be used to enhance their efficiency as biological control agents. This approach has been used successfully for larval parasitoids. The curculionid *Pissodes strobi* attacks the leaders of the Sitka spruce in British Columbia, Oregon and Washington, and both this species and its parasitoids overwinter in the infested leaders. These leaders could be clipped in the autumn and subjected to $-26°C$ before being returned to the field. This temperature kills all overwintering larvae of *P. strobi* while most associated parasitoids survive (Hulme *et al.*, 1986). Such cold treatment reduces the population density of the adult weevils the following spring while preserving most of the parasitoid population. Cultural techniques can also be used to augment the survival of a parasitoid population. Late autumn and midwinter disking of piles of cotton bolls that contain diapause prepupae of *Bracon mellitor*, a larval ectoparasitoid of the boll weevil, increases the survival of female parasitoids, especially during harsh winters (Slosser *et al.*, 1990) because disking buries some of the bolls and protects them from the lowest winter temperatures.

As mentioned earlier, the dormancy characteristics of egg parasitoids can be used to synchronize production and to increase storage time (Voegelé *et al.*, 1986). Dormancy could also be used to induce staggered emergence following diapause. In *C. cinctus*, most non-diapause individuals generally emerge within a few days of each other while diapause individuals emerge over a long period (Jackson, 1963). If used for mass-reared species, such a characteristic could permit a single release, thereby reducing treatment costs, while obtaining a control effect over several weeks. Of course, the cumulative mortality from desiccation and predation should be considered too.

Conclusions

Most of the information available on the overwintering strategies of egg parasitoids deals with their practical impacts such as the use of diapause for the storage of mass-reared parasitoids and the presence of overwintering hosts for the introduction of exotic species. Few studies have been devoted to the cold hardiness of egg parasitoids and even fewer to how these characteristics have influenced the population dynamics of a species and its geographical distribution. In several instances, however, a better knowledge of the overwintering strategies of egg parasitoids would either permit a better exploitation of their possibilities as biological control agents or avoid costly mistakes. Despite numerous introductions in California, the eulophid *Tetrastichus gallerucae*, an egg parasitoid of the elm leaf beetle, *Xanthogaleruca luteola*, has failed to become firmly established (Dreistadt & Dahlsten, 1991). Although these introductions spanned more than 60 years, the overwintering stage of *T. gallerucae* and the habitat where it survives during the winter are still unknown. A detailed study on the overwintering strategy of this species, including whether or not it can enter diapause and if alternative hosts are necessary, would be a prerequisite for a successful introduction.

Egg parasitoids can themselves be limiting factors if their hosts are used in biological control programmes. Their capacity to overwinter then becomes an important factor to consider. The success in North America of the chrysomelid *Longitarsus jacobaeae* as a biological control agent of the tansy ragwort could be endangered by the presence of its egg parasitoid, *Anaphes euryale* (Windig, 1991). The absence of this parasitoid in North America could explain the larger impact of *L. jacobaeae* on its host outside its native habitat. In Europe, *L. jacobaeae* overwinters as adults in the southern part of its distribution but as eggs farther north. If *L. jacobaeae* overwinters as eggs in North America, the cold hardiness of *A. euryale* should be evaluated to assess whether it could become established if accidentally introduced.

Acknowledgements

I wish to express my thanks to Drs S.M. Smith, University of Toronto, and H.V. Danks, Biological Survey of Canada, Canadian Museum of Nature, for their critical reading of this manuscript.

References

Aeschlimann, J.P. (1977) Notes on *Patasson lameerei* (Hym.: Mymaridae), an egg parasitoid of *Sitona* spp. (Col.: Curculionidae) in the Mediterranean region. *Entomophaga* 22, 111–114.

Al-Ghamdi, K.M., Stewart, R.K. & Boivin, G. (1993) Note on overwintering of *Polynema pratensiphagum* (Walley) (Hymenoptera: Mymaridae), in southwestern Quebec. *Canadian Entomologist* 125, 407–408.

Anderson, J.F. & Kaya, H.K. (1974) Diapause induction by photoperiod and temperature in the elm spanworm egg parasitoid, *Ooencyrtus* sp. *Annals of the*

Entomological Society of America 67, 845–849.

Anderson, J.F. & Kaya, H.K. (1975) Influence of temperature on diapause termination in *Ooencyrtus ennomus*, an elm spanworm egg parasitoid. *Annals of the Entomological Society of America* 68, 671–672.

Austin, A.D. (1984) The fecundity, development and host relationships of *Ceratobaeus* spp. (Hymenoptera: Scelionidae), parasites of spider eggs. *Ecological Entomology* 9, 125–138.

Bale, J.S. (1991) Implications of cold hardiness for pest management. In: Lee, R.E. & Denlinger, D.L. (eds) *Insects at Low Temperature*. Chapman and Hall, New York, pp. 461–498.

Barbosa, J.T., Filho, M.L. & Riscado, G.M. (1979) Diapause on eggs of *Mahanarva posticata* (Stal, 1855) and its effects on *Acmopolynema hervali* (Gomes, 1948), an egg parasite. *International Society of Sugarcane Technologists, Entomology Newsletter* 6, 6.

Battisti, A., Ianne, P., Milani, N. & Zanata, M. (1990) Preliminary accounts on the rearing of *Ooencyrtus pityocampae* (Mercet) (Hym., Encyrtidae). *Journal of Applied Entomology* 110, 121–127.

Beck, S.D. (1980) *Insect Photoperiodism*, 2nd edn. Academic Press, New York, 387 pp.

Berg, M.A. van den (1971) Studies on the induction and termination of diapause in *Mesocomys pulchriceps* Cam. (Hymenoptera: Eupelmidae) an egg parasite of Saturniidae (Lepidoptera). *Phytophylactica* 3, 85–88.

Block, W. (1982) Cold hardiness in invertebrate poikilotherms. *Comparative Biochemistry and Physiology* 73, 581–593.

Boivin, G. (1993) Density dependence of *Anaphes sordidatus* (Hymenoptera: Mymaridae) parasitism on eggs of *Listronotus oregonensis* (Coleoptera: Curculionidae). *Oecologia* 93, 73–79.

Bonnemaison, L. (1972) Diapause et superparasitisme chez *Trichogramma evanescens* Westwood (Hym. Trichogrammatidae). *Bulletin de la Société Entomologique de France* 77, 122–132.

Broodryk, S.W. (1969) The biology of *Chelonos* (*Microchelonus*) *curvimaculatus* Cameron (Hymenoptera: Braconidae). *Journal of the Entomological Society of South Africa* 32, 169–189.

Brown, J.R. & Phillips, J.R. (1990) Diapause in *Microplitis croceipes* (Hymenoptera: Braconidae). *Annals of the Entomological Society of America* 83, 1125–1129.

Brown, M.W. & Cameron, E.A. (1982) Spatial distribution of adults of *Ooencyrtus kuvanae* (Hymenoptera: Encyrtidae), an egg parasite of *Lymantria dispar* (Lepidoptera: Lymantriidae). *Canadian Entomologist* 114, 1109–1120.

Burbutis, P.P., Curl, G.D. & Davis, C.P. (1976) Overwintering of *Trichogramma nubilale* in Delaware. *Environmental Entomology* 5, 888–890.

Chantarasa-Ard, S. (1984) Preliminary study on the overwintering of *Anagrus incarnatus* Haliday (Hymenoptera: Mymaridae), an egg parasitoid of the rice planthoppers. *Esakia* 22, 159–162.

Curl, G.D. & Burbutis, P.P. (1977) The mode of overwintering of *Trichogramma nubilale* Ertle and Davis. *Environmental Entomology* 6, 629–632.

Danks, H.V. (1987) Insect dormancy: an ecological perspective. *Biological Survey of Canada, Monograph Series* No. 1, 439 pp.

Doutt, R.L. & Nakata, J. (1965) Overwintering refuge of *Anagrus epos* (Hymenoptera: Mymaridae). *Journal of Economic Entomology* 58, 586.

Dreistadt, S.H. & Dahlsten, D.L. (1991) Establishment and overwintering of *Tetrastichus gallerucae* (Hymenoptera: Eulophidae), an egg parasitoid of the elm leaf beetle (Coleoptera: Chrysomelidae) in northern California. *Environmental Entomology* 20, 1711–1719.

Drooz, A.T. & Solomon, J.D. (1980) Rearing the egg parasite *Ooencyrtus ennomophagus* (Hymenoptera: Encyrtidae) on eggs of *Clostera inclusa* (Lepidoptera: Notodontidae) kept below freezing. *Canadian Entomologist* 112, 739–740.

Dysart, R.J. (1990) The introduction and recovery in the United States of *Anaphes diana* (Hym.: Mymaridae), an egg parasite of *Sitona* weevils (Col.: Curculionidae). *Entomophaga* 35, 307–313.

Gautam, R.D. (1986) Effect of cold storage on the adult parasitoid *Telenomus remus* Nixon (Scelionidae: Hymenoptera) and the parasitised eggs of *Spodoptera litura* (Fabr.) (Noctuidae: Lepidoptera). *Journal of Entomological Research* 10, 125–131.

Griffiths, K.J. & Sullivan, C.R. (1978) The potential for establishment of the egg parasite *Ooencyrtus kuwanai* in Ontario populations of the gypsy moth. *Canadian Entomologist* 110, 633–638.

Grijpma, P. (1984) Host specificity of *Telenomus nitidulus* (Thomson) (Hymenoptera: Scelionidae), egg parasite of the satin moth, *Leucoma salicis* L. *Nederlands Bosbouwtijdschrift* 56, 201–207.

Hafez, M., El-Kifl, A.H. & Fayad, Y.H. (1977) On the bionomics of *Platytelenomus hylas* Nixon, an egg parasite of *Sesamia cretica* Led., in Egypt. *Bulletin of the Society of Entomology of Egypt* 61, 161–178.

Hance, T. & Boivin, G. (1993) Effect of parasitism by *Anaphes* n. sp. (Hymenoptera: Mymaridae) on cold hardiness of *Listronotus oregonensis* eggs (Coleoptera: Curculionidae). *Canadian Journal of Zoology* 71, 759–764.

Henderson, C.F. (1955) Overwintering, spring emergence and host synchronization of two egg parasites of the beet leafhopper in southern Idaho. *United States Department of Agriculture Circular* No. 967, 16 pp.

Hulme, M.A., Dawson, A.F. & Harris, J.W.E. (1986) Exploiting cold-hardiness to separate *Pissodes strobi* (Peck) (Coleoptera: Curculionidae) from associated insects in leaders of *Picea sitchensis* (Bong.) Carr. *Canadian Entomologist* 118, 1115–1122.

Jackson, C.G. (1986) Effects of cold storage of adult *Anaphes ovijentatus* on survival, longevity, and ovipositor. *Southwestern Entomologist* 11, 149–153.

Jackson, C.G. (1987) Biology of *Anaphes ovijentatus* (Hymenoptera: Mymaridae) and its host, *Lygus hesperus* (Hemiptera: Miridae), at low and high temperatures. *Annals of the Entomological Society of America* 80, 367–372.

Jackson, D.J. (1963) Diapause in *Caraphractus cinctus* Walker (Hymenoptera: Mymaridae), a parasitoid of the eggs of Dytiscidae (Coleoptera). *Parasitology* 53, 225–251.

Jalali, S.K. & Singh, S.P. (1992) Differential response of four *Trichogramma* species to low temperatures for short term storage. *Entomophaga* 37, 159–165.

James, D.G. (1988) Fecundity, longevity and overwintering of *Trissolcus biproruli* Girault (Hymenoptera: Scelionidae) a parasitoid of *Biprorulus bibax* Breddin (Hemiptera: Pentatomidae). *Journal of the Australian Entomological Society* 27, 297–301.

Jennings, D.T. & Houseweart, M.W. (1983) Parasitism of spruce budworm (Lepidoptera: Tortricidae) eggs by *Trichogramma minutum* and absence of

overwintering parasitoids. *Environmental Entomology* 12, 535–540.

Keller, M.A. (1986) Overwintering by *Trichogramma exiguum* in North Carolina. *Environmental Entomology* 15, 659–661.

Kido, H., Flaherty, D.L., Bosch, D.F. & Valero, K.A. (1984) French prune trees as overwintering sites for the grape leafhopper egg parasite. *American Journal of Entomology and Viticulture* 35, 156–160.

Kumar, D.A., Divakar, B.J. & Pawar, A.D. (1984) Observation on the storage of life stages of *Telenomus remus* Nixon (Hymenoptera: Scelionidae) under low temperature. *Plant Protection Bulletin* 36, 13–14.

Lee, R.E. (1991) Principles of insect low temperature tolerance. In: Lee, R.E. & Denlinger, D.L. (eds) *Insects at Low Temperature*. Chapman and Hall, New York, pp. 17–46.

Lees, A.D. (1953) Experimental factors controlling the evocation and termination of diapause in the fruit tree red spider mite *Metatetranychus ulmi* Koch (Acarina: Tetranychidae). *Annals of Applied Biology* 40, 449–486.

López, J.D. & Morrison, R.K. (1980a) Overwintering of *Trichogramma pretiosum* in Central Texas. *Environmental Entomology* 9, 75–78.

López, J.D. & Morrison, R.K. (1980b) Susceptibility of immature *Trichogramma pretiosum* to freezing and subfreezing temperatures. *Environmental Entomology* 9, 697–700.

McKenzie, L.M. & Beirne, B.P. (1972) A grape leafhopper, *Erythroneura ziczac* (Homoptera: Cicadellidae), and its mymarid (Hymenoptera) egg-parasite in the Okanagan valley, British Columbia. *Canadian Entomologist* 104, 1229–1233.

Meyerdirk, D.E. & Moratorio, M.S. (1987) Seasonal population density of *Anagrus giraulti* (Hymenoptera: Mymaridae), an egg parasitoid of *Circulifer tenellus* and *Empoasca* spp. (Homoptera: Cicadellidae). *Journal of Economic Entomology* 80, 362–365.

Moore, S.D. (1989) Regulation of host diapause by an insect diapause. *Ecological Entomology* 14, 93–98.

Oatman, E.R. & Platner, G.R. (1972) Colonization of *Trichogramma evanescens* and *Apanteles rubecola* on the imported cabbageworm on cabbage in Southern California. *Environmental Entomology* 1, 347–352.

Obrycki, J.J., Tauber, M.J., Tauber, C.A. & Gollands, B. (1985) *Edovum puttleri* (Hymenoptera: Eulophidae), an exotic egg parasitoid of the Colorado potato beetle (Coleoptera: Chrysomelidae): responses to temperate zone conditions and resistant potato plants. *Environmental Entomology* 14, 48–54.

Overholt, W.A. & Smith, J.W. (1990) Colonization of six exotic parasites (Hymenoptera) against *Diatraea grandiosella* (Lepidoptera: Pyralidae) in corn. *Environmental Entomology* 19, 1889–1902.

Pak, G.A., Buis, H.C.E.M., Heck, I.C.C. & Hermans, M.L.G. (1986) Behavioural variations among strains of *Trichogramma* spp.: host-age selection. *Entomologia Experimentalis et Applicata* 40, 247–258 .

Parker, F.D. & Pinnell, R.E. (1971) Overwintering of some *Trichogramma* spp. in Missouri. *Journal of Economic Entomology* 64, 80–81.

Peterson, A. (1931) Refrigeration of *Trichogramma minutum* Riley and other notes. *Journal of Economic Entomology* 24, 1070–1074.

Pizzol, J. & Voegelé, J. (1987) The diapause of *Trichogramma maidis* Pintureau & Voegelé in relation to some characteristics of its alternative host *Ephestia kuehniella* Zell. In: Boulétreau, M. & Bonnot, G. (eds) *Parasitoid Insects. Les Colloques de l'INRA* 48, 93–94.

Pucci, C. & Dominici, M. (1988) Field evaluation of *Edovum puttleri* Grissel (Hym., Eulophidae) on eggs of *Leptinotarsa decemlineata* Say (Col., Chrysomelidae) in Central Italy. *Journal of Applied Entomology* 106, 465–472.

Reznik, S.Y. & Umarova, T.Y. (1990) The influence of host's age on the selectivity of parasitism and fecundity of *Trichogramma. Entomophaga* 35, 31–37.

Ring, R.A. & Tesar, D. (1981) Adaptations to cold in Canadian Arctic insects. *Cryobiology* 18, 199–211.

Rodriguez-del-Bosque, L.A., Smith, J.W. & Pfannenstiel, R.S. (1989) Parasitization of *Diatraea grandiosella* eggs by Trichogrammatids on corn in Jalisco, Mexico. *Southwestern Entomologist* 14, 179–180.

Ryan, R.B., Mortensen, R.W. & Torgersen, T.R. (1981) Reproductive biology of *Telonomus californicus* Ashmead, an egg parasite of the Douglas-fir tussock moth: laboratory studies. *Annals of the Entomological Society of America* 74, 213–216.

Salt, R.W. (1961) Principles of insect cold hardiness. *Annual Review of Entomology* 6, 55–74.

Saunders, D.S. (1982) *Insect Clocks*, 2nd edn. Pergamon Press, Oxford, 409 pp.

Schaber, B.D. (1981) Recovery of *Patasson luna* (Hymenoptera: Mymaridae) a parasite of the alfalfa weevil, *Hypera postica* (Coleoptera: Curculionidae), in Alberta. *Quaestiones Entomologicae* 17, 169–170.

Seyedoleslami, H. & Croft, B.A. (1980) Spatial distribution of overwintering eggs of the white apple leafhopper, *Typhlocyba pomaria*, and parasitism by *Anagrus epos*. *Environmental Entomology* 9, 624–628.

Singh, H.N., Lakshmi, A.U. & Singh, H.K. (1987) Natural parasitization in eggs of *Lymantria obfuscata* Walk. at Shrinagar (J. & K.). *Bulletin of Entomology (New Delhi)* 28, 70–72.

Slosser, J.E., Price, J.R. & Puterka, G.J. (1990) Manipulation of overwintering survival and spring emergence of *Bracon mellitor* Say (Hymenoptera: Braconidae). *Environmental Entomology* 19, 1110–1114.

Smith, S.M., Hubbes, M. & Carrow, J.R. (1987) Ground releases of *Trichogramma minutum* Riley (Hymenoptera: Trichogrammatidae) against the spruce budworm (Lepidoptera: Tortricidae). *Canadian Entomologist* 119, 251–263.

Sømme, L. (1982) Supercooling and winter survival in terrestrial arthropods. *Comparative Biochemistry and Physiology* 73, 519–543.

Sorokina, A.P. & Maslennikova, V.A. (1988) Optimum temperature for diapause formation in species of *Trichogramma* (Hymenoptera, Trichogrammatidae). *Entomological Review* 67, 40–51.

Stengel, M., Voegelé, J. & Lewis, J.W. (1977) Les Trichogrammes. V.b. Survie hivernale de *Trichogramma evanescens* Westw. souche moldave et découverte de *T. cacoeciae* Mar. sur pontes d'*Ostrinia nubilalis* Hübn. dans les conditions agroclimatiques de l'Alsace. *Annales de Zoologie et d'Ecologie Animale* 9, 313–317.

Stinner, R.E., Ridgway, R.L. & Kinzer, R.E. (1974) Storage, manipulation of emergence, and estimation of numbers of *Trichogramma pretiosum*. *Environmental Entomology* 3, 505–507.

Storey, K.B. & Storey, J.M. (1988) Freeze tolerance in animals. *Physiological Review* 68, 27–84.

Storey, K.B. & Storey, J.M. (1991) Biochemistry of cryoprotectants. In: Lee, R.E. &

244   *G. Boivin*

Denlinger, D.L. (eds) *Insects at Low Temperature*. Chapman and Hall, New York, pp. 64–93.

Sullivan, C.R., Griffiths, K.J. & Wallace, D.R. (1977) Low winter temperatures and the potential for establishment of the egg parasite *Anastatus disparis* (Hymenoptera: Eupelmidae) in Ontario populations of the gypsy moth. *Canadian Entomologist* 109, 215–220.

Takagi, M. (1987) The reproductive strategy of the gregarious parasitoid, *Pteromalus puparum* (Hymenoptera: Pteromalidae). 3. Superparasitism in a field population. *Oecologia* 71, 321–324.

Tauber, M.J., Tauber, C.A., Nechols, J.R. & Obrycki, J.J. (1983) Seasonal activity of parasitoids: control by external, internal and genetic factors. In: Brown, V.K. & Hodek, I. (eds) *Diapause and Life Cycle Strategies in Insects*. Dr. W. Junk Publishers, The Hague, pp. 87–108.

Tauber, M.J., Tauber, C.A. & Masaki, S. (1986) *Seasonal Adaptations of Insects*. Oxford University Press, New York, 411 pp.

Tenow, O. & Nilssen, A. (1990) Egg cold hardiness and topoclimatic limitations to outbreaks of *Epirrita autumnata* in northern fennoscandia. *Journal of Applied Ecology* 27, 723–734.

Torgersen, T.R. & Ryan, R.B. (1981) Field biology of *Telonomus californicus* Ashmead, an important egg parasite of Douglas-fir tussock moth. *Annals of the Entomological Society of America* 74, 185–186.

Voegelé, J. (1976) La diapause et l'hétérogénéité du développement chez les *Aelia* (Heteroptera, Pentatomidae) et les Trichogrammes (Hymenoptera, Trichogrammatidae). *Annales de Zoologie et d'Ecologie Animale* 8, 367–371.

Voegelé, J., Pizzol, J., Raynaud, B. & Hawlitzky, N. (1986) La diapause chez les Trichogrammes et ses avantages pour la production de masse et la lutte biologique. *Mededelingen Faculteit Landbouwwetenschappen Rijksuniversiteit Gent* 51, 1033–1039.

Voegelé, J., Pizzol, J. & Babi, A. (1988) The overwintering of some *Trichogramma* species. In: Voegelé, J., Waage, J.K. & van Lenteren, J.C. (eds) Trichogramma *and Other Egg Parasites. Les Colloques de l'INRA* 43, 275–282.

Walker, T.J. (1980) Migrating Lepidoptera: are butterflies better than moths? *Florida Entomologist* 63, 79–98.

Windig, J.J. (1991) Life cycle and abundance of *Longitarsus jacobaea* (Col.: Chrysomelidae), biocontrol agent of *Senecio jacobaea*. *Entomophaga* 36, 605–618.

Zachariassen, K.E. (1985) Physiology of cold tolerance in insects. *Physiological Review* 65, 799–832.

Zaslavski, V.A. (1988) *Insect Development. Photoperiodic and Temperature Control*. Springer Verlag, Berlin, 187 pp.

Zaslavski, V.A. & Umarova, T.Y. (1982) Photoperiodic and temperature control of diapause in *Trichogramma evanescens* Westw. (Hymenoptera, Trichogrammatidae). *Entomological Review* 60, 1–12.

Zaslavski, V.A. & Umarova, T.Y. (1990) Environmental and endogenous control of diapause in *Trichogramma* species. *Entomophaga* 35, 23–290.

Intra-population Genetic Variation in *Trichogramma* 12

ERIC WAJNBERG
Laboratoire de Biologie des Invertébrés, Unité de Biologie des
Populations, INRA, 37 Bld. du Cap, 06600 Antibes, France

Abstract

As pointed out by several authors, accurate quantification of genetic variation within beneficial organism populations should lead to an improvement in their efficiency in the control of pests. In *Trichogramma*, however, genetic variation in quantitative biological traits has seldom been investigated.

In this chapter, the first section summarizes current knowledge about intra-specific, intra-population genetic variation in this genus. Then, results of experimental analyses of the variation observed in different biological traits of *T. brassicae* Bezdenko, related to their reproductive strategy, will be presented. These traits concern: (i) the way females detect the presence of their hosts and attack them; (ii) the way they distribute their progeny in them; and (iii) the way they determine the optimal sex ratio they have to lay according to their own density on a host patch (local mate competition). For each of these traits, a simulation model, using Monte Carlo procedures, is worked out in order to understand the ecological implications of the observed variation.

Introduction

Species of the hymenopterous genus *Trichogramma* Westwood 1879 have been used more than any other natural enemies for inundative biological control programmes (DeBach, 1974; Stinner, 1977; DeBach & Rosen, 1991) (see Chapter 2 for a recent worldwide survey). This is mainly due to the fact that this genus has a wide range of hosts, especially among the Lepidoptera (Nagarkatti & Nagaraja, 1977). Indeed, this wide range of hosts enables *Trichogramma* to be both potential efficient candidates to control a large

number of phytophagous pests, and to be mass-produced inexpensively on different suitable factitious hosts (Starler & Ridgway, 1977). Since their first use at the beginning of this century in the USA and in eastern Europe, these egg parasitoids have been mass-produced and released in several countries on to pests attacking a large variety of crops, including: corn (Hassan, 1981; Voegelé, 1981; Bigler, 1986); cotton (King et al., 1986); tomato (Voegelé et al., 1990); cabbage (Hassan & Rost, 1985); vines (Babi, 1990); and many others.

The genus presents a high specific and geographic diversity (Pintureau, 1990; see Chapter 1), which should make it possible to select the optimal species and strains to control future pests, as long as the egg is the vulnerable stage. But, in order to be as efficient as possible in this process, several ecological mechanisms must first be accurately understood (Pak, 1988; see Chapter 3). More precisely, three main steps must to be studied:

1. The way the different *Trichogramma* species and/or populations interact with each other, with their hosts and with environmental factors has to be analysed thoroughly in natural conditions. Such analysis, which has to be handled from a population point of view (Wajnberg, 1991a), should result both in a reliable quantification of the mechanisms involved in the host specificity process, and in the accurate description of the different populations' biological characteristics within each species studied. This first step, based on both an inter-specific and an intra-specific, inter-population analysis, should optimize the selection of the appropriate species and population to control a given pest on a given crop and in a given environment.

2. Once the most suitable candidate has been chosen, it has to be efficiently mass-produced, usually on a factitious host and under highly standardized conditions. Such a mass-rearing procedure usually induces severe selection pressures that may lead both to a loss in the population genetic variability (Boller, 1972; Mackauer, 1972; Boller & Chambers, 1977; Joslyn, 1984), and to the uncontrolled selection for undesirable biological characteristics (Schmidt, 1991; Wajnberg, 1991a). Methods have to be developed to reduce as much as possible such negative mechanisms, and the corresponding analysis is now based on an intra-specific, intra-population approach. Genetic variability in the biological traits of the wasp produced has to be quantified, and an accurate study has to be done to estimate to what extent the mass-produced population is able to evolve in response to the rearing conditions and how long it takes.

3. Finally, the ecological consequences of inundative releases of the wasp have to be quantitatively predicted. More particularly, the pest control level has to be accurately estimated. This analysis should lead to the identification of the biological traits of the wasp that is involved in controlling the target pest. Therefore, an artificial selection procedure could be started to increase the pest control ability (Wajnberg, 1991a). The possible negative environmental impact of such field treatments should also be studied (Howarth, 1991). For instance, we have to know to what extent the released wasp population competes with endemic parasitic species and how such mechanisms could change the pest control level. This is thus an intra-specific, inter- and intra-population analysis.

In this chapter, only the intra-specific, intra-population level of variability in different biological traits of *Trichogramma brassicae* Bezdenko (= *T. maidis* Pintureau & Voegelé) related to their reproductive strategy will be quantified in a given population, and its ecological meaning will be estimated. This level of variability will be discussed here because analyses at this level of genetic variation may: (i) lead to an accurate estimation of adaptive responses of *Trichogramma* to the environmental conditions encountered during the mass-production process; and (ii) provide the necessary means to start artificial selection programmes in order to improve wasp efficiency.

The *T. brassicae* population studied here originated from the Moldavian population which has been efficiently used in France since 1974 against the European corn borer, *Ostrinia nubilalis* Hübner (Pintureau & Voegelé, 1980; Voegelé, 1981). Among the different biological features of the *Trichogramma* reproductive strategy, those that have been analysed are likely to be connected with both our ability to mass-produce them on a factitious host (in our case the Mediterranean flour moth, *Ephestia kuehniella* Zeller (Daumal *et al.*, 1975)), and with their efficiency as biocontrol agents. More precisely, they concern: (i) the way females detect the presence of their hosts and attack them; (ii) the way they distribute their progeny in them (progeny allocation strategy); and (iii) the way they determine the optimal sex ratio they have to lay (for arrhenotokous species) according to the different environmental characteristics encountered (sex allocation strategy). Thus, they concern the mechanisms involved in both the host selection process (Vinson, 1976; Weseloh, 1981; see Chapter 9) and family planning (Waage, 1986).

The first section will summarize what is already known about the intra-specific, intra-population genetic variability in quantitative traits in *Trichogramma*. Then, results of genetic analysis of the variation observed in the different biological characteristics studied will be presented. Finally, results of simulation models, based on Monte Carlo procedures, will be given in order to understand the ecological meanings of such genetic variability and how it may be manipulated in order to improve the efficiency of the mass-produced and released population.

Intra-population Genetic Variation in the *Trichogramma* Genus

Despite the fact that genetic variations in *Trichogramma* have been described both between species and between populations (see Pak, 1988, for a review),

genetic variability within natural populations has seldom been investigated. Some of the corresponding works were done to describe the variability in selectively neutral traits: at the molecular level, either for enzymatic proteins (Pintureau, 1988) or for mitochondrial DNA (Vanlerberghe, 1991). For other biological characteristics, genetic variations were studied at most of the steps of the parasitization process. Chronologically, eight types of traits were analysed:

1. Walking behaviour. For *Trichogramma*, the last step of the host-finding behaviour is predominantly performed by walking (see below for a detailed experimental analysis). Limburg & Pak (1991) have shown that this walking behaviour presents a significant genetic variability in populations of *T. evanescens* Westwood and *T. dendrolimi* Matsumura. Using a family analysis (see Wajnberg, 1991a, for a general review of methods available to estimate genetic variability in quantitative traits) on a drawing of the video-recorded females' tracks, genetic variation was observed in both species, for various parameters used to describe the female's walking path: walking speed, turning angle and pause duration.

2. Spatial distribution of parasitizations. The average distance between all the hosts attacked by an isolated female also shows a significant genetic variability in populations of *T. brassicae*, *T. cacoeciae* Marchal (Chassain & Boulétreau, 1987; Chassain, 1988; Chassain *et al.*, 1988c) and *T. voegelei* Pintureau (Mimouni, 1990): some females aggregate their eggs in adjacent hosts whereas others scatter them amongst distant hosts. This trait is correlated with the linear speed of the females' displacement: females that present the higher linear speed present higher dispersion of their attacks among hosts (Chassain, 1988). Moreover, in the *T. brassicae* population, an artificial selection programme has made it possible to separate different strains according to the spatial distribution of host attacks (Chassain *et al.*, 1988b).

3. Handling time. Once a host is found, the female has to decide whether it is suitable for the development of her progeny before attacking it. This step involves three major behaviours: antennal drumming, ovipositor drilling and oviposition (see Klomp *et al.*, 1980, for a detailed description). Using mother–daughter regression analysis, a significant genetic variation has been shown in the duration of the oviposition phase within a *T. brassicae* population (Wajnberg, 1988, 1989). However, in the same population, the variation in the duration of both the drumming and the drilling phases does not seem to be under genetic control.

4. Capacity of parasitization. The number of hosts attacked by isolated females, usually estimated by the number of hosts that turn black after a minimal development, also presents a significant genetic variability within populations of *T. cacoeciae* (Babi, 1990), *T. bourarachae* Pintureau & Babault and *T. voegelei* (Mimouni, 1990, 1991).

5. Superparasitism. Based on histological staining which makes it possible to count the number of eggs deposited in each host (*E. kuehniella*), a clear genetic variability in superparasitism intensity has been demonstrated in a *T. brassicae* population (Wajnberg & Pizzol, 1989; Wajnberg *et al.*, 1989). This was observed both for the average number of eggs laid in each host, and for the frequency distribution of wasp eggs among hosts. More recently, such a result was confirmed in a population of *T. evanescens*, by counting the number of adult wasps emerging from *Manduca sexta* L. eggs attacked by isolated females (Schmidt, 1991).

6. Sex ratio. The proportion of sons (or daughters) produced by mated arrhenotokous females also shows a significant genetic variation in a *T. brassicae* population (Chassain, 1988; Chassain & Boulétreau, 1991). Moreover, this intra-population variability is observed not only in the sex ratio laid by the females, but also in the way they produce it accurately. Indeed, using a non-parametric, multivariate statistical method that makes it possible to describe oviposition sequences, a strong genetic variability was observed in the way that *T. brassicae* females organize the sequence of the sexes in their progeny (Wajnberg, 1991b,c, 1993).

7. Rate of development. A significant intra-population genetic variability was also observed in the rate of preimaginal development in a *T. voegelei* population (Mimouni, 1990). However, this result, based on statistical quantification of the blackening kinetic of parasitized hosts, only concerns the development of the first larval stages.

8. Preimaginal viability. Finally, a significant genetic variation also exists in the rate of success of the preimaginal development in populations of *T. brassicae* (Chassain, 1988; Chassain & Boulétreau, 1991) and *T. voegelei* (Mimouni, 1990).

All these intra-population genetic variabilities concern important biological and ecological *Trichogramma* features. However, at least for some of them, their real meaning remains to be understood. Therefore, in the following sections, some of these traits will be analysed simultaneously in the same population. Stochastic models that will be worked out will lead to the estimation of how such genetic variations enable wasp populations to evolve in response to the biotic and abiotic environmental characteristics encountered.

Genetic Variation in the Strategy of Progeny Distribution

In fact, genetic variations in the spatial dispersion of parasitizations mentioned above and in the intensity of superparasitism are likely to correspond to a global variability in the different components of a whole strategy of progeny distribution by *Trichogramma* females among available hosts. Therefore, a genetic analysis was carried out on these two biological features simultaneously in the same *T. brassicae* population. Twenty-three families

Table 12.1. Analysis of variance used to compare the average values of the spatial distribution of the attacks by *T. brassicae* females, and the superparasitism intensity between the 23 families analysed.

Source of variation	Distribution of the attacks			Superparasitism intensity		
	d.f.	Variance	F	d.f.	Variance	F
Families	22	15.727×10^{-3}	3.78*	22	43.171	3.07*
Error	112	4.161×10^{-3}		110	14.051	
Total	134	6.052×10^{-3}		132	18.904	

*$P < 0.01$.

were found, each from a single couple taken at random from recently trapped individuals. For each of them, the distribution of the attacks was estimated in a Petri dish by providing single, newly emerged, mated females with an 'artificial' patch of 500 UV-killed *E. kuehniella* eggs, for 24 h. After 5 days the attacked hosts turned black and the patch was digitalized with video equipment. The average distance between all the parasitized eggs, taken two by two, was then computed. The estimation of the superparasitism intensity, for each replicate and for each family, entailed providing ten mated females (less than 24 h old) with 20 UV-killed *E. kuehniella* eggs for 24 h. Then, for each replicate, using the histological staining method described in Wajnberg *et al.* (1989), the number of *T. brassicae* eggs laid in each host was counted. The main difference between these two experimental procedures concerns the female : host ratio. In the analysis of the distribution of attacks among hosts, this ratio equals 1 : 500; for the estimation of the superparasitism intensity, this ratio reaches 1 : 2. This difference has been decided in order to have, in each case, the best experimental conditions to quantify the trait under study.

Table 12.1 shows that a strongly significant difference between the average values of the 23 families exist for both traits. This result confirms those obtained by Chassain & Boulétreau (1987) and by Wajnberg *et al.* (1989): the way the females distribute their attacks among hosts is a family feature; this is also the case for the intensity of superparasitism. Therefore, variations in these two traits appear to be under genetic control in the *T. brassicae* population analysed.

Furthermore, Fig. 12.1 shows that there is a significant negative relationship between these two biological characteristics. Thus, within the analysed population, some genotypes lead the females to disperse intensively their attacks with a low level of superparasitization, while others lead them to lump spatially their parasitization with a higher average number of eggs laid in each host.

Such a result seems to indicate the simultaneous existence, within the same population, of different strategies used by the females to colonize host

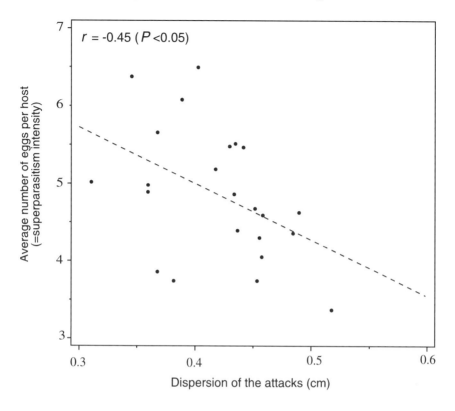

Fig. 12.1. Correlation between the spatial distribution of the attacks and the superparasitism intensity (estimated by the average number of eggs laid in each host) by 23 *T. brassicae* families.

patches and to distribute their progeny. In order to lay eggs in healthy hosts, some females would allocate their time to displacement, so that the probability of encountering already attacked hosts will progressively decrease. On the other hand, other females would bet on the superparasitization of spatially lumped hosts with a sufficient probability to give rise to a healthy progeny. The fact that these different strategies can simultaneously be found in the same population probably plays an important part in the ability of this population to respond to variations in the environmental characteristics encountered. This is particularly true for the species studied here which is known to be polyphagous and thus to experience high spatial and temporal variations in the characteristics of their hosts (i.e. variations in the hosts' density and spatial distribution).

   However, these results are in contradiction with the theoretical results obtained by Charnov & Skinner (1984, 1985) which predict that the optimal clutch size laid by females increases with an increase in the time allocated to travel between hosts. Such a contradiction seems to indicate that, at least in

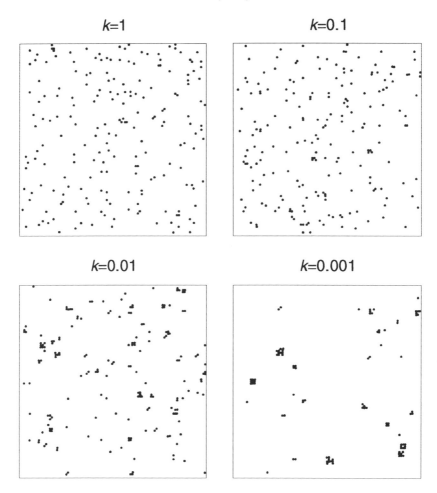

Fig. 12.2. Simulated spatial distribution of 200 hosts on a 100 × 100 grid, as a function of k, the parameter of a negative binomial distribution.

the present case, selective constraints differ from those related to the way the females manage their time. More precisely, it could be interesting to understand the real adaptive meaning of the negative genetic correlation presented Fig. 12.1. What are the real selective values of the different strategies observed in relation to the variation in the characteristics of the different types of hosts attacked?

In order to answer this question, a simple stochastic model has been worked out to simulate the evolution, in the course of generations, of the genetic structure of a hypothetical *Trichogramma* population, according to different environmental parameters. In this model, the spatial distribution of the hosts was simulated using a negative binomial distribution. As shown in

Table 12.2. Four hypothetical genotypes and the corresponding phenotypes used to describe the genetic structure of a *Trichogramma* population in a simulation model. Values for both traits are chosen to simulate the experimental result given in Fig. 12.1. Dispersion ability is given in arbitrary units (see text).

	A	B	C	D
Eggs laid per host	1	2	3	4
Dispersion ability	32	16	8	4

Fig. 12.2, this enables the generation of any kind of spatial pattern, from random (i.e. Poisson) to very aggregated distributions, by changing only the value of k, a parameter measuring the degree of aggregation.

The model works as follows. At the beginning, the wasp population consists of 20 thelytokous females, divided into four equal genotypic classes whose phenotypic characteristics are described in Table 12.2, and each of them has a fixed number of eggs to lay in her life.

Each host cannot contain more than four wasp eggs. So, for a given generation, all females repeat continuously the two following steps, until all eggs are laid:

1. In the attacked host (which is randomly chosen only at the first step), each female lays either a number of eggs corresponding to her own genotype (see Table 12.2), or, if this is not possible, adds one or several wasp eggs into the host to get a total of four.
2. The females then move to another host, the minimal travelling distance being determined by genotype (see Table 12.2), and the direction being randomly drawn from a uniform distribution between 0 and 2π.

The number and genotype of adults wasps emerging from all the attacked hosts were then computed using the simple following equation (Skinner, 1985):

$$w = \frac{(5 - n)^s}{4}$$

which is illustrated in Fig. 12.3.

This equation relates the average fitness (w) of wasps (here, the probability of emergence) to the number (n) developing in each host. s defines the shape of the relationship and serves as a measure of the level of competition (for food and/or space) within the host. Using this equation, the frequency of each genotype is computed, and the procedure is run again for the next generation which is made up of 20 new females; this is done up to

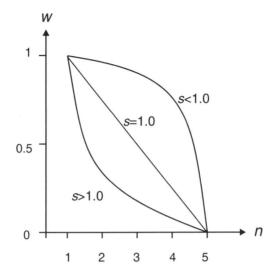

Fig. 12.3. Different possible relationships between the number of eggs developing in a host (n) and their probability of emergence (w). The parameter s measures the level of competition within the host (after Skinner, 1985).

the 50th generation. Results are presented Fig. 12.4.

When there is an intense level of competition within hosts ($s > 1$), and whatever their spatial distribution or females' fecundity, the only remaining genotype after a few generations is the one that presents the lowest level of superparasitism. However, when the level of competition decreases ($s < 1$), the situation appears to be rather different and depends on the hosts' spatial pattern. More precisely, when hosts are strongly aggregated ($k = 0.001$), and especially when females' fecundity is high, a genetic polymorphism seems to be maintained, even after 50 generations of simulation. Therefore, the negative correlation presented Fig. 12.1 could be a characteristic of a *Trichogramma* population attacking spatially aggregated hosts such as the naturally occurring *O. nubilalis* egg masses in the field. Thus, the corresponding genetic variability could be advantageous when this population experiences spatial and temporal variations in the characteristics of their hosts.

Finally, superparasitism is usually considered as disadvantageous (Fiske, 1910; Lenteren, 1981; but see also Alphen & Nell, 1982; Bakker *et al.*, 1985; Alphen *et al.*, 1987; Alphen & Visser, 1990). However, in the present model, it is worth noting that, under some particular conditions, superparasitism, when correlated with other biological traits (here, females' dispersion ability), can be maintained in an evolving population.

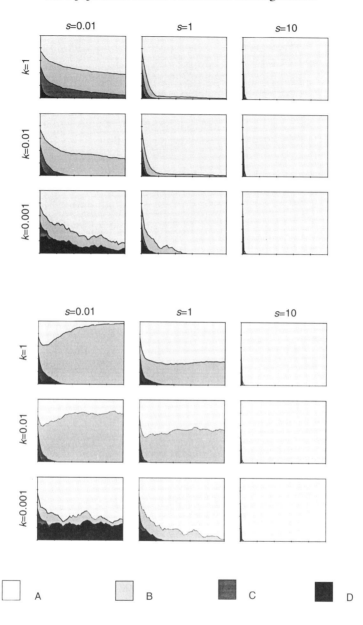

Fig. 12.4. Evolution of the genotype frequency during 50 generations of simulation. In the model, the spatial distribution of hosts range from random ($k = 1$) up to very aggregated patterns ($k = 0.001$) (see Fig. 12.2). s measures the level of competition within hosts (see Fig. 12.3). At each generation, wasp females can lay a maximum of either 20 (top) or 40 (bottom) eggs in hosts. The characteristics of each genotype are described in Table 12.2. Each graph is the average of ten simulations.

Genetic Variation in the Ability of *Trichogramma* Females to Perceive Their Hosts

The fact that *Trichogramma* females may respond to spatial distribution and, therefore, to the presence of their hosts strongly suggests that they are able to perceive them at a distance, using both chemical and visual cues. Laing (1937, 1938) was the first to try to estimate the maximal distance from which *T. evanescens* females perceive their hosts. According to an indirect method based on percentage parasitism, Laing concluded that *T. evanescens* visually perceived eggs of *Sitotroga cereallela* Olivier from a distance of 1.8 mm. More recently, more accurate methods were developed to estimate this trait, all in an indirect way, some of them using video equipment (Edwards, 1961; Yano, 1978; Glas *et al.*, 1981; Pak *et al.*, 1991).

Because this maximal distance, now called 'reactive distance' (Roitberg, 1985; Pak *et al.*, 1991), is likely to be correlated with wasp searching efficiency, it is usually assumed to be an important factor determining parasitic wasps' success as biological control agents. Thus, an increase, through artificial selection, of the average value of this trait in the population released would improve pest control efficiency.

In order to analyse such a hypothesis, a stochastic model, simulating the walking path of isolated *Trichogramma* females during their searching behaviour, was developed. This model uses the one described by Yano (1978) after an experimental analysis of *T. dendrolimi* walking behaviour. In this model, a path is generated using the following assumptions (using Yano's notation):

• The path orientation of a female leaving the host she has just attacked is drawn from a uniform distribution between $-\pi$ and π, the corresponding walking speed being fixed to its minimal value, V_{min}.
• Then, the female's walking speed increases linearly up to its maximal value, V_{max}, in relation to the distance to the last attacked host, the path direction for the nth step being drawn from a normal distribution with the direction in the $n-1$th step as mean and SD as standard deviation. This last procedure leads to an autocorrelation in the successive path directions, but this autocorrelation is not taken into account if the absolute value of the last path direction is greater that a given value: TH_{max} (see Yano, 1978, for a more detailed explanation).
• When the distance to the last attacked host gets greater than RSA (using Yano's notation), or if the time before this last attack becomes greater than GT (i.e. 'giving-up time'), the female is supposed to walk more rapidly and straighter: the walking speed is turned to V_{max}, and the successive directions of the path are drawn from a normal distribution with always the same mean and SD.
• Finally, when the female enters within a circle of a certain radius RD (i.e.

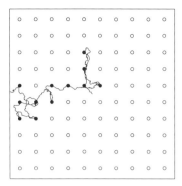
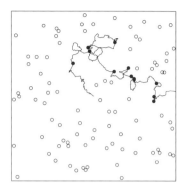
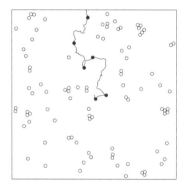

Fig. 12.5. Examples of the simulation of the walking behaviour of isolated *Trichogramma* females using the model published by Yano (1978). In each case there are 100 hosts, and their spatial distribution is either regular (top left), random (top right) or aggregated (bottom). The aggregated spatial pattern of the hosts is simulated using a negative binomial distribution (see Fig. 12.2). The female is always released in the centre of the graph, and the simulation stopped as soon as the females reached the outer square. In all cases, the parameters' values are: SD = 40 degrees; TH_{max} = 30 degrees; V_{min} = 1.5; V_{max} = 7.5; RSA = 80; RD = 30; GT = 300 (arbitrary units).

reactive distance; *RP* in Yano's notation) with any healthy host as centre, she moves straight to this host and attacks it.

These assumptions were used with hosts showing different spatial patterns, ranging from regular to aggregated distributions. Fig. 12.5 gives some examples of the results obtained.

Whatever the host spatial distribution, an increase in the reactive distance clearly leads to an increase in the number of hosts attacked (Fig. 12.6). This trait is thus likely to be correlated with wasp efficiency in biological control release programmes.

Moreover, according to this result, females that have to be close to their

258 E. Wajnberg

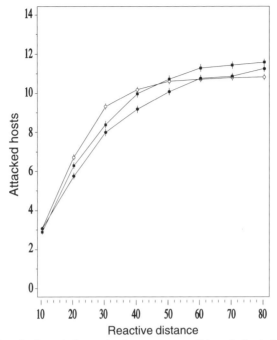

Fig. 12.6. Effect of a change in the reactive distance on wasp efficiency (estimated by the number of hosts attacked) obtained by simulation. Each point is the average (± 2 SD) of 2000 simulations, each run as explained in Fig. 12.5 with hosts showing regular (○), random (■) or aggregated (●) spatial patterns.

hosts to perceive them (i.e. low reactive distance) seem to be more efficient when their hosts are regularly distributed. On the other hand, females that perceive their hosts from a longer distance (i.e. high reactive distance) appear more efficient on randomly distributed hosts. Thus, owing to the fact that wasp females may experience a variation in the spatial pattern of their hosts, a genetic variation in the reactive distance in a *Trichogramma* population would be advantageous.

Therefore, the genetic variability in this trait was estimated in the *T. brassicae* population analysed in the previous section. Because both smell and visual cues are likely to be involved in this trait, results obtained on three types of hosts were compared: (i) eggs of *E. kuehniella*, oval-shaped, about 0.5–0.6 mm long and 0.3–0.4 mm wide; (ii) eggs of *Mamestra brassicae* L., spherical, about 0.6 mm in diameter; and (iii) odourless, transparent spherical glass beads with a diameter of 0.45 mm.

For each of these hosts, respectively 30, 19 and 29 less than 24-h-old experienced females taken at random from the mass-reared population, were observed. For all of them, three of their daughters were also observed at the next generation. For each experiment, an isolated female was released in a Petri dish (9 cm diameter) in which 12 hosts were placed in a regular pattern,

Table 12.3. Overall average (± 1.96 SE) reactive distance (mm) of
T. brassicae females for each host tested.

	n	Reactive distance*
E. kuehniella	422	3.69 ± 0.10 b
M. brassicae	291	4.01 ± 0.15 a
Glass beads	396	3.67 ± 0.10 b

*Values followed by the same letter do not differ significantly at the 5% level.

the distance between two neighbouring hosts being 15 mm for rows and columns. Then, the walking path of the female was recorded with a video camera up to the moment when the female touched a host. On average, about five measurements were made for each female.

Using the system described by Coulon *et al.* (1983) and by Chassain *et al.* (1988a), the path of each female was then automatically transformed into *x-y* coordinates with an accuracy of 25 points s^{-1}. Finally, the reactive distance was estimated from these coordinates using the method described by Pak *et al.* (1991): the angle between the path and the female–host direction was computed in relation to the wasp–host distance. As soon as she perceived a host, the female went straight to it. Therefore, the reactive distance was estimated by the female–host distance from which the computed angle dropped to zero.

As shown in Table 12.3, *T. brassicae* females perceived the *M. brassicae* eggs from a longer distance than the *E. kuehniella* eggs. Although the odour of these eggs might be different, the difference in reactive distance might mainly be explained by the difference in their size: *M. brassicae* eggs are much bigger. The small, smelling *E. kuehniella* eggs are perceived from the same distance as the bigger, odourless glass beads. An eventual chemical perception of these eggs may have compensated for the probably smaller visual perception of them.

Moreover, with glass beads, a significant relationship can be observed between the values obtained for the mothers and their daughters (Fig. 12.7). This suggests that, in this case, there is a significant genetic variability in this trait in the *T. brassicae* population analysed. For the two other hosts, however, no significant relationship can be observed. The odour of these two hosts may have hidden some genetic variation in visual host perception mechanisms only.

Table 12.4 shows that, for both *M. brassicae* and *E. kuehniella* as hosts, there are significant differences between the average values of each family. Thus, for these two hosts, the reactive distance seems to be a family feature, which also suggests that the observed variability is under a genetic control. With glass beads, however, no family effect is observed.

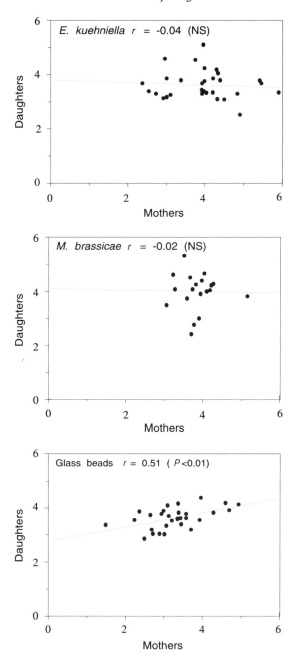

Fig. 12.7. Mothers/daughters regression analysis of the reactive distance (mm) of *T. brassicae* females for three types of hosts (NS, not significant).

Table 12.4. Analysis of variance used to compare the average values of the reactive distance to three types of hosts between different *T. brassicae* families. In this analysis, mothers were not taken into account, a family being constituted by their different daughters. The 'female' effect is nested within the 'family' effect.

Source of variation	d.f.	Variance	F
E. kuehniella			
Families	29	2.6056	1.65*
Females.families	60	1.5751	2.54[†]
Error	222	0.6208	
Total	311	0.9900	
M. brassicae			
Families	18	4.5735	2.41[†]
Females.families	37	1.8949	1.35 (NS)
Error	164	1.4009	
Total	219	1.7451	
Glass beads			
Families	28	1.7138	1.15 (NS)
Females.families	58	1.4868	1.89[†]
Error	224	0.7851	
Total	310	1.0003	

* $P < 0.05$.
[†] $P < 0.01$.
NS, not significant.

These results suggest that, in the analysed population, there is a genetic variability in the maximal distance from which *T. brassicae* females perceive their hosts. Within this population, some females have to be close to their hosts to perceive them, others are able to identify the presence of hosts from a longer distance. According to the simulation model described above, this genetic variability could be useful when there is a variation in the spatial distribution patterns of the hosts.

Genetic Variation in Sex Allocation

Once an arrhenotokous mated *Trichogramma* female has perceived healthy hosts and decided to parasitize them, she has to decide accurately what sex ratio to lay. This decision, which is achieved through the control of the proportion of fertilized eggs she oviposits, is considered to be strongly determined by the different environmental characteristics encountered (Flanders, 1946). Among others, such characteristics could be 'host quality' (e.g.

size) (Charnov, 1979; Charnov *et al.*, 1981; Werren, 1984); encounter rate with hosts (Strand, 1988; Colazza *et al.*, 1991); internal factors such as egg load (Putters, 1988), etc.

One of these characteristics also has a particular effect: the density of conspecific females on a host patch. Indeed, in non-random mating populations, where females mate as soon as they emerge and then disperse to colonize new host patches, it can be shown that the optimal sex ratio (the one that provides the highest fitness to the laying mother) depends on the number of females colonizing a host patch and, in fact, on the female : host ratio on that patch. This is the Hamilton's local mate competition (LMC) model (Hamilton, 1967; Taylor & Bulmer, 1980; Waage, 1986): when the female density increases, the sex ratio produced progressively increases from a female-biased sex ratio to a 50 : 50 sex ratio. Several experimental results, including some on *Trichogramma*, are in accordance with this model which is expected when the host population shows a significant aggregated spatial distribution (Taylor & Bulmer, 1980; Waage, 1982; Werren, 1983; Waage & Lane, 1984; Strand, 1988).

What are the mechanisms used by female wasps to change the sex ratio they produce in response to this particular constraint? Several ways to do this have been reported so far; a shift in sex allocation can be induced in response to: (i) the frequency of physical contacts with other females while ovipositing (Wylie, 1976); (ii) the perception of chemical traces left by previous females on the host patch (Viktorov & Kochetova, 1973); or (iii) the frequency of encounters with already parasitized hosts (Wylie, 1973). However, according to Waage & Lane (1984), the most efficient and simple mechanism females could use lies in their laying their son and daughter eggs in a particular order. Such sequence patterns would enable female wasps to produce accurate sex ratios without being obliged to 'count' the number of hosts in a patch and to calculate what fraction of their progeny should be male and female (Green *et al.*, 1982; Waage, 1986).

Sequences of oviposition have been recorded on *T. chilonis* Ishii (Suzuki *et al.*, 1984) and on *T. evanescens* (Waage & Lane, 1984; Waage & Ng, 1984; Dijken & Waage, 1987). In all cases, females tend to produce proportionately more male offspring early in an oviposition bout. Using a proper mathematical descriptive procedure, such a 'male-first' strategy has been statistically verified on *T. brassicae* also (Wajnberg, 1991a,b, 1993).

As suggested by Waage & Lane (1984), as long as females tend to lay fewer eggs to avoid superparasitism, such male-first strategy would simply explain how mothers can increase the sex ratio in their progeny under an increase in their density on a host patch (i.e. decreasing LMC). This hypothesis has been analysed with a simulation model. In this model, a patch of 16 healthy hosts is attacked by 1, 2, 4, 8 or 16 females. Only one egg is oviposited in each host. So, out of a sequence of 16 eggs, each female lays respectively its first 16, 8, 4, 2 or 1 eggs only. The oviposition sequence for each female

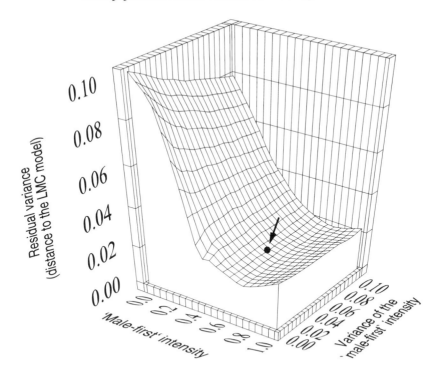

Fig. 12.8. Surface response of a simulation model run to test if an egg laying sequence effect may enable *Trichogramma* females to change their sex ratio in response to their own density on a host patch (LMC). The 'male-first' intensity (=(1−SMR); see Appendix) ranges from 0 (all males are laid at the end of the sequence) to 1.0 (all males are laid at the beginning of the sequence). The distance to the LMC model is estimated by the residual variance to the model published by Taylor & Bulmer (1980): when N arrhenotokous females colonize a host patch, the optimal sex ratio they have to produce is $((N-1)(2N-1))/(N(4N-1))$. Each point of this surface is the average response of 1000 simulations. The black dot (arrowed) corresponds to the values obtained on real *T. brassicae* females (see Appendix).

is drawn randomly using the method described in the Appendix. The variation in the sex ratio obtained, in response to changes in the intensity and accuracy of the male-first strategy, is shown Fig. 12.8.

This result shows that, whatever the accuracy of such a strategy, laying males at the beginning of the sequence seems sufficient to explain how females change the sex ratio in their progeny in response to their own density on a host patch (LMC).

Such a sex sequence pattern shows a significant genetic variation in the *T. brassicae* population studied (Wajnberg, 1991b,c, 1993). Therefore, according to this simulation model, a significant genetic variability in the change in sex ratio in response to females' density should be observed in this population. Such a hypothesis has been verified experimentally: 12 families were found,

Table 12.5. Variation in the average (± SE) sex ratio produced by mated *T. brassicae* females under different females' density on a host patch.

Females	n	Sex ratio*
1	49	0.1791 ± 0.0548 a
2	56	0.2525 ± 0.0581 b
4	82	0.2735 ± 0.0492 bc
6	90	0.2974 ± 0.0482 c
10	96	0.2543 ± 0.0444 bc
20	87	0.2613 ± 0.0471 bc

*Values followed by the same letter do not differ significantly at the 5% level.

Table 12.6. Analysis of variance used to compare 12 families of *T. brassicae* according to changes in the sex ratio they produced under different females' density on a host patch (i.e. different level of LMC). The computation was done after arcsin(sqrt(p)) transformation.

Source of variation	d.f.	Variance	F
Females' density (1)	5	0.20468	3.14*
Families (2)	11	0.06441	2.48*
Interaction (1)–(2)	53	0.06520	2.51*
Error	390	0.02593	
Total	459	0.03331	

*$P < 0.01$.

each from a single couple taken at random from the population studied. For each of them, 200 *E. kuehniella* eggs were offered, during 1 h at 25°C, to 1, 2, 4, 6, 10 or 20 mated, less than 24-h-old females. A preliminary analysis, comparing experiments done with virgin and mated females shows that, under such conditions, no differential mortality is observed between the developing males and females. Therefore, the sex ratio laid in each case was estimated by counting adults emerging from each replicate, after a proper development at 25°C. Table 12.5 gives the overall sex ratio produced by the females as a function of their own density.

There is a clear significant change in the sex ratio produced by the females in response to their density on a host patch. Moreover, the comparison between the results obtained for each family (Table 12.6) shows that, whatever the females' density, there is also a significant variation in the overall sex ratio produced by the different tested families.

This result confirms those previously obtained by Chassain (1988) and Chassain & Boulétreau (1991): there is a significant intra-population genetic variability in the sex ratio produced by *T. brassicae* mated females. Finally, a strongly significant interaction is also observed between the 'females' density' and the 'family' effects. This last result shows that the way females respond to their own density on a host patch differs between the different families analysed, and this strongly indicates that there is a genetic variation in this trait. Such a genetic variability, which is likely to be the consequence of the genetic variation in the egg laying sequence produced by the females, has probably been implicated in the progressive evolution to the production of a female-biased sex ratio under LMC.

Conclusion

As pointed out by Roitberg (1990), studies done to quantify genetic variation among individuals within beneficial organism populations should lead to improved results in the biological control of pests. Concerning *Trichogramma*, genetic variability in quantitative biological traits has seldom been investigated. According both to the few studies available and to the results presented here, significant variations can be observed in basic biological features of *Trichogramma* reproductive strategy. Such a variability provides the fundamental basis from which artificial selection programmes can be started in order to improve the efficiency of released *Trichogramma* to control target pests in the field.

Using stochastic modelization, this chapter provides a way of estimating how each biological characteristic studied, and its variability, is related to the success of *Trichogramma* as a biological control agent. Such an approach, which is novel in this field, could lead rapidly to valuable information. More precisely, it could provide a way of concentrating our efforts on the most important biological characteristics of wasps released in the field, aiming both to quantify their efficiency and to improve it. Moreover, it could provide a way of optimizing mass-production methods in order to produce beneficial organisms without losing their biological potentialities. Finally, by providing a way of comparing rapidly different potential populations and/or species, it could give a means of accelerating the very time-consuming process involved in the selection of biological control agents (Waage, 1990). This process, which is generally considered to be the most critical phase for the success of a biological control programme (Pak, 1988), is usually at the centre of conflicting ideas between theoretical ecologists and practitioners of biological control (Harris, 1973; Lenteren, 1980). The present approach lies between these two extremities: models developed are only simulating biological characteristics that are likely to occur in real conditions, and the simulated situations do not go beyond the range of the experimental results obtained.

E. Wajnberg

Acknowledgements

I am indebted to C. Chassain, M. Boulétreau and E. Bruins for allowing me to present unpublished data, and I thank C. Curty for her valuable help in experimental work, J. Daumal and H. Boinel for providing the *E. kuehniella* eggs, and M. Gigout and J.P. Merlet for technical computer advice. C. Caquineau, from Thinking Machines Corporation, is also thanked for computer assistance, on a parallel computer, for the simulation model on sex-sequence patterns. Finally, B. Legerstee-Pernelle is thanked for careful reading of the English version of the manuscript.

References

Alphen, J.J.M. Van & Nell, H.W. (1982) Superparasitism and host discrimination by *Asobara tabida* Nees (Braconidae; Alysiinae), a larval parasitoid of Drosophilidae. *Netherlands Journal of Zoology* 32, 232–260.

Alphen, J.J.M. & Visser, M.E. (1990) Superparasitism as an adaptive strategy for insect parasitoids. *Annual Review of Entomology* 35, 59–79.

Alphen, J.J.M. Van, Van Dijken, M.J. & Waage, J.K. (1987) A functional approach to superparasitism: host discrimination needs not be learnt. *Netherlands Journal of Zoology* 37, 167–179.

Babi, A. (1990) Bioécologie de *Trichogramma cacoeciae* Marchal et *T. daumalae* Dugast & Voegelé (Hym.; Trichogrammatidae). Utilisation en lutte biologique contre *Lobesia botrana* Den. & Schiff. (Lep.; Tortricidae). PhD Thesis, Aix-Marseille, France, 143 pp.

Bakker, K., Van Alphen, J.J.M., Van Batenburg, F.H.D., Van der Hoeven, N., Nell, H.W., Van Strien-Van Liempt, W.T.F.H. & Turling, T.C. (1985) The function of host discrimination and superparasitization in parasitoids. *Oecologia* 67, 572–576.

Bigler, F. (1986) Mass-production of *Trichogramma maidis* Pint. & Voeg. and its field application against *Ostrinia nubilalis* Hbn. in Switzerland. *Journal of Applied Entomology* 101, 23–29.

Boller, E.F. (1972) Behavioral aspects of mass-rearing of insects. *Entomophaga* 17, 9–25.

Boller, E.F. & Chambers, D.L. (1977) Quality aspects of mass-reared insects. In: Ridgway, R.L. & Vinson, S.B. (eds) *Biological Control by Augmentation of Natural Enemies*. Plenum Press, New York, pp. 219–236.

Charnov, E.L. (1979) The genetical evolution of patterns of sexuality: Darwinian fitness. *American Naturalist* 113, 465–480.

Charnov, E.L. & Skinner, S.W. (1984) Evolution of host selection and clutch size in parasitoid wasps. *Florida Entomologist* 67, 5–21.

Charnov, E.L. & Skinner, S.W. (1985) Complementary approaches to the understanding of parasitoid oviposition decisions. *Environmental Entomology* 14, 383–391.

Charnov, E.L., Los-den Hartogh, R.L., Jones, W.T. & Van den Assem, J. (1981) Sex ratio evolution in a variable environment. *Nature* 289, 27–33.

Chassain, C. (1988) Reproduction et comportements d'infestation des hôtes chez les trichogrammes: facteurs de variations génétiques et épigénétiques. Doctorate Thesis, University of Lyon I, France, 125 pp.

Chassain, C. & Boulétreau, M. (1987) Genetic variability in the egg-laying behaviour of *Trichogramma maidis*. *Entomophaga* 32, 149–157.

Chassain, C. & Boulétreau, M. (1991) Genetic variability in quantitative traits of host exploitation in *Trichogramma* (Hymenoptera: Trichogrammatidae). *Genetica* 83, 195–202.

Chassain, C., Clément, P., Chassé, J.L., Fouillet, P. & Boulétreau, M. (1988a) Quantitative features of displacement by *Trichogramma* females. In: Boulétreau, M. & Bonnot, G. (eds) *Parasitoid Insects. Les Colloques de l'INRA* 48, 47–49.

Chassain, C., Fleury, F., Fouillet, P. & Boulétreau, M. (1988b) Variabilité génétique du comportement de ponte chez les trichogrammes (Hymenoptera). *Annales de la Société Entomologique de France* 24, 458.

Chassain, C., Fouillet, P. & Boulétreau, M. (1988c) Variability in foraging behaviour in *Trichogramma* females: genetic basis and comparison of local populations. In: Boulétreau, M. & Bonnot, G. (eds) *Parasitoid Insects. Les Colloques de l'INRA* 48, 51–53.

Colazza, S., Vinson, S.B., Li, T.Y. & Bin, F. (1991) Sex ratio strategies of the egg parasitoid *Trissolcus basalis* (Woll.) (Hymenoptera Scelionidae): influence of the host egg path size. *Redia* 74, 279–286.

Coulon, P.Y., Charras, J.P., Chassé, J.L., Clément, J.L., Cornillac, P., Luciani, A. & Wurdak, E. (1983) An experimental system for the automatic tracking and analysis of rotifer swimming behaviour. *Hydrobiologia* 104, 197–202.

Daumal, J., Voegelé, J., & Brun, P. (1975) Les trichogrammes. II. Unité de production massive et quotidienne d'un hôte de substitution *Ephestia kuehniella* Zell (Lepidoptera, Pyralidae). *Annales de Zoologie, Ecologie Animales* 7, 45–59.

DeBach, P. (1974) *Biological Control by Natural Enemies*. Cambridge University Press, Cambridge.

DeBach, P. & Rosen, D. (1991) *Biological Control by Natural Enemies*, 2nd edn. Cambridge University Press, Cambridge, 440 pp.

Dijken, M.J. Van & Waage, J.K. (1987) Self and conspecific superparasitism by the egg parasitoid *Trichogramma evanescens*. *Entomologia Experimentalis et Applicata* 43, 183–192.

Edwards, R.L. (1961) The area of discovery of two insect parasites, *Nasonia vitripennis* (Walker) and *Trichogramma evanescens* (Westwood) in an artificial environment. *Canadian Entomologist* 93, 475–481.

Fiske, W.F. (1910) Superparasitism: an important factor in the natural control of insects. *Journal of Economic Entomology* 3, 88–97.

Flanders, S.E. (1946) Control of sex and sex-limited polymorphism in the Hymenoptera. *Quarterly Review of Biology* 21, 635–655.

Glas, P.C., Smits, P.H., Vlaming, P. & Van Lenteren, J.C. (1981) Biological control of lepidopteran pests in cabbage crops by means of inundative releases of *Trichogramma* species (*T. evanescens* Westwood and *T. cacoeciae* March): a combination of field and laboratory experiments. *Mededelingen van de Faculteit Landbouwwetenschappen Rijksuniversiteit Gent* 46, 487–497.

Green, R.F., Gordh, G. & Hawkins, B.A. (1982) Precise sex ratio in highly inbred parasitic wasp. *American Naturalist* 120, 635–655.

Hamilton, W.D. (1967) Extraordinary sex ratio. *Science* 156, 477–488.

Harris, P. (1973) The selection of effective agents for the biological control of weeds. *Canadian Entomologist* 105, 1495–1503.

Hassan, S.A. (1981) Mass-production and utilization of *Trichogramma*. 2. Four years successful biological control of the European corn borer. *Mededelingen van de Faculteit Landbouwwetenschappen Rijksuniversiteit Gent* 46, 417–427.

Hassan, S.A. & Rost, W.M. (1985) Mass-production and utilization of *Trichogramma*. 6. Studies towards the use against cabbage lepidopterous pests. *Mededelingen van de Faculteit Landbouwwetenschappen Rijksuniversiteit Gent* 50, 389–398.

Howarth, F.G. (1991) Environmental impacts of classical biological control. *Annual Review of Entomology* 36, 485–509.

Joslyn, D.J. (1984) Maintenance of genetic variability in reared insects. In: King, E.G. & Leppla, N.C. (eds) *Advances and Challenges in Insect Rearing*. USDA, Agricultural Research Service, pp. 20–29.

King, E.G., Bouse, L.F., Bull, R.L., Coleman, W.A., Dickerson, W.A., Lewis, W.J., Lopez, J.D., Morrison, R.K. & Phillips, J.R. (1986) Management of *Heliothis* spp. in cotton by augmentative releases of *Trichogramma pretiosum*. *Journal of Applied Entomology* 101, 2–10.

Klomp, H., Teerink, B.J. & Ma, W.C. (1980) Discrimination between parasitized and unparasitized hosts in the egg parasite *Trichogramma embryophagum* (Hym.: Trichogrammatidae): a matter of learning and forgetting. *Netherlands Journal of Zoology* 30, 254–277.

Laing, J. (1937) Host finding by insect parasites. I. Observations on the finding of hosts by *Alysia manducator*, *Mormoniella vitripennis* and *Trichogramma evanescens*. *Journal of Animal Ecology* 6, 298–317.

Laing, J. (1938) Host finding by insect parasites. II. The chance of *Trichogramma evanescens* finding its hosts. *Journal of Experimental Biology* 15, 281–302.

Lenteren, J.C. van (1980) Evaluation of control capabilities of natural enemies: does art have to become science? *Netherlands Journal of Zoology* 30, 369–381.

Lenteren, J.C. van (1981) Host discrimination by parasitoids. In: Nordlund, D.A., Jones, R.L. & Lewis, W.J. (eds) *Semiochemicals, Their Role in Pest Control*. J. Wiley & Sons, New York, pp. 153–179.

Limburg, H. & Pak, G.A. (1991) Genetic variation in the walking behaviour of the egg parasite *Trichogramma*. In: Bigler, F. (ed.) *Fifth Workshop of the IOBC Global Working Group. Quality Control of Mass Reared Arthropods*. Wageningen, The Netherlands, pp. 47–55.

Mackauer, M. (1972) Genetic aspects of insect control. *Entomophaga* 17, 27–48.

Mimouni, F. (1990) Caractérisation biologique et comportementale de deux espèces de trichogrammes marocains: étude de facteurs génétiques et épigénétiques. Doctorate Thesis, University of Lyon I, France, 89 pp.

Mimouni, F. (1991) Genetic variations in host infestation efficiency in two *Trichogramma* species from Morocco. *Redia* 74, 393–400.

Nagarkatti, S. & Nagaraja, H. (1977) Biosystematics of *Trichogramma* and *Trichogrammatidae* species. *Annual Review of Entomology* 22, 157–176.

Pak, G.A. (1988) Selection of *Trichogramma* for inundative biological control. PhD Thesis, Wageningen Agricultural University, The Netherlands, 224 pp.

Pak, G.A., Berkhout, H. & Klapwijk, J. (1991) Do *Trichogramma* look for hosts? In: Wajnberg, E. & Vinson, S.B. (eds) Trichogramma *and Other Egg Parasitoids. 3rd International Symposium. Les Colloques de l'INRA* 56, 77–80.

Pintureau, B. (1988) Genetic variation in the genus *Trichogramma* Westwood (Hym. Trichogrammatidae). In: Boulétreau, M. & Bonnot, G. (eds) *Parasitoid Insects. Les Colloques de l'INRA* 48, 111–113.

Pintureau, B. (1990) Polymorphisme, biogéographie et spécificité parasitaire des trichogrammes européens (Hym.; Trichogrammatidae). *Bulletin de la Société Entomologique de France* 95, 17–38.

Pintureau, B. & Voegelé, J. (1980) Une nouvelle espèce proche de *Trichogramma evanescens: T. maidis* (Hym.; Trichogrammatidae). *Entomophaga* 26, 431–440.

Putters, F.A. (1988) Proximate factors in sex ratio shift in gregarious parasitoid wasps. In: Boulétreau, M. & Bonnot, G. (eds) *Parasitoid Insects. Les Colloques de l'INRA* 48, 115–116.

Roitberg, B.D. (1985) Search dynamics in fruit-parasitic insects. *Journal of Insect Physiology* 31, 865–872.

Roitberg, B.D. (1990) Variation in behaviour of individual parasitic insects: bane or boon? In: Mackauer, M., Ehler, L.E. & Roland, J. (eds) *Critical Issues in Biological Control.* Intercept, Andover, UK, pp. 25–39.

Schmidt, J.M. (1991) The inheritance of clutch size regulation in *Trichogramma* species (Hymenoptera: Chalcidoidea: Trichogrammatidae). In: Bigler, F. (ed.) *Fifth Workshop of the IOBC Global Working Group. Quality Control of Mass Reared Arthropods.* Wageningen, The Netherlands, pp. 26–37.

Skinner, S.W. (1985) Clutch size as an optimal foraging problem for insects. *Behavioral Ecology and Sociobiology* 17, 231–238.

Starler, N.H. & Ridgway, R.L. (1977) Economic and social considerations for the utilization of augmentation of natural enemies. In: Ridgway, R.L. & Vinson, S.B. (eds) *Biological Control by Augmentation of Natural Enemies.* Plenum Press, New York, pp. 431–450.

Stinner, R.E. (1977) Efficacy of inundative releases. *Annual Review of Entomology* 22, 515–531.

Strand, M.R. (1988) Variable sex ratio strategy of *Telenomus heliothidis* (Hymenoptera: Scelionidae): adaptation to host and conspecific density. *Oecologia* 77, 219–224.

Suzuki, Y., Tsuji, H. & Sasakaw, M. (1984) Sex allocation and effects of superparasitism on secondary sex ratios in the gregarious parasitoid, *Trichogramma chilonis* (Hymenoptera: Trichogrammatidae). *Animal Behavior* 32, 478–484.

Taylor, P.D. & Bulmer, M.G. (1980) Local mate competition and the sex ratio. *Journal of Theoretical Biology* 86, 409–419.

Vanlerberghe, F. (1991) Mitochondrial DNA polymorphism and relevance for *Trichogramma* population genetics. In: Wajnberg, E. & Vinson, S.B. (eds) Trichogramma *and Other Egg Parasitoids. 3rd International Symposium. Les Colloques de l'INRA* 56, 123–126.

Viktorov, G.A. & Kochetova, N.I. (1973) The role of trace pheromones in regulating the sex ratio in *Trissolcus grandis* (Hymenoptera, Scelionidae). *Zhurnal Obshchei Biologii* 34, 559–562.

Vinson, S.B. (1976) Host selection by insect parasitoids. *Annual Review of Entomology* 21, 109–133.

Voegelé, J. (1981) Lutte biologique contre *Ostrinia nubilalis* à l'aide de trichogrammes. *Bulletin OEPP* 11, 91–95.

Voegelé, J., Pizzol, J., Ciociola, A., Marro, J.P., Millot, P. & Guillomard, G. (1990) Tomatoes noctuids control in France by *Trichogramma* inundative releases. In:

International Symposium on Biological Control, Turkey, November 1989. pp. 42–46.

Waage, J.K. (1982) Sib-mating and sex ratio strategies in scelionid wasps. *Ecological Entomology* 7, 103–113.

Waage, J.K. (1986) Family planning in parasitoids: adaptive patterns of progeny and sex allocation. In: Waage, J.K. & Greathead, D. (eds) *Insect Parasitoids*. Academic Press, London, pp. 63–95.

Waage, J.K. (1990) Ecological theory and the selection of biological control agents. In: Mackauer, M., Ehler, L.E. & Roland, J. (eds) *Critical Issues in Biological Control*. Intercept, Andover, UK, pp. 135–157.

Waage, J.K. & Lane, J.A. (1984) The reproductive strategy of a parasitic wasp. II. Sex allocation and local mate competition in *Trichogramma evanescens*. *Journal of Animal Ecology* 53, 417–426.

Waage, J.K. & Ng, S.M. (1984) The reproductive strategy of a parasitic wasp. I. Optimal progeny and sex allocation in *Trichogramma evanescens*. *Journal of Animal Ecology* 53, 401–415.

Wajnberg, E. (1988) Analysis of variations of handling time in *Trichogramma maidis* (Hym.; Chalc.). In: Boulétreau, M. & Bonnot, G. (eds) *Parasitoid Insects. Les Colloques de l'INRA* 48, 69–70.

Wajnberg, E. (1989) Analysis of variations of handling-time in *Trichogramma maidis*. *Entomophaga* 34, 397–407.

Wajnberg, E. (1991a) Quality control of mass-reared arthropods: a genetical and statistical approach. In: Bigler, F. (ed.) *Fifth Workshop of the IOBC Global Working Group. Quality Control of Mass Reared Arthropods*. Wageningen, The Netherlands, pp. 15–25.

Wajnberg, E. (1991b) A new statistical method for quantifying sex patterns within sequences of oviposition produced by parasitic wasps. *Redia* 74, 401–405.

Wajnberg, E. (1991c) Genetic variation in sex allocation in *Trichogramma maidis*: variation in the sex pattern within sequences of oviposition. In: Wajnberg, E. & Vinson, S.B. (eds) Trichogramma *and Other Egg Parasitoids. 3rd International Symposium. Les Colloques de l'INRA* 56, 127–129.

Wajnberg, E. (1993) Genetic variation in sex allocation in a parasitic wasp. Variation in sex pattern within sequences of oviposition. *Entomologia Experimentalis et Applicata* 69, 221–229.

Wajnberg, E. & Pizzol, J. (1989) The problem of superparasitism in the production of natural enemies for inundative biological control: a genetical approach. In: Cavalloro, R. & Delucchi, V. (eds) *Parasitis 88. Proceedings of a Scientific Congress*. Ministerio de Agricultura Pesca y Alimentation, Madrid, pp. 437–444.

Wajnberg, E., Pizzol, J. & Babault, M. (1989) Genetic variation in progeny allocation in *Trichogramma maidis. Entomologia Experimentalis et Applicata* 53, 177–187.

Weseloh, R.M. (1981) Host location by parasitoids. In: Nordlund, D.A. & Lewis, W.J. (eds) *Semiochemicals: Their Role in Pest Control*. John Wiley, New York, pp. 79–95.

Werren, J.H. (1983) Sex ratio evolution under local mate competition in a parasitic wasp. *Evolution* 37, 116–124.

Werren, J.H. (1984) Brood size and sex ratio regulation in the parasitic wasp *Nasonia vitripennis* (Walker) (Hymenoptera: Pteromalidae). *Netherlands Journal of Zoology* 34, 123–143.

Westwood, J.O. (1879) Descriptions of some minute hymenopterous insects. *Transactions of the Linnean Society of London, Zoology* 1, 583–593.

Wylie, H.G. (1973) Control of egg fertilization by *Nasonia vitripennis* (Walk.) (Hymenoptera: Pteromalidae) reared from super-parasitized housefly pupae. *Canadian Entomologist* 98, 645–653.

Wylie, H.G. (1976) Interference among females of *Nasonia vitripennis* (Hymenoptera: Pteromalidae) and its effect on the sex ratio of their progeny. *Canadian Entomologist* 108, 655–661.

Yano, E. (1978) A simulation model of searching behaviour of a parasite. *Research on Population Ecology* 20, 105–122.

Appendix

A statistical method for a random drawing of likely sequences of oviposition for a simulation model

Wajnberg (1991b, 1993) provided a non-parametric, multivariate method for describing sequences of oviposition by parasitic female wasps. Briefly, each sequence is quantified by a set of five non-parametric statistics describing different sequential features of the oviposition sequences. These statistics are: the sum and variance of the males' rank position (respectively SMR and VMR), the 'centre-group' of males (CGM) or females (CGF), and the number of runs (NR) of males or females. These statistics were built in order to see if males are laid at the beginning (for SMR) or in the middle (for VMR) of the sequence, if there is some pooling of males (CGM) or females (CGF) within the sequence, and if there is any autocorrelation of males or females within each sequence described. Moreover, in order to describe sequences differing in length or composed of different sex ratios, these five parameters were first standardized to all be included in the [0;1] interval (see Wajnberg, 1993, for a more detailed description).

A sample of 113 sequences of oviposition, produced by mated *T. brassicae* females, was described with this method. **E** and **V** were respectively the mean vector and the variance matrix obtained (see black dot in Fig. 12.8). The overall sex ratio was 0.2454 ± 0.0065. According to these results, the procedure for drawing randomly a sequence of 16 eggs works as follows: using the Cholesky's decomposition of **V**, a vector **X** of five components is drawn from a 5D multinormal distribution with **E** as mean vector and **V** as variance matrix. Out of the 16 eggs, the number M of males was estimated by drawing the corresponding sex ratio from a normal distribution with the same parameters as those computed on the sample described. Then, sequences corresponding to all the possible combinations of placing M males in a sequence of 16 eggs were quantified with the method described above, and this quantification was used to compute the Euclidian distance between each of them and **X**. The sequence corresponding to the lower distance obtained was used in the simulation model corresponding to Fig. 12.8. This procedure was repeated with different values of SMR and of its variance.

Index